Stochastic Processes

Wolfgang Paul · Jörg Baschnagel

Stochastic Processes

From Physics to Finance

Second Edition

Wolfgang Paul
Institut für Physik
Martin-Luther-Universität
Halle (Saale), Germany

Jörg Baschnagel
Institut Charles Sadron
Université de Strasbourg
Strasbourg, France

ISBN 978-3-319-00326-9   ISBN 978-3-319-00327-6 (eBook)
DOI 10.1007/978-3-319-00327-6
Springer Heidelberg New York Dordrecht London

Library of Congress Control Number: 2013944566

© Springer International Publishing Switzerland 2000, 2013
This work is subject to copyright. All rights are reserved by the Publisher, whether the whole or part of the material is concerned, specifically the rights of translation, reprinting, reuse of illustrations, recitation, broadcasting, reproduction on microfilms or in any other physical way, and transmission or information storage and retrieval, electronic adaptation, computer software, or by similar or dissimilar methodology now known or hereafter developed. Exempted from this legal reservation are brief excerpts in connection with reviews or scholarly analysis or material supplied specifically for the purpose of being entered and executed on a computer system, for exclusive use by the purchaser of the work. Duplication of this publication or parts thereof is permitted only under the provisions of the Copyright Law of the Publisher's location, in its current version, and permission for use must always be obtained from Springer. Permissions for use may be obtained through RightsLink at the Copyright Clearance Center. Violations are liable to prosecution under the respective Copyright Law.
The use of general descriptive names, registered names, trademarks, service marks, etc. in this publication does not imply, even in the absence of a specific statement, that such names are exempt from the relevant protective laws and regulations and therefore free for general use.
While the advice and information in this book are believed to be true and accurate at the date of publication, neither the authors nor the editors nor the publisher can accept any legal responsibility for any errors or omissions that may be made. The publisher makes no warranty, express or implied, with respect to the material contained herein.

Printed on acid-free paper

Springer is part of Springer Science+Business Media (www.springer.com)

Preface to the Second Edition

Thirteen years have passed by since the publication of the first edition of this book. Its favorable reception encouraged us to work on this second edition. We took advantage of this opportunity to update the references and to correct several mistakes which (inevitably) occurred in the first edition. Furthermore we added several new sections in Chaps. 2, 3 and 5. In Chap. 2 we give an introduction to Jaynes' treatment of probability theory as a form of logic employed to judge rational expectations and to his famous maximum entropy principle. Additionally, we now also discuss limiting distributions for extreme values. In Chap. 3 we added a section on the Caldeira-Leggett model which allows to derive a (generalized) Langevin equation starting from the deterministic Newtonian description of the dynamics. Furthermore there is now also a section about the first passage time problem for unbounded diffusion as an example of the power of renewal equation techniques and a discussion of the extreme excursions of Brownian motion. Finally we extended the section on Nelson's stochastic mechanics by giving a detailed discussion on the treatment of the tunnel effect. Chapter 5 of the first edition contained a discussion of credit risk, which was based on the commonly accepted understanding at that time. This discussion has been made obsolete by the upheavals on the financial market occurring since 2008, and we changed it accordingly. We now also address a problem that has been discussed much in the recent literature, the (possible) non-stationarity of financial time series and its consequences. Furthermore, we extended the discussion of microscopic modeling approaches by introducing agent based modeling techniques. These models allow to correlate the behavior of the agents—the microscopic 'degrees of freedom' of the financial market—with the features of the simulated financial time series, thereby providing insight into possible underlying mechanisms. Finally, we also augmented the discussion about the description of extreme events in financial time series.

We would like to extend the acknowledgments of the first edition to thank M. Ebert, T. Preis and T. Schwiertz for fruitful collaborations on the modeling of financial markets. Without their contribution Chap. 5 would not have its present form.

Halle and Strasbourg, Wolfgang Paul
February 2013 Jörg Baschnagel

Preface to the First Edition

Twice a week, the condensed matter theory group of the University of Mainz meets for a coffee break. During these breaks we usually exchange ideas about physics, discuss administrative duties, or try to tackle problems with hardware and software maintenance. All of this seems quite natural for physicists doing computer simulations.

However, about two years ago a new topic arose in these routine conversations. There were some Ph.D. students who started discussions about the financial market. They had founded a 'working group on finance and physics'. The group met frequently to study textbooks on 'derivatives', such as 'options' and 'futures', and to work through recent articles from the emerging 'econophysics' community which is trying to apply well-established physical concepts to the financial market. Furthermore, the students organized special seminars on these subjects. They invited speakers from banks and consultancy firms who work in the field of 'risk management'. Although these seminars took place in the late afternoon and sometimes had to be postponed at short notice, they were better attended than some of our regular group seminars. This lively interest evidently arose partly for the reason that banks, insurance companies, and consultancy firms currently hire many physicists. Sure enough, the members of the 'working group' found jobs in this field after graduating.

It was this initiative and the professional success of our students that encouraged us to expand our course on 'Stochastic Processes' to include a part dealing with applications to finance. The course, held in the winter semester 1998/1999, was well attended and lively discussions throughout all parts gave us much enjoyment. This book has accrued from these lectures. It is meant as a textbook for graduate students who want to learn about concepts and 'tricks of the trade' of stochastic processes and to get an introduction to the modeling of financial markets and financial derivatives based on the theory of stochastic processes. It is mainly oriented towards students with a physics or chemistry background as far as our decisions about what constituted 'simple' and 'illustrative' examples are concerned. Nevertheless, we tried to keep our exposition so self-contained that it is hopefully also interesting and helpful to students with a background in mathematics, economics or engineering. The book is also meant as a guide for our colleagues who may plan to teach a similar course.

The selection of applications is a very personal one and by no means exhaustive. Our intention was to combine classical subjects, such as random walks and Brownian motion, with non-conventional themes.

One example of the latter is the financial part and the treatment of 'geometric Brownian motion'. Geometric Brownian motion is a viable model for the time evolution of stock prices. It underlies the Black-Scholes theory for option pricing, which was honored by the Nobel Prize for economics in 1997.

An example from physics is Nelson's 'stochastic mechanics'. In 1966, Nelson presented a derivation of non-relativistic quantum mechanics based on Brownian motion. The relevant stochastic processes are energy-conserving diffusion processes. The consequences of this approach still constitute a field of active research.

A final example comprises stable distributions. In the 1930s the mathematicians Lévy and Khintchine searched for all possible limiting distributions which could occur for sums of random variables. They discovered that these distributions have to be 'stable', and formulated a generalization of the central limit theorem. Whereas the central limit theorem is intimately related to Brownian diffusive motion, stable distributions offer a natural approach to anomalous diffusion, i.e., subdiffusive or superdiffusive behavior. Lévy's and Khintchine's works are therefore not only of mathematical interest; progressively they find applications in physics, chemistry, biology and the financial market.

All of these examples should show that the field of stochastic processes is copious and attractive, with applications in fields as diverse as physics and finance. The theory of stochastic processes is the 'golden thread' which provides the connection. Since our choice of examples is naturally incomplete, we have added at the end of each chapter references to the pertinent literature from which we have greatly profited, and which we believe to be excellent sources for further information. We have chosen a mixed style of referencing. The reference section at the end of the book is in alphabetical order to group the work of a given author and facilitate its location. To reduce interruption of the text we cite these references, however, by number.

In total, the book consists of five chapters and six appendices, which are structured as follows. Chapter 1 serves as an introduction. It briefly sketches the history of probability theory. An important issue in this development was the problem of the random walk. The solution of this problem in one dimension is given in detail in the second part of the chapter. With this, we aim to provide an easy stepping stone onto the concepts and techniques typical in the treatment of stochastic processes.

Chapter 2 formalizes many of the ideas of the previous chapter in a mathematical language. The first part of the chapter begins with the measure theoretic formalization of probabilities, but quickly specializes to the presentation in terms of probability densities over $\mathbb{R}^d$. This presentation will then be used throughout the remainder of the book. The abstract definitions may be skipped on first reading, but are included to provide a key to the mathematical literature on stochastic processes. The second part of the chapter introduces several levels of description of Markov processes (stochastic processes without memory) and their interrelations, starting from the Chapman-Kolmogorov equation. All the ensuing applications will be Markov processes.

Chapter 3 revisits Brownian motion. The first three sections cover classical applications of the theory of stochastic processes. The chapter begins with random walks on a d-dimensional lattice. It derives the probability that a random walker will be at a lattice point r after N steps, and thereby answers 'Polya's question': What is the probability of return to the origin on a d-dimensional lattice? The second section discusses the original Brownian motion problem, i.e., the irregular motion of a heavy particle immersed in a fluid of lighter particles. The same type of motion can occur in an external potential which acts as a barrier to the motion. When asking about the time it takes the particle to overcome that barrier, we are treating the so-called 'Kramers problem'. The solution of this problem is given in the third section of the chapter. The fourth section treats the mean field approximation of the Ising model. It is chosen as a vehicle to present a discussion of the static (probabilistic) structure as well as the kinetic (stochastic) behavior of a model, using the various levels of description of Markov processes introduced in Chap. 2. The chapter concludes with Nelson's stochastic mechanics to show that diffusion processes are not necessarily dissipative (consume energy), but can conserve energy. We will see that one such process is non-relativistic quantum mechanics.

Chapter 4 leaves the realm of Brownian motion and of the central limit theorem. It introduces stable distributions and Lévy processes. The chapter starts with some mathematical background on stable distributions. The distinguishing feature of these distributions is the presence of long-ranged power-law tails, which might lead to the divergence of even the lowest-order moments. Physically speaking, these lower-order moments set the pertinent time and length scales. For instance, they define the diffusion coefficient in the case of Brownian motion. The divergence of these moments therefore implies deviations from normal diffusion. We present two examples, one for superdiffusive behavior and one for subdiffusive behavior. The chapter closes with a special variant of a Lévy process, the truncated Lévy flight, which has been proposed as a possible description of the time evolution of stock prices.

The final chapter (Chap. 5) deals with the modeling of financial markets. It differs from the previous chapters in two respects. First, it begins with a fairly verbose introduction to the field. Since we assume our readers are not well acquainted with the notions pertaining to financial markets, we try to compile and explain the terminology and basic ideas carefully. An important model for the time evolution of asset prices is geometric Brownian motion. Built upon it is the Black-Scholes theory for option pricing. As these are standard concepts of the financial market, we discuss them in detail. The second difference to previous chapters is that the last two sections have more of a review character. They do not present well-established knowledge, but rather current opinions which are at the moment strongly advocated by the physics community. Our presentation focuses on those suggestions that employ methods from the theory of stochastic processes. Even within this limited scope, we do not discuss all approaches, but present only a selection of those topics which we believe to fit well in the context of the previous chapters and which are extensively discussed in the current literature. Among those topics are the statistical analysis of financial data and the modeling of crashes.

Finally, some more technical algebra has been relegated to the appendices and we have tried to provide a comprehensive subject index to the book to enable the reader to quickly locate topics of interest.

One incentive for opening or even studying this book could be the hope that it holds the secret to becoming rich. We regret that this is (probably) an illusion. One does not necessarily learn how to make money by reading this book, at least not if this means how to privately trade successfully in financial assets or derivatives. This would require one to own a personal budget which is amenable to statistical treatment, which is true neither for the authors nor probably for most of their readers. However, although it does not provide the 'ABC' to becoming a wizard investor, reading this book can still help to make a living. One may acquire useful knowledge for a prospective professional career in financial risk management. Given the complexity of the current financial market, which will certainly still grow in the future, it is important to understand at least the basic parts of it.

This will also help to manage the largest risk of them all which has been expressed in a slogan-like fashion by the most successful investor of all time, Warren Buffet [23]:

Risk is not knowing what you're doing.

In order to know what one is doing, a thorough background in economics, a lot of experience, but also a familiarity with the stochastic modeling of the market are important. This book tries to help in the last respect.

To our dismay, we have to admit that we cannot recall all occasions when we obtained advice from students, colleagues and friends. Among others, we are indebted to C. Bennemann, K. Binder, H.-P. Deutsch, H. Frisch, S. Krouchev, A. Schäcker and A. Werner. The last chapter on the financial market would not have its present form without the permission to reproduce artwork from the research of J.-P. Bouchaud, M. Potters and coworkers, of R.N. Mantegna, H.E. Stanley and coworkers, and of A. Johansen, D. Sornette and coworkers. We are very grateful that they kindly and quickly provided the figures requested. Everybody who has worked on a book project knows that the current standards of publishing can hardly be met without the professional support of an experienced publisher like Springer. We have obtained invaluable help from C. Ascheron, A. Lahee and many (unknown) others. Thank you very much.

Mainz,　　　　　　　　　　　　　　　　　　　　　　　　　　　　　　　　Wolfgang Paul
October 1999　　　　　　　　　　　　　　　　　　　　　　　　　　　　　　Jörg Baschnagel

Contents

1 A First Glimpse of Stochastic Processes 1
  1.1 Some History 1
  1.2 Random Walk on a Line 3
      1.2.1 From Binomial to Gaussian 6
      1.2.2 From Binomial to Poisson 11
      1.2.3 Log–Normal Distribution 13
  1.3 Further Reading 15

2 A Brief Survey of the Mathematics of Probability Theory 17
  2.1 Some Basics of Probability Theory 17
      2.1.1 Probability Spaces and Random Variables 18
      2.1.2 Probability Theory and Logic 21
      2.1.3 Equivalent Measures 29
      2.1.4 Distribution Functions and Probability Densities ... 30
      2.1.5 Statistical Independence and Conditional Probabilities ... 31
      2.1.6 Central Limit Theorem 33
      2.1.7 Extreme Value Distributions 35
  2.2 Stochastic Processes and Their Evolution Equations 38
      2.2.1 Martingale Processes 41
      2.2.2 Markov Processes 43
  2.3 Itô Stochastic Calculus 53
      2.3.1 Stochastic Integrals 53
      2.3.2 Stochastic Differential Equations and the Itô Formula ... 57
  2.4 Summary 58
  2.5 Further Reading 59

3 Diffusion Processes 63
  3.1 The Random Walk Revisited 63
      3.1.1 Polya Problem 66
      3.1.2 Rayleigh-Pearson Walk 69
      3.1.3 Continuous-Time Random Walk 72
  3.2 Free Brownian Motion 75

		3.2.1 Velocity Process	77

        3.2.1 Velocity Process 77
        3.2.2 Position Process 81
  3.3 Caldeira-Leggett Model 84
        3.3.1 Definition of the Model 85
        3.3.2 Velocity Process and Generalized Langevin Equation . . . 86
  3.4 On the Maximal Excursion of Brownian Motion 90
  3.5 Brownian Motion in a Potential: Kramers Problem 92
        3.5.1 First Passage Time for One-dimensional Fokker-Planck
                Equations 94
        3.5.2 Kramers Result 97
  3.6 A First Passage Problem for Unbounded Diffusion 98
  3.7 Kinetic Ising Models and Monte Carlo Simulations 101
        3.7.1 Probabilistic Structure 102
        3.7.2 Monte Carlo Kinetics 102
        3.7.3 Mean-Field Kinetic Ising Model 105
  3.8 Quantum Mechanics as a Diffusion Process 110
        3.8.1 Hydrodynamics of Brownian Motion 110
        3.8.2 Conservative Diffusion Processes 114
        3.8.3 Hypothesis of Universal Brownian Motion 115
        3.8.4 Tunnel Effect 118
        3.8.5 Harmonic Oscillator and Quantum Fields 122
  3.9 Summary 126
  3.10 Further Reading 127

4 Beyond the Central Limit Theorem: Lévy Distributions 131
  4.1 Back to Mathematics: Stable Distributions 132
  4.2 The Weierstrass Random Walk 136
        4.2.1 Definition and Solution 137
        4.2.2 Superdiffusive Behavior 143
        4.2.3 Generalization to Higher Dimensions 147
  4.3 Fractal-Time Random Walks 150
        4.3.1 A Fractal-Time Poisson Process 151
        4.3.2 Subdiffusive Behavior 154
  4.4 A Way to Avoid Diverging Variance: The Truncated Lévy Flight . 155
  4.5 Summary 159
  4.6 Further Reading 161

5 Modeling the Financial Market 163
  5.1 Basic Notions Pertaining to Financial Markets 164
  5.2 Classical Option Pricing: The Black-Scholes Theory 173
        5.2.1 The Black-Scholes Equation: Assumptions and Derivation . 174
        5.2.2 The Black-Scholes Equation: Solution and Interpretation . 179
        5.2.3 Risk-Neutral Valuation 184
        5.2.4 Deviations from Black-Scholes: Implied Volatility 189
  5.3 Models Beyond Geometric Brownian Motion 191
        5.3.1 Statistical Analysis of Stock Prices 192

		5.3.2	The Volatility Smile: Precursor to Gaussian Behavior? . . .	205
		5.3.3	Are Financial Time Series Stationary?	209
		5.3.4	Agent Based Modeling of Financial Markets	214
	5.4	Towards a Model of Financial Crashes		221
		5.4.1	Some Empirical Properties	222
		5.4.2	A Market Model: From Self-organization to Criticality . . .	225
	5.5	Summary		233
	5.6	Further Reading		234

Appendix A Stable Distributions Revisited 237
  A.1 Testing for Domains of Attraction 237
  A.2 Closed-Form Expressions and Asymptotic Behavior 239

Appendix B Hyperspherical Polar Coordinates 243

Appendix C The Weierstrass Random Walk Revisited 247

Appendix D The Exponentially Truncated Lévy Flight 253

Appendix E Put–Call Parity 259

Appendix F Geometric Brownian Motion 261

References 265

Index 273

Chapter 1
A First Glimpse of Stochastic Processes

In this introductory chapter we will give a short overview of the history of *probability theory* and *stochastic processes*, and then we will discuss the properties of a simple example of a stochastic process, namely the *random walk* in one dimension. This example will introduce us to many of the typical questions that arise in situations involving *randomness* and to the tools for tackling them, which we will formalize and expand on in subsequent chapters.

1.1 Some History

Let us start this historical introduction with a quote from the superb review article *On the Wonderful World of Random Walks* by E.W. Montroll and M.F. Shlesinger [145], which also contains a more detailed historical account of the development of probability theory:

> *Since traveling was onerous (and expensive), and eating, hunting and wenching generally did not fill the 17th century gentleman's day, two possibilities remained to occupy the empty hours, praying and gambling; many preferred the latter.*

In fact, it is in the area of gambling that the theory of probability and stochastic processes has its origin. People had always engaged in gambling, but it was only through the thinking of the Enlightenment that the outcome of a gambling game was no longer seen as a divine decision, but became amenable to rational thinking and speculation. One of these 17th century gentlemen, a certain Chevalier de Méré, is reported to have posed a question concerning the odds at a gambling game to Pascal (1623–1662). The ensuing exchange of letters between Pascal and Fermat (1601–1665) on this problem is generally seen as the starting point of probability theory.

The first book on probability theory was written by Christiaan Huygens (1629–1695) in 1657 and had the title *De Ratiociniis in Ludo Aleae* (On Reasoning in the Game of Dice). The first mathematical treatise on probability theory in the modern sense was Jakob Bernoulli's (1662–1705) book *Ars Conjectandi* (The Art of Conjecturing), which was published posthumously in 1713. It contained

- a critical discussion of Huygens' book,
- *combinatorics*, as it is taught today,
- probabilities in the context of gambling games, and
- an application of probability theory to daily problems, especially in *economics*.

We can see that even at the beginning of the 18th century the two key ingredients responsible for the importance of stochastic ideas in economics today are already discernable: combine a probabilistic description of economic processes with the control of risks and odds in gambling games, and you get the risk-management in financial markets that has seen such an upsurge of interest in the last 30 years.

A decisive step in the stochastic treatment of the price of financial assets was made in the Ph.D. thesis of Louis Bachelier (1870–1946), *Théorie de la Spéculation*, for which he obtained his Ph.D. in mathematics on March 19, 1900. His advisor was the famous mathematician Henri Poincaré (1854–1912), who is also well known for his contributions to theoretical physics. The thesis is very remarkable at least for two reasons:

- It already contained many results of the theory of stochastic processes as it stands today which were only later mathematically formalized.
- It was so completely ignored that even Poincaré forgot that it contained the solution to the *Brownian motion* problem when he later started to work on that problem.

Brownian motion is the archetypical problem in the theory of stochastic processes. In 1827 the Scottish botanist Robert Brown had reported the observation of a very irregular motion displayed by a pollen particle immersed in a fluid. It was the *kinetic theory of gases*, dating back to the book *Hydrodynamica, sive de viribus et motibus fluidorum commentarii* (Hydrodynamics, or commentaries on the forces and motions of fluids) by Daniel Bernoulli (1700–1782), published in 1738, which would provide the basis for Einstein's (1879–1955) and Smoluchowski's (1872–1917) successful treatments of the Brownian motion problem in 1905 and 1906, respectively. Through the work of Maxwell (1831–1879) and Boltzmann (1844–1906), *Statistical Mechanics*, as it grew out of the kinetic theory of gases, was the main area of application of probabilistic concepts in theoretical physics in the 19th century.

In the Brownian motion problem and all its variants—whether in physics, chemistry and biology or in finance, sociology and politics—one deals with a phenomenon (motion of the pollen particle, daily change in a stock market index) that is the outcome of many unpredictable and sometimes unobservable events (collisions with the particles of the surrounding liquid, buy/sell decisions of the single investor) which individually contribute a negligible amount to the observed phenomenon, but collectively lead to an observable effect. The individual events cannot sensibly be treated in detail, but their statistical properties may be known, and, in the end, it is these that determine the observed macroscopic behavior.

1.2 Random Walk on a Line

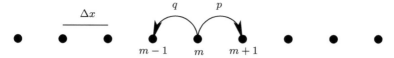

Fig. 1.1 Random walker on a 1-dimensional lattice of sites that are a fixed distance Δx apart. The walker jumps to the right with probability p and to the left with $q = 1 - p$

Another problem, which is closely related to Brownian motion and which we will examine in the next section, is that of a *random walker*. This concept was introduced into science by Karl Pearson (1857–1936) in a letter to Nature in 1905:

> *A man starts from a point 0 and walks l yards in a straight line: he then turns through any angle whatever and walks another l yards in a straight line. He repeats this process n times. I require the probability that after these n stretches he is at a distance between r and r + δr from his starting point 0.*

The solution was provided in the same volume of Nature by Lord Rayleigh (1842–1919), who pointed out that he had solved this problem 25 years earlier when studying the superposition of sound waves of equal frequency and amplitude but with random phases.

The random walker, however, is still with us today and we will now turn to it.

1.2 Random Walk on a Line

Let us assume that a walker can sit at regularly spaced positions along a line that are a distance Δx apart (see Fig. 1.1); so we can label the positions by the set of whole numbers, $\mathbb{Z}$. Furthermore we require the walker to be at position 0 at time 0. After fixed time intervals Δt the walker either jumps to the right with probability p or to the left with probability $q = 1 - p$; so we can work with discrete time points, labeled by the natural numbers including zero, $\mathbb{N}_0$.

Our aim is to answer the following question: What is the probability $p(m, N)$ that the walker will be at position m after N steps?

For $m < N$ there are many ways to start at 0, go through N jumps to nearest-neighbor sites, and end up at m. But since all these possibilities are independent of each other we have to add up their probabilities. For all these ways we know that the walker must have made $m + l$ jumps to the right and l jumps to the left; and since $m + 2l = N$, the walker must have made

- $(N + m)/2$ jumps to the right and
- $(N - m)/2$ jumps to the left.

So whenever N is even, so is m. Furthermore we know that the probability for the next jump is always p to the right and q to the left, irrespective of what the path of the walker up to that point was. The probability for a sequence of left and right jumps is the product of the probabilities of the individual jumps. Since the probability of the individual jumps does not depend on their position in the sequence, all paths starting at 0 and ending at m have the same overall probability. The probability for making exactly $(N+m)/2$ jumps to the right and exactly $(N-m)/2$ jumps to the left is

$$p^{\frac{1}{2}(N+m)} q^{\frac{1}{2}(N-m)}.$$

To finally get the answer to our question, we have to find out how many such paths there are. This is given by the number of ways to make $(N+m)/2$ out of N jumps to the right (and consequently $N - (N+m)/2 = (N-m)/2$ jumps to the left), where the order of the jumps does not matter (and repetitions are not allowed):

$$\frac{N!}{(\frac{N+m}{2})!(\frac{N-m}{2})!}.$$

The probability of being at position m after N jumps is therefore given as

$$p(m, N) = \frac{N!}{(\frac{N+m}{2})!(\frac{N-m}{2})!} p^{\frac{1}{2}(N+m)} (1-p)^{\frac{1}{2}(N-m)}, \tag{1.1}$$

which is the binomial distribution. If we know the probability distribution $p(m, N)$, we can calculate all the moments of m at fixed time N. Let us denote the number of jumps to the right as $r = (N+m)/2$ and write

$$p(m, N) = p_N(r) = \frac{N!}{r!(N-r)!} p^r q^{N-r} \tag{1.2}$$

and calculate the moments of $p_N(r)$. For this purpose we use the property of the binomial distribution that $p_N(r)$ is the coefficient of u^r in $(pu+q)^N$. With this trick it is easy, for instance, to convince ourselves that $p_N(r)$ is properly normalized to one

$$\sum_{r=0}^{N} p_N(r) = \left[\sum_{r=0}^{N} \binom{N}{r} u^r p^r q^{N-r}\right]_{u=1} = \left[(pu+q)^N\right]_{u=1} = 1. \tag{1.3}$$

The first moment or *expectation value* of r is:

$$\langle r \rangle = \sum_{r=0}^{N} r p_N(r)$$

$$= \left[\sum_{r=0}^{N} r \binom{N}{r} u^r p^r q^{N-r}\right]_{u=1} = \left[\sum_{r=0}^{N} \binom{N}{r} u \frac{d}{du}(u^r p^r q^{N-r})\right]_{u=1}$$

$$= \left[u \frac{d}{du} \sum_{r=0}^{N} \binom{N}{r} u^r p^r q^{N-r}\right]_{u=1} = \left[u \frac{d}{du}(pu+q)^N\right]_{u=1}$$

$$= \left[Nup(pu+q)^{N-1}\right]_{u=1}$$

1.2 Random Walk on a Line

leading to

$$E[r] \equiv \langle r \rangle = Np. \tag{1.4}$$

In the same manner, one can derive the following for the second moment:

$$E[r^2] \equiv \langle r^2 \rangle = \left[\left(u\frac{d}{du}\right)^2 (pu+q)^N\right]_{u=1} = Np + N(N-1)p^2. \tag{1.5}$$

From this one can calculate the *variance* or *second central moment*

$$\text{Var}[r] \equiv \sigma^2 := \langle (r - \langle r \rangle)^2 \rangle = \langle r^2 \rangle - \langle r \rangle^2 \tag{1.6}$$

of the distribution, which is a measure of the width of the distribution

$$\sigma^2 = Npq. \tag{1.7}$$

The relative width of the distribution

$$\frac{\sigma}{\langle r \rangle} = \sqrt{\frac{q}{p}} N^{-1/2} \tag{1.8}$$

goes to zero with increasing number of performed steps, N. Distributions with this property are called (strongly) *self-averaging*. This term can be understood in the following way: The outcome for r after N measurements has a statistical error of order σ. For self-averaging systems this error may be neglected relative to $\langle r \rangle$, if the system size (N) becomes large. In the large-N limit the system thus 'averages itself' and r behaves as if it was a non-random variable (with value $\langle r \rangle$). This self-averaging property is important, for instance, in statistical mechanics.

Figure 1.2 shows a plot of the binomial distribution for $N = 100$ and $p = 0.8$. As one can see, the distribution has a bell-shaped form with a maximum occurring around the average value $\langle r \rangle = Np = 80$, and for this choice of parameters it is almost symmetric around its maximum.

When we translate the results for the number of steps to the right back into the position of the random walker we get the following results

$$\langle m \rangle = 2N\left(p - \frac{1}{2}\right) \tag{1.9}$$

and

$$\langle m^2 \rangle = 4Np(1-p) + 4N^2\left(p - \frac{1}{2}\right)^2, \tag{1.10}$$

$$\sigma^2 = \langle m^2 \rangle - \langle m \rangle^2 = 4Npq. \tag{1.11}$$

In the case of symmetric jump rates, this reduces to

$$\langle m \rangle = 0 \quad \text{and} \quad \langle m^2 \rangle = N. \tag{1.12}$$

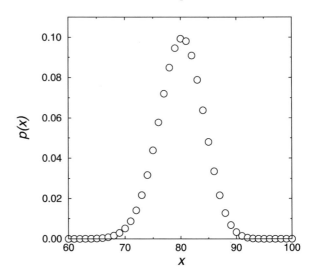

Fig. 1.2 Plot of the binomial distribution for a number of steps $N = 100$ and the probability of a jump to the right $p = 0.8$

This behavior, in which the square of the distance traveled is proportional to time, is called *free diffusion*.

1.2.1 From Binomial to Gaussian

The reader may be familiar with the description of particle diffusion in the context of partial differential equations, i.e., Fick's equation. To examine the relation between our jump process and Fickian diffusion, we will now study approximations of the binomial distribution which will be valid for large number of jumps ($N \to \infty$), i.e., for long times.

Assuming $N \gg 1$ we can use Stirling's formula to approximate the factorials in the binomial distribution

$$\ln N! = \left(N + \frac{1}{2}\right) \ln N - N + \frac{1}{2} \ln 2\pi + O(N^{-1}). \tag{1.13}$$

Using Stirling's formula, we get

$$\begin{aligned}\ln p(m, N) = &\left(N + \frac{1}{2}\right) \ln N - \left(\frac{N+m}{2} + \frac{1}{2}\right) \ln\left[\frac{N}{2}\left(1 + \frac{m}{N}\right)\right] \\ &- \left(\frac{N-m}{2} + \frac{1}{2}\right) \ln\left[\frac{N}{2}\left(1 - \frac{m}{N}\right)\right] \\ &+ \frac{N+m}{2} \ln p + \frac{N-m}{2} \ln q - \frac{1}{2} \ln 2\pi.\end{aligned}$$

1.2 Random Walk on a Line

Now we want to derive an approximation to the binomial distribution close to its maximum, which is also close to the expectation value $\langle m \rangle$. So let us write

$$m = \langle m \rangle + \delta m = 2Np - N + \delta m,$$

which leads to

$$\frac{N+m}{2} = Np + \frac{\delta m}{2} \quad \text{and} \quad \frac{N-m}{2} = Nq - \frac{\delta m}{2}.$$

Using these relations, we get

$$\ln p(m, N) = \left(N + \frac{1}{2}\right) \ln N - \frac{1}{2} \ln 2\pi$$
$$+ \left(Np + \frac{\delta m}{2}\right) \ln p + \left(Nq - \frac{\delta m}{2}\right) \ln q$$
$$- \left(Np + \frac{\delta m}{2} + \frac{1}{2}\right) \ln\left[Np\left(1 + \frac{\delta m}{2Np}\right)\right]$$
$$- \left(Nq - \frac{\delta m}{2} + \frac{1}{2}\right) \ln\left[Nq\left(1 - \frac{\delta m}{2Nq}\right)\right]$$
$$= -\frac{1}{2} \ln(2\pi Npq) - \left(Np + \frac{\delta m}{2} + \frac{1}{2}\right) \ln\left(1 + \frac{\delta m}{2Np}\right)$$
$$- \left(Nq - \frac{\delta m}{2} + \frac{1}{2}\right) \ln\left(1 - \frac{\delta m}{2Nq}\right).$$

Expanding the logarithm

$$\ln(1 \pm x) = \pm x - \frac{1}{2}x^2 + O(x^3)$$

yields

$$\ln p(m, N) \simeq -\frac{1}{2} \ln(2\pi Npq) - \frac{1}{2}\frac{(\delta m)^2}{4Npq} - \frac{\delta m(q-p)}{4Npq} + \frac{(\delta m)^2(p^2+q^2)}{16(Npq)^2}.$$

We should remember that the variance (squared width) of the binomial distribution was $\sigma^2 = 4Npq$. When we want to approximate the distribution in its center and up to fluctuations around the mean value of the order $(\delta m)^2 = O(\sigma^2)$, we find for the last terms in the above equation:

$$\frac{\delta m(q-p)}{4Npq} = O\left((Np)^{-1/2}\right) \quad \text{and} \quad \frac{(\delta m)^2(p^2+q^2)}{16(Npq)^2} = O\left((Np)^{-1}\right).$$

These terms can be neglected if $Np \to \infty$. We therefore finally obtain

$$p(m, N) \to \frac{2}{\sqrt{2\pi 4Npq}} \exp\left[-\frac{1}{2}\frac{(\delta m)^2}{4Npq}\right], \tag{1.14}$$

which is the Gaussian (C.F. Gauss (1777–1855)) or normal distribution. The factor 2 in front of the exponential comes from the fact that for fixed N (odd or even) only every other m (odd or even, respectively) has a non-zero probability, so $\Delta m = 2$.

This distribution is called 'normal' distribution because of its ubiquity. Whenever one adds up random variables, x_i, with finite first and second moments, so $\langle x_i \rangle < \infty$ and $\langle x_i^2 \rangle < \infty$ (in our case the jump distance of the random walker is such a variable with first moment $(p-q)\Delta x$ and second moment $(p+q)(\Delta x)^2 = (\Delta x)^2$), then the sum variable

$$S_N := \frac{1}{N}\sum_{i=1}^{N} x_i$$

is distributed according to a normal distribution for $N \to \infty$. This is the gist of the *central limit theorem*, to which we will return in the next chapter.

There are, however, also cases where either $\langle x_i^2 \rangle$ or even $\langle x_i \rangle$ does not exist. In these cases the limiting distribution of the sum variable is not a Gaussian distribution but a so-called *stable* or *Lévy* distribution, named after the mathematician Paul Lévy (1886–1971), who began to study these distributions in the 1930s. The Gaussian distribution is a special case of these stable distributions. We will discuss the properties of these distributions, which have become of increasing importance in all areas of application of the theory of stochastic processes, in Chap. 4.

We now want to leave the discrete description and perform a continuum limit. Let us write

$$x = m\Delta x, \quad \text{i.e.,} \quad \langle x \rangle = \langle m \rangle \Delta x,$$
$$t = N\Delta t, \tag{1.15}$$
$$D = 2pq\frac{(\Delta x)^2}{\Delta t},$$

so that we can interpret

$$p(m\Delta x, N\Delta t) = \frac{2\Delta x}{\sqrt{2\pi 2Dt}}\exp\left[-\frac{1}{2}\frac{(x-\langle x \rangle)^2}{2Dt}\right]$$

as the probability of finding our random walker in an interval of width $2\Delta x$ around a certain position x at time t. We now require that

$$\Delta x \to 0, \quad \Delta t \to 0, \quad \text{and} \quad 2pq\frac{(\Delta x)^2}{\Delta t} = D = \text{const.} \tag{1.16}$$

Here, D, with the units length^2/time, is called the *diffusion coefficient* of the walker. For the probability of finding our random walker in an interval of width dx around the position x, we get

$$p(x,t)dx = \frac{1}{\sqrt{2\pi 2Dt}}\exp\left[-\frac{1}{2}\frac{(x-\langle x \rangle)^2}{2Dt}\right]dx. \tag{1.17}$$

1.2 Random Walk on a Line

When we look closer at the definition of $\langle x \rangle$ above, we see that another assumption was buried in our limiting procedure:

$$\langle x \rangle(t) = \Delta x \langle m \rangle = 2\left(p - \frac{1}{2}\right) N \Delta x = 2\left(p - \frac{1}{2}\right) \frac{\Delta x}{\Delta t} t.$$

So our limiting procedure also has to include the requirement

$$\Delta x \to 0, \quad \Delta t \to 0 \quad \text{and} \quad \frac{2(p - \frac{1}{2})\Delta x}{\Delta t} = v = \text{const.} \quad (1.18)$$

As we have already discussed before, when $p = 1/2$ the average position of the walker stays at zero for all times and the velocity of the walker vanishes. Any asymmetry in the transition rates ($p \neq q$) produces a net velocity of the walker. However, when $v = 0$, we have $\langle x \rangle = 0$ and $\langle x^2 \rangle = 2Dt$. Finally, we can write down the probability density for the position of the random walker at time t,

$$p(x, t) = \frac{1}{\sqrt{2\pi 2Dt}} \exp\left[-\frac{1}{2} \frac{(x - vt)^2}{2Dt}\right], \quad (1.19)$$

with starting condition

$$p(x, 0) = \delta(x)$$

and boundary conditions

$$p(x, t) \xrightarrow{x \to \pm\infty} 0.$$

By substitution one can convince oneself that (1.19) is the solution of the following partial differential equation:

$$\frac{\partial}{\partial t} p(x, t) = -v \frac{\partial}{\partial x} p(x, t) + D \frac{\partial^2}{\partial x^2} p(x, t), \quad (1.20)$$

which is Fick's equation for diffusion in the presence of a constant drift velocity.

To close the loop, we now want to derive this evolution equation for the probability density starting from the discrete random walker. For this we have to rethink our treatment of the random walker from a slightly different perspective, the perspective of *rate equations*.

How does the probability of the discrete random walker being at position m at time N change in the next time interval Δt? Since our walker is supposed to perform one jump in every time interval Δt, we can write

$$p(m, N + 1) = p p(m - 1, N) + q p(m + 1, N). \quad (1.21)$$

The walker has to jump to position m either from the position to the left or to the right of m. This is an example of a *master equation* for a stochastic process. In the next chapter we will discuss for which types of stochastic processes this evolution equation is applicable.

In order to introduce the drift velocity, v, and the diffusion coefficient, D, into this equation let us rewrite the definition of D:

$$D = 2pq\frac{(\Delta x)^2}{\Delta t}$$

$$= (2p-1)(1-p)\frac{(\Delta x)^2}{\Delta t} + (1-p)\frac{(\Delta x)^2}{\Delta t}$$

$$= v(1-p)\Delta x + (1-p)\frac{(\Delta x)^2}{\Delta t}$$

$$= vq\Delta x + q\frac{(\Delta x)^2}{\Delta t}.$$

We therefore can write

$$q = (D - vq\Delta x)\frac{\Delta t}{(\Delta x)^2}, \tag{1.22}$$

$$p = (D + vp\Delta x)\frac{\Delta t}{(\Delta x)^2}. \tag{1.23}$$

Inserting this into (1.21) and subtracting $p(m, N)$ we get

$$\frac{p(m, N+1) - p(m, N)}{\Delta t} = \frac{vp}{\Delta x}p(m-1, N) - \frac{vq}{\Delta x}p(m+1, N)$$

$$+ D\frac{p(m+1, N) - 2p(m, N) + p(m-1, N)}{(\Delta x)^2}$$

$$+ \left(\frac{2D}{(\Delta x)^2} - \frac{1}{\Delta t}\right)p(m, N)$$

and from this

$$\frac{p(m, N+1) - p(m, N)}{\Delta t} = -vp\frac{p(m, N) - p(m-1, N)}{\Delta x}$$

$$- vq\frac{p(m+1, N) - p(m, N)}{\Delta x}$$

$$+ D\frac{p(m+1, N) - 2p(m, N) + p(m-1, N)}{(\Delta x)^2}$$

$$+ \left(\frac{2D}{(\Delta x)^2} - \frac{1}{\Delta t} + \frac{vp}{\Delta x} - \frac{vq}{\Delta x}\right)p(m, N).$$

Reinserting the definitions of v and D into the last term, it is easy to show that it identically vanishes. When we now perform the continuum limit of this equation

1.2 Random Walk on a Line

keeping v and D constant, we again arrive at (1.20). The Fickian diffusion equation therefore can be derived via a prescribed limiting procedure from the rate equation (1.21). Most important is the unfamiliar requirement $(\Delta x)^2/\Delta t = \text{const}$ which does not occur in deterministic motion and which captures the diffusion behavior, $x^2 \propto t$, of the random walker.

1.2.2 From Binomial to Poisson

Let us now turn back to analyzing the limiting behavior of the binomial distribution. The Gaussian distribution is not the only limiting distribution we can derive from it. In order to derive the Gaussian distribution, we had to require that $Np \to \infty$ for $N \to \infty$. Let us ask the question of what the limiting distribution is for

$$N \to \infty, \quad p \to 0, \quad Np = \text{const}.$$

Again we are only interested in the behavior of the distribution close to its maximum and expectation value, i.e., for $r \approx Np$; however, now $r \ll N$, and

$$p_N(r) = \frac{N!}{r!(N-r)!} p^r q^{N-r}$$

$$= \frac{N(N-1)\cdots(N-r+1)(N-r)!}{r!(N-r)!} p^r (1-p)^{N-r}$$

$$\approx \frac{(Np)^r}{r!} (1-p)^N,$$

where we have approximated all terms $(N-1)$ up to $(N-r)$ by N. So we arrive at

$$\lim_{\substack{N \to \infty, p \to 0 \\ Np=\text{const}}} \frac{(Np)^r}{r!} \left(1 - \frac{Np}{N}\right)^N = \frac{\langle r \rangle^r}{r!} e^{-\langle r \rangle}. \quad (1.24)$$

This is the Poisson distribution which is completely characterized by its first moment. To compare the two limiting regimes for the binomial distributions which we have derived, take a look at Figs. 1.3 and 1.4.

For the first figure we have chosen $N = 1000$ and $p = 0.8$, so that $\langle r \rangle = Np = 800$. The binomial distribution and the Gaussian distribution of the same mean and width are already indistinguishable. The Poisson distribution with the same mean is much broader and not a valid approximation for the binomial distribution in this parameter range. The situation is reversed in Fig. 1.4, where again $N = 1000$ but now $p = 0.01$, so that $\langle r \rangle = Np = 10 \ll N$. Now the Poisson distribution is the better approximation to the binomial distribution, capturing especially the fact that for these parameters the distribution is not symmetric around the maximum, as a comparison with the Gaussian distribution (which is symmetric by definition)

Fig. 1.3 Plot of the binomial distribution for a number of steps $N = 1000$ and the probability of a jump to the right $p = 0.8$ (*open circles*). This is compared with the Gaussian approximation with the same mean and width (*solid curve*) and the Poisson distribution with the same mean (*dashed curve*)

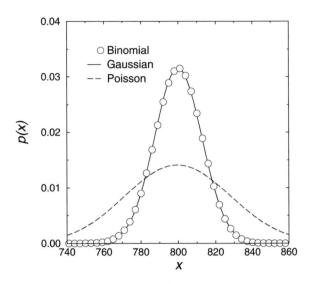

Fig. 1.4 Plot of the binomial distribution for a number of steps $N = 1000$ and the probability of a jump to the right $p = 0.01$ (*open circles*). This is compared with the Gaussian approximation with the same mean and width (*solid curve*) and the Poisson distribution with the same mean (*dashed curve*)

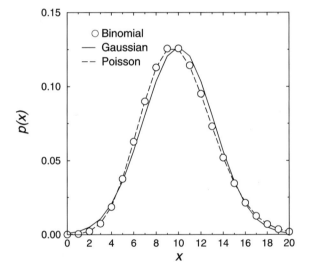

shows. For large r the binomial and Poisson distributions show a larger probability than the Gaussian distribution, whereas the situation is reversed for r close to zero.

To quantify such asymmetry in a distribution, one must look at higher moments. For the Poisson distribution one can easily derive the following recurrence relation between the moments,

$$\langle r^n \rangle = \lambda \left(\frac{\mathrm{d}}{\mathrm{d}\lambda} + 1 \right) \langle r^{n-1} \rangle \quad (n \geq 1), \tag{1.25}$$

1.2 Random Walk on a Line

where $\lambda = Np = \langle r \rangle$. So we have for the first four moments

$$\langle r^0 \rangle = \sum_{r=0}^{\infty} p_N(r) = 1,$$

$$\langle r^1 \rangle = \lambda,$$

$$\langle r^2 \rangle = \lambda(1+\lambda) \Rightarrow \sigma^2 = \lambda,$$

$$\langle r^3 \rangle = \lambda^3 + 3\lambda^2 + \lambda,$$

$$\langle r^4 \rangle = \lambda^4 + 6\lambda^3 + 7\lambda^2 + \lambda.$$

We saw that the second central moment, σ^2, is a measure of the width of the distribution and in a similar way the normalized third central moment, or *skewness*, is used to quantify the asymmetry of the distribution,

$$\hat{\kappa}_3 = \frac{\langle (r - \langle r \rangle)^3 \rangle}{\sigma^3} = \lambda^{-1/2}. \tag{1.26}$$

The normalized fourth central moment, or *kurtosis*, measures the 'fatness' of the distribution or the excess probability in the tails,

$$\hat{\kappa}_4 = \frac{\langle (r - \langle r \rangle)^4 \rangle - 3\langle (r - \langle r \rangle)^2 \rangle^2}{\sigma^4} = \lambda^{-1}. \tag{1.27}$$

Excess has to be understood in relation to the Gaussian distribution,

$$p_G(x) = \frac{1}{\sqrt{2\pi\sigma^2}} \exp\left[-\frac{(x - \langle x \rangle)^2}{2\sigma^2}\right], \tag{1.28}$$

for which due to symmetry all odd central moments $\langle (x - \langle x \rangle)^n \rangle$ with $n \geq 3$ are zero. Using the substitution $y = (x - \langle x \rangle)^2/2\sigma^2$ we find for the even central moments

$$\langle (x - \langle x \rangle)^{2n} \rangle_G = \int_{-\infty}^{+\infty} (x - \langle x \rangle)^{2n} p_G(x) dx = \frac{(2\sigma^2)^n}{\sqrt{\pi}} \int_0^{\infty} y^{(n+1/2)-1} e^{-y} dy$$

$$= \frac{(2\sigma^2)^n}{\sqrt{\pi}} \Gamma\left(n + \frac{1}{2}\right), \tag{1.29}$$

where $\Gamma(x)$ is the Gamma function [1]. From this one easily derives that $\hat{\kappa}_4$ is zero for the Gaussian distribution. In fact, for $\lambda = \langle r \rangle \to \infty$, we again recover the Gaussian distribution from the Poisson distribution.

1.2.3 Log–Normal Distribution

As a last example of important distribution functions, we want to look at the so-called log–normal distribution, which is an example of a strongly asymmetric distribution. Starting from the Gaussian distribution $p_G(x)$ and focusing for $\langle x \rangle \gg 1$ on $x > 0$, we can define

$$x = \ln y, \quad \langle x \rangle = \ln y_0, \quad dx = \frac{dy}{y} \qquad (1.30)$$

and get the log–normal distribution

$$p_G(x)dx = p_{LN}(y)dy = \frac{1}{\sqrt{2\pi\sigma^2}} \exp\left[-\frac{[\ln(y/y_0)]^2}{2\sigma^2}\right]\frac{dy}{y}. \qquad (1.31)$$

The log–normal distribution is normalized in the following way,

$$\int_0^\infty p_{LN}(y)dy = \int_{-\infty}^{+\infty} p_G(x)dx = 1. \qquad (1.32)$$

The moments are given by

$$\langle y^n \rangle = \int_0^\infty y^n p_{LN}(y)dy$$

$$= \int_{-\infty}^{+\infty} e^{nx} p_G(x)dx$$

$$= y_0^n e^{n^2\sigma^2/2} \int_{-\infty}^{+\infty} p_G(x - n\sigma^2)dx,$$

leading to

$$\langle y^n \rangle = y_0^n e^{n^2\sigma^2/2}. \qquad (1.33)$$

The maximum of the distribution is at

$$y_{max} = y_0 e^{-\sigma^2}. \qquad (1.34)$$

A further quantity to characterize the location of a distribution is its median, which is the point where the cumulative probability is equal to one-half,

$$F(y_{med}) = \int_0^{y_{med}} p_{LN}(y)dy = \frac{1}{2}. \qquad (1.35)$$

With $x = \ln y$ and $x_{med} = \ln y_{med}$, this is equivalent to

$$\int_{-\infty}^{x_{med}} p_G(x)dx = \frac{1}{2},$$

or, because of the symmetry of the Gaussian distribution,

$$x_{med} = \langle x \rangle = \ln y_0 \Rightarrow y_{med} = y_0. \qquad (1.36)$$

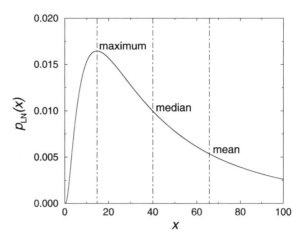

Fig. 1.5 Plot of the log–normal distribution, with *vertical dashed lines* indicating the positions of the maximum, median and mean of the distribution

We therefore find the relative order

$$y_{\max} = y_0 e^{-\sigma^2} < y_{\text{med}} = y_0 < \langle y \rangle = y_0 e^{\sigma^2/2}.$$

In Fig. 1.5 we show the log–normal distribution for $y_0 = 40$ and $\sigma = 1$. This distribution shows a pronounced skewness around the maximum and a fat-tailed behavior for large argument.

1.3 Further Reading

This section was split into two parts. The first one gave a short introduction to the history of stochastic processes and the second one was devoted to the discussion of the one-dimensional random walk.

Section 1.1

- The history of probability theory and the theory of stochastic processes, as we have shortly sketched it, is nicely described in the review article by Montroll and Shlesinger [145]. Those interested in more background information may want to have a look at the books by Pearson [165] or Schneider [183] (in German).

Section 1.2

- The treatment of the random walk on the one-dimensional lattice can be found in many textbooks and review articles on stochastic processes, noticeably that of Chandrasekhar [26]. An elucidating treatment in the context of One-Step or Birth-and-Death processes can be found in the book by van Kampen [101]. Further literature for the random walk can be found at the end of Chap. 3.

- The Fickian diffusion equation is a special case of a Fokker-Planck equation, which we will discuss in the second part of Chap. 2. Its applications in the natural sciences are widespread. All transport phenomena obeying the diffusion approximation $\Delta x^2/\Delta t = \text{const.}$, can be modeled by this equation. Examples include energy transport (following a temperature gradient), mass transport (following a density gradient) and interdiffusion (following a concentration gradient).
- The Gaussian limit of probability distributions is of central importance in statistical physics and is therefore discussed in many textbooks, e.g. in Reichl [176].
- The log–normal distribution is of importance in finance [18, 217], as will be seen in Chap. 5. A detailed account of its properties as well as further applications may be found in [99].

Chapter 2
A Brief Survey of the Mathematics of Probability Theory

In this chapter we want to give a short introduction to the mathematical structure of probability theory and stochastic processes. We do not intend this to be a mathematical treatise, but will strive for mathematical exactness in the definitions and theorems presented. For proof of the theorems the reader is referred to the relevant mathematical literature. We will also try to motivate the mathematical structures from intuition and *empirical wisdom*, drawing on examples from physics to show how the mathematical objects are realized in actual situations. The chapter presents some aspects in a more general setting than will be needed later on in the book. In this way, we hope to enable the reader to put problems outside the realm of those we chose for this book into a helpful mathematical form. We also hope to provide a starting point from which the mathematical literature on stochastic processes in general and on mathematical finance in particular is more easily accessible.

If you are not mathematically inclined it may be a good idea, on first reading, to skip over the first parts and start with the subsection on statistical independence and conditional probabilities. The first part of the chapter may later be consulted to see the connection between the abstract mathematical formulation and the description with probability densities over $\mathbb{R}$ or $\mathbb{R}^d$ one usually encounters in physics texts.

2.1 Some Basics of Probability Theory

In the introductory example in the last chapter we studied the properties of a process that involved some randomness in each single step: jumping to the right with probability p and to the left with probability $q = 1 - p$. Let us call the outcome of an experiment involving randomness an *event*. In general, the number of possible outcomes of such an experiment will be larger than the two possibilities above. The collection of all possible outcomes of an experiment involving randomness is called the *sample space*. There are certain events among these that are indecomposable and are called *simple events*.

As an example, let us look at the finite and discrete sample space corresponding to the experiment of rolling a die:

- The sample space is $\Omega = \{1, 2, 3, 4, 5, 6\}$, the collection of all possible outcomes of rolling a die.
- The simple (indecomposable) events are $\{1\}, \{2\}, \{3\}, \{4\}, \{5\}, \{6\}$, corresponding to rolling a one, two, etc.
- Examples for possible events are:
  - A_1: the result is an even number.
  - A_2: the result is larger than three.

We see that $A_1 = \{2, 4, 6\}$ and $A_2 = \{4, 5, 6\}$ are unions of the simple events; $A_1 = \{2\} \cup \{4\} \cup \{6\}$ and $A_2 = \{4\} \cup \{5\} \cup \{6\}$. For both of them we can also define their complement, i.e. A_1': 'rolling an odd number' and A_2': 'rolling a number smaller than or equal to three'. Clearly $A_1' = \Omega \setminus A_1$ and $A_2' = \Omega \setminus A_2$. We can also define the event 'rolling an even number larger than three', which is the simultaneous occurrence of A_1 and A_2, $A_1 \cap A_2 = \{4, 6\}$, and is given by the intersection of the sets A_1 and A_2. Equally, we can define the event 'rolling a number larger than three or rolling an even number', which is the union of A_1 and A_2, $A_1 \cup A_2 = \{2, 4, 5, 6\}$.

In order to find out what the probabilities for all these events are, we first have to know what probabilities to assign to the simple events. Within the standard framework of probability theory (but see Sect. 2.1.2) these cannot be deduced but have to be postulated. As with the choice of the sample space, the probabilities of the simple events are input into a probabilistic treatment of a given problem. They are either obtained from empirical knowledge, by intuition or on the basis of some different theoretical description of the problem. Probability theory and the theory of stochastic processes are then only concerned with rules to manipulate these basic probabilities in order to calculate the probabilities of the more complex events under consideration [51].

In the case of the die our intuition and our hope for a fair game lead us to assign a probability of $p = 1/6$ for all the six possible outcomes of rolling a single die once. From this we would deduce $p(A_1) = p(A_2) = 1/2$, $p(A_1 \cap A_2) = 1/3$, $p(A_1 \cup A_2) = 2/3$ and, of course, $p(\Omega) = 1$.

Most of the situations we will encounter, however, are not describable within the simple framework of a discrete and finite sample space. Consider, for instance, the probabilistic treatment of a velocity component of a particle in an ideal gas. A priori we would allow any real number for this quantity, $v_i \in \mathbb{R}$, leading to an infinite continuum of sample points. What is the adequate mathematical structure to deal with events and their probabilities in this case? The axiomatic answer to this question was formulated by A. Kolmogorov (1903–1987) [107] in 1933.

2.1.1 Probability Spaces and Random Variables

Following Kolmogorov, probability theory can be formalized based on three axioms for a (real) function p over a set Ω of possible outcomes A of an experiment involving randomness:

2.1 Some Basics of Probability Theory

- *non-negativity:* $p(A) \geq 0$;
- *normalization:* $p(\Omega) = 1$;
- *additivity:* $p(A \cup B) = p(A) + p(B)$ when $A \cap B = \emptyset$, where $\emptyset$ is the empty set.

These basic requirements are ideally fulfilled by identifying probabilities as measures over some sample space.

Definition 2.1 The sample space is a probability space, i.e., a measure space $(\Omega, \mathcal{F}, \mu)$, given through a set Ω, a σ-algebra $\mathcal{F}$ of subsets of Ω and a measure μ on $\mathcal{F}$. An event is a set $A \in \mathcal{F}$ and the probability for this event is the measure $\mu(A)$. The measure has the property: $\mu(\Omega) = 1$.

A σ-*algebra* has the following properties that correspond to the intuitively required properties of events, as discussed for the rolling of a die:

- $\Omega \in \mathcal{F}$ and $\emptyset \in \mathcal{F}$.
- $A \in \mathcal{F} \Rightarrow A' = \Omega \setminus A \in \mathcal{F}$.
- $A, B \in \mathcal{F} \Rightarrow A \cap B \in \mathcal{F}$.
- $A_i \in \mathcal{F}, i = 1, 2, \ldots \Rightarrow \bigcup_{i=1}^{\infty} A_i \in \mathcal{F}$; if furthermore $A_i \cap A_j = \emptyset$ for $i \neq j$ then $\mu(\bigcup_{i=1}^{\infty} A_i) = \sum_{i=1}^{\infty} \mu(A_i)$.

The role of the simple events now has to be taken over by a collection of *simple sets*, which includes the empty set (needed for an algebraic formalization of our simple example with the die) and the whole space, and which generates a σ-algebra through the operations of building the complement, finite intersections and countable unions. Again this does not specify the simple events uniquely. The choice of simple events, and the corresponding σ-algebra, is an input into the theory and depends on our prior information about the problem that we want to treat probabilistically. The most simple σ-algebra definable on Ω is $\mathcal{F}_0 = \{\emptyset, \Omega\}$, where all we can tell is that the event Ω occurs. We do not know of any subsets of Ω and since Ω occurs with probability one, there is also no randomness in our experiment. It is deterministic and not amenable to probabilistic treatment. If we are able to distinguish one subset $A \in \Omega$ in our experiment, we can define the σ-algebra $\mathcal{F}_1 = \{\emptyset, A, A', \Omega\}$. To complete the construction of a probability space, however, we also have to be able to assign some probability to the set A, i.e., to measure A.

This is the same situation as with the construction of the Lebesgue integral over $\mathbb{R}$. We have an intuitive feeling to take the value $b - a$ as the length, that is, as the measure, of the open interval (a, b) (the interval could also be closed or half open). From this starting point, integration can be defined on the measure space $(\mathbb{R}, \mathcal{B}, \mu_L)$, where $\mathcal{B}$ is the Borel algebra, i.e., the σ-algebra generated by the open intervals, and μ_L is the Lebesgue measure, which is $\mu_L(A) = b - a$ for $A = (a, b)$. This is not a probability space, since $\mu_L(\mathbb{R}) = \infty$, but can be made into one by a different choice of the measure. We can now proceed to define what is meant by a *random variable (stochastic variable)*.

Definition 2.2 A random variable, f, is a μ-integrable function on the probability space $(\Omega, \mathcal{F}, \mu)$. Its value, when the event A occurs, is

$$f(A) = \int_A f(\omega) d\mu(\omega).$$

The expectation value of the random variable f is defined as

$$\langle f \rangle \equiv \mathrm{E}[f] := \int_\Omega f(\omega) d\mu(\omega). \tag{2.1}$$

This means that f is in $L^1(\Omega, \mathcal{F}, \mu)$, i.e., in the space of μ-integrable functions on Ω. If f is also in $L^n(\Omega, \mathcal{F}, \mu)$, we can define the nth moment of f as

$$\langle f^n \rangle \equiv \mathrm{E}[f^n] := \int_\Omega f^n(\omega) d\mu(\omega). \tag{2.2}$$

In this way we can formalize the meaning of, e.g., expectation value, variance, skewness and kurtosis already used in the introductory example.

To familiarize ourselves with these concepts, let us translate two examples from physics into this language. The first example is drawn from statistical physics (of course). Consider a system of N indistinguishable particles (of mass m) in a volume $V \subseteq \mathbb{R}^3$, interacting through a pair potential U. The Hamiltonian of this system is

$$H(x) = \sum_{i=1}^{N} \frac{p_i^2}{2m} + \sum_{i=1}^{N} \sum_{j>i} U(r_i, r_j), \tag{2.3}$$

where $x = (r_1, \ldots, r_N, p_1, \ldots, p_N)$ stands for the collection of $3N$ spatial coordinates and $3N$ momenta specifying the state of the system. Furthermore, let the system be in contact with a heat bath of temperature T. In statistical physics this system in its specific state is said to be in the canonical ensemble, specified by prescribing the values of the macroscopic variables (N, V, T). All observables of this thermodynamic ensemble are then given by integrable functions on the probability space $(\Omega = V^N \otimes \mathbb{R}^{3N}, \mathcal{B}, \mu)$, where $\mathcal{B}$ is the Borel algebra on Ω and the measure μ is given by the Gibbs measure

$$d\mu(x) = \frac{1}{Z} \exp\left[-\frac{H(x)}{k_B T}\right] \frac{d^{6N} x}{N! h^{3N}}. \tag{2.4}$$

Here k_B is the Boltzmann constant, h is the Planck constant, and we have written $d^{6N} x$ for the standard Lebesgue measure $d\mu_L(x)$. The $N!$ factor accounts for the indistinguishability of the particles and h^{3N} is the quantum-mechanical minimal phase space volume. The normalization constant Z is the canonical partition function of the system,

$$Z = \int_\Omega \exp\left[-\frac{H(x)}{k_B T}\right] \frac{d^{6N} x}{N! h^{3N}}. \tag{2.5}$$

2.1 Some Basics of Probability Theory

From this we can calculate, for instance, the mean energy of the system as

$$\langle E \rangle \equiv \mathrm{E}[H] = \int_{\Omega} H(x) \mathrm{d}\mu(x), \tag{2.6}$$

and the specific heat at constant volume as

$$C_V := \left(\frac{\partial \langle E \rangle}{\partial T}\right)_V = \frac{1}{k_\mathrm{B} T^2} \left(\mathrm{E}[H^2] - \mathrm{E}^2[H]\right). \tag{2.7}$$

Our second example for the occurrence of such probability measures in physics is non-relativistic quantum mechanics. Consider the stationary solution of the one-dimensional Schrödinger equation

$$\left[-\frac{\hbar^2}{2m} \frac{\mathrm{d}^2}{\mathrm{d}x^2} + U(x) - E\right] \psi(x) = 0. \tag{2.8}$$

The solution $\psi(x)$ of this equation is normalized in the L^2-norm

$$\int_{-\infty}^{+\infty} |\psi(x)|^2 \mathrm{d}x = 1, \tag{2.9}$$

which means that $|\psi(x)|^2 \mathrm{d}x$ is a probability measure. In quantum mechanics we interpret it as the probability of finding a quantum particle in the interval $(x, x+\mathrm{d}x)$. Observables (operators), which are only functions of the position coordinates, have expectation values

$$\langle f(x) \rangle = \langle \psi | f | \psi \rangle = \int_{-\infty}^{+\infty} f(x) |\psi(x)|^2 \mathrm{d}x = \mathrm{E}[f]. \tag{2.10}$$

For the particle momentum and functions thereof we can make the same argument in the momentum representation.

In quantum mechanics, every solution of the Schrödinger equation defines a probability space $(\mathbb{R}, \mathcal{B}, |\psi(x)|^2 \mathrm{d}x)$, where $\mathcal{B}$ is the Borel algebra on $\mathbb{R}$. We will see in Chap. 3 how the connection between quantum mechanics and probability theory and stochastic processes can be further expanded.

2.1.2 Probability Theory and Logic

Our introduction to probability theory followed the axiomatization given to this field by Kolmogorov. In this form it deals with sets of random events and measure theory on these sets. This builds on basic intuition and reflects one aspect of a definition of probability exposed in the ground breaking work on probability theory by

P.S. Laplace (1749–1827), *Théorie Analytique des Probabilités* (1812). After [93] we quote from this work:

> The probability for an event is the ratio of the number of cases favorable to it to the number of all cases possible, when nothing leads us to expect that any one of these cases should occur more than any other, which renders them, for us, equally possible.

The first part of this statement tells us that in an experiment involving randomness we get an approximation for the probability of an event by measuring its frequency of occurrence. This part is axiomatized in Kolmogorov's theory. The second part of the sentence evoking our expectations based on prior information ("...nothing leads us to expect...") is ignored in the standard probability framework, although, considering the aforementioned 17th century gentleman Chevalier de Méré interested in the odds at gambling, the question of rational expectations was the starting point of probability theory. One consequence of this reduction in scope is the fact that we have to assign probabilities to simple events from considerations lying outside the scope of Kolmogorov probability theory as discussed above.

Extending the scope of probability theory is, however, possible when one sees it as a form of applied logic. The mathematical formalization of this approach, which accounts for the second part of the quoted sentence by Laplace, was mostly advanced by E.T. Jaynes (1922–1998) [93]. Deductive logic derives conclusions about the truth value of propositions based on their relations to other propositions. Given two propositions, A and B, and the information "if A is true then B is true" we can conclude:

- if we find A to be true then B must be true;
- if we find B to be false then A must be false.

These are the so-called strong *syllogisms* (word of Greek origin, meaning deductive reasoning from the general to the particular). Drawing on the concept of strong syllogism, Jaynes argues that our everyday reasoning and evaluation of situations—as well as the reasoning followed by scientists in their work—is based on exploring weaker syllogisms like, e.g.,

- if B is true then A becomes more plausible;
- if A is false then B becomes less plausible.

While deductive reasoning based on strong syllogisms deals with propositions which are either true (1) or false (0), Jaynes argues that probability theory can be based on an extended logic dealing with the plausibilities of propositions. The mathematical formalization of the plausibility of statements, i.e., of a rational expectation that they are true, is then based on the following 'desiderata':

1. Degrees of plausibility are represented by real numbers (which can be chosen to lie between zero and one).
2. The assignment of plausibilities to propositions agrees qualitatively with common sense.
3. The assignment of plausibilities is done consistently, i.e.:
    - Alternative routes to assign a plausibility to a proposition lead to the same numerical value.

2.1 Some Basics of Probability Theory

- No prior information is ignored.
- Equivalent states of knowledge lead to equivalent plausibility assignments.

From these requirements—which, in a sense, incorporate common sense or a rational way of human thinking—the rules of probability theory can be derived. We will employ the notation of Boolean algebra as it is also used to perform deductive reasoning:

- Let A and B be propositions.
- Then A' is the denial (complement) of A,
- AB is the conjunction: A and B,
- $A + B$ is the disjunction: A or B, and
- $A|B$ (A given B) indicates the plausibility that A is true, given that B is true.

Then there exists a function $0 \leq p(A) \leq 1$ of the plausibility of statements—their probability to be true—which obeys the following rules:

$$p(AB|C) = p(A|BC)p(B|C)$$
$$= p(B|AC)p(A|C) \quad \text{(product rule)}, \qquad (2.11)$$

$$p(A|C) + p(A'|C) = 1 \quad \text{(sum rule)}. \qquad (2.12)$$

From this an extended sum rule for $A + B|C$ can be derived. To this end, we note that the denial of $A + B$ is $A'B'$. Applying then the sum and product rules repeatedly, we get

$$\begin{aligned} p(A + B|C) &= 1 - p(A'B'|C) = 1 - p(A'|B'C)p(B'|C) \\ &= 1 - \left[1 - p(A|B'C)\right]p(B'|C) \\ &= p(B|C) + p(A|B'C)p(B'|C) \\ &= p(B|C) + p(B'|AC)p(A|C) \\ &= p(B|C) + \left[1 - p(B|AC)\right]p(A|C), \end{aligned}$$

and so

$$p(A + B|C) = p(A|C) + p(B|C) - p(AB|C). \qquad (2.13)$$

One way to visualize the extended sum rule is to think of probability in terms of statements about sets of events (see Fig. 2.1), which lies at the heart of Kolmogorov's axiomatization of probability theory by way of measure theory. However, this offers only one possible way of visualizing the logical statement included in the extended sum rule and the logic framework employed here goes beyond what can be derived within the measure theoretic framework.

To exemplify this let us look at a given prior information B and some propositions A_1, A_2, A_3. By applying the extended sum rule we can derive [93]

$$p(A_1 + A_2 + A_3|B) = p(A_1 + A_2|B) + p(A_3|B) - p(A_1A_3 + A_2A_3|B) \quad (2.14)$$

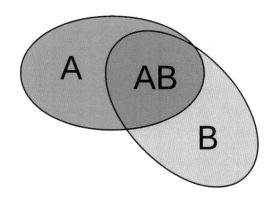

Fig. 2.1 Venn diagram for two not mutually exclusive propositions A and B. The disjunction $A + B$ is given by the union of both sets representing A and B, and the plausibility of the union is obtained from the extended sum rule: $p(A + B) = p(A) + p(B) - p(AB)$, where AB is the overlap (conjunction) between the sets

where we considered $A_1 + A_2$ as one proposition. Applying the extended sum rule once more to this result one obtains

$$p(A_1 + A_2 + A_3|B) = p(A_1|B) + p(A_2|B) + p(A_3|B)$$
$$- p(A_1 A_2|B) - p(A_1 A_3|B) - p(A_2 A_3|B)$$
$$+ p(A_1 A_2 A_3|B). \quad (2.15)$$

When we now assume that the propositions are mutually exclusive given the prior information B, i.e., $p(A_i A_j|B) \sim \delta_{ij}$ (and similarly for $p(A_1 A_2 A_3|B)$), we have

$$p(A_1 + A_2 + A_3|B) = p(A_1|B) + p(A_2|B) + p(A_3|B). \quad (2.16)$$

This is easily extended to N mutually exclusive propositions

$$p(A_1 + A_2 + \cdots + A_N|B) = \sum_{i=1}^{N} p(A_i|B). \quad (2.17)$$

If we furthermore know that these N propositions are exhaustive given the prior information B, we must have

$$\sum_{i=1}^{N} p(A_i|B) = 1. \quad (2.18)$$

If we now assume that information B is indifferent about the truth values of propositions $A_1, \ldots, A_N$, the only consistent (see desideratum number 3 above) assignment of probabilities for the different propositions A_i is

$$p(A_i|B) = \frac{1}{N} \quad (i = 1, \ldots, N). \quad (2.19)$$

Thus, application of the extended sum rule and consistency requirements in assigning probabilities (plausibilities) is sufficient to derive the probability values we have

2.1 Some Basics of Probability Theory

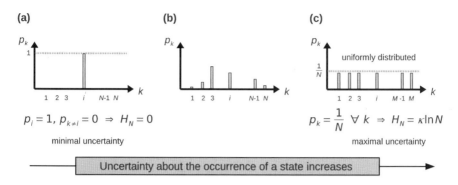

Fig. 2.2 Three possible cases of probabilities p_k for a system with $k = 1, \ldots, N$ states: (**a**) $p_i = 1$, $p_k = 0$ $\forall k \neq i$ ("minimal uncertainty" about the occurrence of a state), (**b**) $p_k \geq 0$ $\forall k$ ("intermediate uncertainty"), (**c**) $p_k = 1/N$ $\forall k$ ("maximal uncertainty")

to assign to propositions which are equally plausible given our prior information. For the general case, however, we need a further concept, the maximum entropy approach [93].

Maximum Entropy Approach

Let us start our argument with the observation that every additional prior information about a system increases the information content of the probabilities we assign to propositions given this prior knowledge. In order to get a qualitative feeling about the information contained in the probabilities, let us consider the following cases for a system with a discrete number of propositions (see Fig. 2.2) which we will call states in the following, anticipating the application to statistical physics. Imagine the system to occupy only one state with probability one. Then all about the system is known; the information is complete and the uncertainty about the occurrence of a particular state vanishes. On the other hand, if the system adopts all states with equal probability, the lack of knowledge about the actual state of the system culminates. Between these two extremes of minimum or maximum uncertainty there are many other distributions conceivable depending on the prior information which can be supplied for the system. To quantify this information content we assume the following (very much in line with the desiderata for probabilities discussed above):

- There exists a real-valued function $H_N(p_1, \ldots, p_N)$ quantifying the amount of uncertainty connected with a given probability assignment.
- H_N is a continuous function of the p_i, so that small changes in the p_i correspond to small changes in the information content.
- The function

$$h(N) = H_N\left(\frac{1}{N}, \ldots, \frac{1}{N}\right)$$

is monotonically increasing with N. That is, the more equally likely states there are, the larger is the amount of uncertainty.
- H_N has to be consistent, i.e., every way to work out its value for a given set of p_i has to lead to the same answer.

It turns out [93, 185] that there is only one function fulfilling these requirements which can be written as[1]

$$H_N(p_1,\ldots,p_N) = -\kappa \sum_{i=1}^{N} p_i \ln p_i \quad (\kappa > 0), \tag{2.20}$$

where κ is a constant. This function is called 'information entropy' by Shannon [185]. We will see in the application to statistical mechanics that it can be linked to the thermodynamic entropy (whence its name). The maximally unbiased assignment of probabilities given all prior information can now be done by maximizing H_N under the constraints imposed by the prior information. This prescription is called 'maximum entropy principle'.

Let us exemplify this principle for the case where we only know that the states (propositions) are exhaustive. So the only prior information consists in the normalization condition of the probabilities, i.e., in

$$\sum_{i=1}^{N} p_i = 1.$$

To find the maximum of H_N we employ the method of Lagrange multipliers. Introducing a new variable (Lagrange multiplier) α we obtain

$$\frac{\partial H_N}{\partial p_k} = \frac{\partial}{\partial p_k}\left\{-\kappa \sum_{i=1}^{N} p_i \ln p_i - \alpha\left(\sum_{i=1}^{N} p_i - 1\right)\right\} = 0,$$

leading to

$$p_k = e^{-(1+\alpha/\kappa)} = \text{constant } \forall k. \tag{2.21}$$

Inserting this result into the normalization condition allows for the determination of α and yields the uniform distribution $p_k = 1/N$ for all k, as derived before.

Maximum Entropy Approach to Statistical Physics

We want to discuss the application of Jaynes maximum entropy principle in the general case for the example of statistical physics. Here the equilibrium probabilities

---

[1] Actually when information (or lack thereof) is measured in bits, the logarithm to basis 2 instead of the natural logarithm needs to be used, but this just leads to a constant shift in (2.20).

2.1 Some Basics of Probability Theory

specify the weight by which each microstate contributes to the equilibrium properties of the system. These probabilities contain all available macroscopic knowledge about the system and should be designed in such a way that they reflect this information. In the case of the isolated system (microcanonical ensemble) this information only consists of the normalization condition of the probabilities, since the thermodynamic variables determining the macroscopic state of the system—the total (internal) energy E, the volume V, and the number of particles N—are all fixed, leading to the uniform probability distribution discussed above (cf. (2.21)). However, if the system is surrounded by a heat bath (canonical ensemble), its internal energy will no longer be constant, but fluctuate around the ensemble average of the Hamiltonian. Therefore measuring this mean value provides additional information which should be incorporated into the determination of the probabilities.

The measure of uncertainty employed in statistical physics, the *statistical entropy* [91, 92], is proportional to the information entropy H_N of (2.20). Suppose we have a physical system with a discrete and finite state space composed of the microstates $k = 1, \ldots, N$ that occur with probability p_k. Let each microstate k be a (non-degenerate) energy eigenstate with eigenvalue E_k. The statistical entropy of the system is then given by

$$S(p_1, \ldots, p_N) := -k_B \sum_{i=1}^{N} p_i \ln p_i, \tag{2.22}$$

where k_B is the Boltzmann constant. So the only difference between H_N and S is that we measure the statistical entropy in units of k_B (i.e., $\kappa = k_B$).

In the canonical ensemble the number (N) of particles (or spins or ...) and the volume (V) of the system are fixed. This determines the possible energy eigenvalues E_k. Furthermore, the system is in contact with a heat bath of temperature T which determines the average energy $\langle E \rangle$ of the system. To obtain p_k we thus have to maximize (2.22) subject to the following prior information:

- The probabilities are normalized:

$$\langle 1 \rangle = \sum_{i=1}^{N} p_i = 1. \tag{2.23}$$

- The expectation value of the energy is fixed:

$$\langle E \rangle = \sum_{i=1}^{N} p_i E_i. \tag{2.24}$$

To find the maximum of S we again employ the method of Lagrange multipliers. Introducing two Lagrange multipliers, α and λ, one for each constraint, we get

$$\frac{\partial S}{\partial p_k} = \frac{\partial}{\partial p_k} \left\{ -k_B \sum_{i=1}^{N} p_i \ln p_i - \lambda \left(\sum_{i=1}^{N} p_i E_i - \langle E \rangle \right) - \alpha \left(\sum_{i=1}^{N} p_i - 1 \right) \right\} = 0,$$

leading to

$$p_k = e^{-(1+\alpha/k_B)} \exp(-\beta E_k), \quad \beta := \frac{\lambda}{k_B}.$$

The Lagrange multipliers can again be determined from the constraints. From (2.23) we see that α is given by

$$Z(N, V, \beta) = e^{1+\alpha/k_B} = \sum_{k=1}^{N} \exp(-\beta E_k) \quad (2.25)$$

so that

$$p_k = \frac{1}{Z} \exp(-\beta E_k). \quad (2.26)$$

Equations (2.25) and (2.26) agree respectively with the canonical partition function and the canonical probability distribution, if $\beta = 1/k_B T$. This equality can be shown as follows: From (2.24) and (2.26) we find

$$\langle E \rangle = -\frac{\partial \ln Z}{\partial \beta} \quad \text{and} \quad \frac{\partial \langle E \rangle}{\partial \beta} = -\left(\langle E^2 \rangle - \langle E \rangle^2\right) < 0. \quad (2.27)$$

The latter results means that $\langle E \rangle$ is a monotonously decreasing function of β. Thus, it can be inverted: $\beta = \beta(\langle E \rangle, N, V)$. But this implies that the entropy becomes a function of $\langle E \rangle$, N and V, since insertion of (2.26) into (2.22) yields

$$S = -k_B \sum_{k=1}^{N} p_k \ln p_k = k_B \beta \langle E \rangle + k_B \ln Z = S(\langle E \rangle, N, V). \quad (2.28)$$

Now we invoke the thermodynamic limit: $N \to \infty$, $V \to \infty$, but $N/V =$ fixed [176, 210]. In this limit, the fluctuations of the internal energy can be neglected because E is a self-averaging (see p. 5) property of the system, and so $S(\langle E \rangle, N, V)$ may be interpreted as the microcanonical entropy of a system with the externally prescribed (non-fluctuating) energy $\langle E \rangle$. Then, we may use the thermodynamic definition of temperature and write

$$\frac{1}{T} := \left(\frac{\partial S}{\partial \langle E \rangle}\right)_{N,V}$$

$$= k_B \beta + k_B \langle E \rangle \frac{\partial \beta}{\partial \langle E \rangle} + k_B \frac{\partial \ln Z}{\partial \beta} \frac{\partial \beta}{\partial \langle E \rangle} \stackrel{(2.27)}{=} k_B \beta, \quad (2.29)$$

which completes the proof that $\beta = 1/k_B T$.

Remarkably, the previous discussion also shows that the statistical entropy—introduced in (2.22) as a function proportional to the information entropy (2.20)—is equal to the thermodynamic entropy introduced by Clausius in the middle of the 19th century. On the one hand, this provides the justification for the use of the word entropy for the function defined in (2.20). On the other hand, it also raises the question of why a physical system populates its possible microstates in such a way that this agrees with what we would conjecture based on our prior information about the system. The debate about the interpretation of entropy in statistical physics has been going on for almost 150 years now, and is still far from being closed [56].

All other ensembles of statistical physics can of course be derived in the same way employing the maximum entropy principle with constraints corresponding to the respective ensemble. Furthermore, maximum entropy methods have proven extremely powerful in many areas where unbiased data analysis is important, ranging from the interpretation of nuclear magnetic resonance spectra to image reconstruction and more [69]. In the remainder of this book, however, we will mostly be concerned with the first half of Laplace's sentence, i.e. how to determine (time dependent) probabilities of events, on the basis that the prior probabilities of the simple events are given. In the following, we will therefore return to the treatment of probabilities as measures in a sample space.

2.1.3 Equivalent Measures

In the examples from statistical physics and quantum mechanics at the end of Sect. 2.1.1 we saw the measure on our probability space expressed as the product of a *probability density* and the Lebesgue measure. Looking at the first example, we furthermore note the following property: if the Lebesgue measure of some set $A \in \mathcal{B}$ is zero then the Gibbs measure of this set is zero, and vice versa. This is an example of the following general definition.

Definition 2.3 A measure μ is called absolutely continuous with respect to another measure ν (written $\mu \ll \nu$) if for all $A \in \mathcal{F}$ we have $\nu(A) = 0 \Rightarrow \mu(A) = 0$. μ and ν are equivalent when $\mu \ll \nu$ and $\nu \ll \mu$.

Equivalent measures have the same sets of measure zero and (for probability measures) also the same sets of measure unity. The importance of this concept lies in the following theorem.

Theorem 2.1 (Radon-Nikodým) *If $\mu \ll \nu$, then there exists an $\mathcal{F}$-measurable function p, such that for every $A \in \mathcal{F}$*

$$\mu(A) = \int_A p \, d\nu.$$

The function $p = d\mu/d\nu$ is called the Radon-Nikodým derivative of μ with respect to ν. If μ and ν are equivalent we have

$$\frac{d\mu}{d\nu} = \left(\frac{d\nu}{d\mu}\right)^{-1}.$$

From the statistical mechanics example, we see that the Boltzmann weight is the Radon-Nikodým derivative of the Gibbs measure with respect to the Lebesgue measure [210]. The probability measures that we will encounter in the applications discussed in this book are all absolutely continuous with respect to the Lebesgue measure. We will be able to write them as $d\mu(x) = p(x)dx$ and often will only use the so-called probability density p to denote the relevant probability measure.

2.1.4 Distribution Functions and Probability Densities

The sample spaces we have considered previously were $\mathbb{R}$ or $\mathbb{R}^d$, which are also the most relevant for the examples we have in mind. These are also the most relevant spaces for the values of the typical random variables we will study, which will be functions $X : \Omega \to \mathbb{R}(\mathbb{R}^d), \omega \mapsto X(\omega)$. For a random variable $X : \Omega \to \mathbb{R}$, let us look at the *cumulative probability distribution*

$$F(x) := \text{Prob}\big[X(\omega) \leq x\big] = \mu\big(\omega \in \Omega : X(\omega) \leq x\big). \tag{2.30}$$

Clearly we have $F(-\infty) = 0$ and $F(+\infty) = 1$, and $F(x)$ is the probability for the open interval $(-\infty, x]$. We also have

$$F(b) - F(a) = \mu\big(\omega \in \Omega : a \leq X(\omega) \leq b\big), \tag{2.31}$$

from which we can construct a probability measure on the Borel algebra of open sets in $\mathbb{R}$. Therefore, the probabilistic structure of $(\Omega, \mathcal{F}, \mu)$ is translated by the random variable X into the probabilistic structure of $(\mathbb{R}, \mathcal{B}, dF)$, and instead of discussing the probabilistic properties of some measurement in terms of the former we can use the latter, more familiar structure, as well. This is the description usually adopted in physical applications, and for our cases we will be able to write $dF(x) = p(x)dx$ and limit ourselves to the probability spaces:

$$\big(\mathbb{R} \text{ (or } \mathbb{R}^d\text{)}, \mathcal{B}, p(x)dx\big).$$

To calculate, for instance, the expectation value of a random variable we will evaluate

$$E[X] = \int_\Omega X(\omega) d\mu(\omega) = \int_\mathbb{R} x p(x) dx. \tag{2.32}$$

The sets $A \subset \Omega$ which are mapped by the random variable X onto open sets Γ in the Borel algebra of $\mathbb{R}$ give rise to a construction that we will need for discussing conditional expectations. If we consider the σ-algebra generated by all the sets A mapped by X onto the open sets in $\mathcal{B}(\mathbb{R})$, denoted by $\sigma(X)$, then this is the smallest σ-algebra with respect to which X is measurable, and in general $\sigma(X) \subset \mathcal{F}$. $\sigma(X)$ is called the σ-algebra generated by X.

In the same way as we did for a random variable with values in $\mathbb{R}$, we can look at the joint distribution of d real-valued random variables, $X_1, \ldots, X_d$, since $\mathbb{R}^d$ is

2.1 Some Basics of Probability Theory

'the same' as $\mathbb{R} \times \mathbb{R} \times \cdots \times \mathbb{R}$, regarding the d random variables as the components of a d-dimensional vector. Through

$$p(x_1, \ldots, x_d)\mathrm{d}x_1 \cdots \mathrm{d}x_d$$
$$= \mu\big(\omega \in \Omega : x_1 \leq X_1(\omega) \leq x_1 + \mathrm{d}x_1, \ldots, x_d \leq X_d(\omega) \leq x_d + \mathrm{d}x_d\big), \quad (2.33)$$

we again transfer the probability structure from Ω to $\mathbb{R}^d$.

2.1.5 Statistical Independence and Conditional Probabilities

For the probability distribution of a random vector $X \in \mathbb{R}^d$ we define its Fourier transform

$$G(\mathbf{k}) := \int_{\mathbb{R}^d} e^{i\mathbf{k}\cdot \mathbf{x}} p(\mathbf{x}) \mathrm{d}^d \mathbf{x}. \quad (2.34)$$

This is called the *characteristic function* of X and one can generate all moments of X (assuming they all exist, i.e., assuming $p(x)$ is sufficiently regular) through its Taylor expansion

$$G(\mathbf{k}) = \int_{\mathbb{R}^d} \mathrm{d}^d \mathbf{x}\, p(\mathbf{x}) \sum_{n=0}^{\infty} \frac{i^n}{n!} \left[\sum_{\alpha=1}^{d} k_\alpha x_\alpha\right]^n$$

$$= \int_{\mathbb{R}^d} \mathrm{d}^d \mathbf{x}\, p(\mathbf{x}) \sum_{n=0}^{\infty} \frac{i^n}{n!} \sum_{m_1+\cdots+m_d=n} \frac{n!}{m_1!\cdots m_d!}$$
$$\times (k_1 x_1)^{m_1} \cdots (k_d x_d)^{m_d}$$

$$= \int_{\mathbb{R}^d} \mathrm{d}^d \mathbf{x}\, p(\mathbf{x}) \sum_{n=0}^{\infty} \sum_{m_1=0}^{\infty}$$
$$\cdots \sum_{m_d=0}^{\infty} \delta_{\sum_{\alpha=1}^{d} m_\alpha, n} \frac{(ik_1 x_1)^{m_1} \cdots (ik_d x_d)^{m_d}}{m_1! \cdots m_d!}.$$

We therefore have

$$G(\mathbf{k}) = \sum_{|m|=0}^{\infty} \frac{(ik_1)^{m_1} \cdots (ik_d)^{m_d}}{m_1! \cdots m_d!} \langle x_1^{m_1} \cdots x_d^{m_d}\rangle, \quad (2.35)$$

where $m = (m_1, \ldots, m_d)$ and $|m|$ is the l^1-norm of m. Similarly, the logarithm of the characteristic function generates all the *cumulants* (central moments) of X:

$$\ln G(\mathbf{k}) = \sum_{|m|\neq 0}^{\infty} \frac{(ik_1)^{m_1} \cdots (ik_d)^{m_d}}{m_1! \cdots m_d!} \langle\!\langle x_1^{m_1} \cdots x_d^{m_d}\rangle\!\rangle, \quad (2.36)$$

where $\langle\langle x_i^{m_i}\rangle\rangle = \langle (x_i - \langle x_i\rangle)^{m_i}\rangle$. Of special interest are the second moments, which define the *covariance matrix*

$$\text{Cov}[x_i, x_j] \equiv \sigma_{ij}^2 \equiv \langle\langle x_i x_j\rangle\rangle = \langle x_i x_j\rangle - \langle x_i\rangle\langle x_j\rangle \tag{2.37}$$

because they serve to define when two random variables are uncorrelated.

Definition 2.4 Two random variables X_1 and X_2 are uncorrelated iff[2]

$$\text{Cov}[x_i, x_j] = 0.$$

We have to distinguish this from the much stronger property of statistical independence.

Definition 2.5 Two random variables X_1 and X_2 are statistically independent iff

$$p(x_1, x_2) = p_1(x_1) p_2(x_2),$$

which is equivalent to the following properties:

- $G(k_1, k_2) = G_1(k_1) G_2(k_2)$,
- $\langle x_1^{m_1} x_2^{m_2}\rangle = \langle x_1^{m_1}\rangle\langle x_2^{m_2}\rangle \,\forall m_1, m_2$,
- $\langle\langle x_1^{m_1} x_2^{m_2}\rangle\rangle = 0 \,\forall m_1 \neq 0, m_2 \neq 0$.

Whenever two random variables are not statistically independent, there is some information about the possible outcome of a measurement of one of the random variables contained in a measurement of the other. Suppose we know the joint distribution of the two random variables $p(x_1, x_2)$. Then we can define their *marginal distributions*

$$p_1(x_1) = \int_{\mathbb{R}} p(x_1, x_2) \mathrm{d}x_2$$

and

$$p_2(x_2) = \int_{\mathbb{R}} p(x_1, x_2) \mathrm{d}x_1.$$

The *conditional probability* density function for x_1 given x_2 is then defined as

$$p(x_1|x_2) = \frac{p(x_1, x_2)}{p_2(x_2)}, \tag{2.38}$$

where $p_2(x_2) \neq 0$ is assumed. We could define $p(x_2|x_1)$ in a symmetric manner. For each fixed x_2, $p(x_1|x_2)$ is a regular probability density and we can calculate the

---

[2] Generally, 'iff' is taken to mean 'if and only if', a necessary and sufficient condition.

2.1 Some Basics of Probability Theory

expectation value of x_1 with this probability density, which is called a *conditional expectation*,

$$E[x_1|x_2] = \int_{\mathbb{R}} x_1 p(x_1|x_2) dx_1. \tag{2.39}$$

The conditional expectation of x_1 with respect to x_2 is itself still a random variable through its dependence on the random variable x_2, so we can determine its expectation value. Of course, we have

$$E[E[x_1|x_2]] = E[x_1] = \int_{\mathbb{R}} \int_{\mathbb{R}} x_1 p(x_1, x_2) dx_1 dx_2, \tag{2.40}$$

as we can easily convince ourselves by using the definitions.

2.1.6 Central Limit Theorem

Many important phenomena that can be addressed with probability theory and the theory of stochastic processes, which we will be concerned with in the next section, have to be described as the compound effect of many small random influences. The observable quantities are most often the sum of a very large number of random events. Take for example the pressure exerted by an ideal gas on a piston, which is a thermodynamic variable made up from all the myriad collisions of the gas particles with the piston. A central question is, therefore, what the distribution of a sum of random variables will ultimately be. The first answer we want to present to this question is given by the central limit theorem, the second—more refined one—will be given in the discussion of the Lévy-stable distributions in Chap. 4.

We consider N statistically independent and identically distributed (iid) random variables $x_1, \ldots, x_N$, i.e.,

$$p(\mathbf{x}) = p(x_1, \ldots, x_N) = p(x_1) p(x_2) \cdots p(x_N).$$

For simplicity we require

$$\langle x_1 \rangle = \cdots = \langle x_N \rangle = 0.$$

Furthermore, let us denote

$$\langle x_1^2 \rangle = \cdots = \langle x_N^2 \rangle = \sigma^2.$$

What is the probability distribution, p_N, of

$$\hat{S}_N = \frac{1}{\sqrt{N}} \sum_{n=1}^{N} x_n ?$$

We can see directly that $\langle \hat{S}_N \rangle = 0$ and $\sigma_N^2 = \langle (\hat{S}_N)^2 \rangle = \sigma^2$. Formally, we have

$$p_N(z) = \int d^N x \, \delta(z - \hat{S}_N) p(x). \tag{2.41}$$

From this we can calculate the characteristic function of the sum variable

$$G_N(k) = \int_{-\infty}^{+\infty} dz \, e^{ikz} p_N(z) = \int_{-\infty}^{+\infty} dz \, e^{ikz} \int d^N x \, \delta(z - \hat{S}_N) p(x)$$

$$= \int d^N x \, p(x) \int_{-\infty}^{+\infty} dz \, e^{ikz} \delta(z - \hat{S}_N)$$

$$= \int_{-\infty}^{+\infty} dx_1 \cdots \int_{-\infty}^{+\infty} dx_N \, p(x_1) \cdots p(x_N) e^{ik\hat{S}_N}$$

$$= \left[\int_{-\infty}^{+\infty} dx_1 \, p(x_1) e^{ikx_1/\sqrt{N}} \right]^N$$

because we assumed iid random variables. We therefore find

$$G_N(k) = \left[G\left(\frac{k}{\sqrt{N}} \right) \right]^N. \tag{2.42}$$

Using now (2.35) and the moments of the individual random variable, we can write this as

$$G_N(k) = \left[1 - \frac{1}{N} \frac{(k\sigma)^2}{2} + O\left(\frac{k^3}{N^{3/2}} \right) \right]^N,$$

and for $N \gg 1$ we get

$$G_N(k) \approx \left[1 - \frac{1}{N} \frac{(k\sigma)^2}{2} \right]^N \xrightarrow{N \to \infty} \exp\left[-\frac{(k\sigma)^2}{2} \right]. \tag{2.43}$$

Ultimately, the characteristic function therefore approaches a Gaussian, which means that the probability distribution also is a Gaussian

$$p_N(z) = \frac{1}{2\pi} \int_{-\infty}^{+\infty} e^{-ikz} G_N(k) dk \xrightarrow{N \to \infty} \frac{1}{\sqrt{2\pi\sigma^2}} \exp\left[-\frac{z^2}{2\sigma^2} \right]. \tag{2.44}$$

This result lies at the heart of the ubiquitous appearance of the Gaussian distribution in statistical phenomena. Whatever the exact form of the underlying distribution of the individual random variables is, as long as the first two moments exist, the sum variable always obeys a Gaussian distribution in the large-N limit. Mathematically, the requirements on the individual distributions can be somewhat weaker than the existence of the first two moments [52]. A criterion for whether a probability distribution is attracted to the Gaussian distribution—in the sense that the sum variable is Gaussian distributed—is given by Gnedenko and Kolmogorov [62]:

2.1 Some Basics of Probability Theory

Theorem 2.2 *A probability density $p(x)$ is attracted to the Gaussian probability density iff*

$$\lim_{y \to \infty} \frac{y^2 \int_{|x|>y} p(x)\,dx}{\int_{|x|<y} x^2 p(x)\,dx} = 0.$$

This condition can be paraphrased by saying that, for probability densities which are attracted to the Gaussian distribution, the normalized sum variable is the sum of *many individually negligible components* [52] (see also Appendix A).

For practical applications, it is also important to know what the deviations from the asymptotic behavior are, when one sums up a large but finite number of random variables. For iid random variables with a sufficiently regular distribution ensuring the existence of higher moments, the answer was derived by Tschebyschew (see [62]). The difference between the cumulative distributions of the finite N sum variable and that of the asymptotic Gaussian distribution can be written as

$$\int_x^\infty dz \left(p_N(z) - \frac{1}{\sqrt{2\pi}} e^{-z^2/2} \right) = \frac{1}{\sqrt{2\pi}} e^{-x^2/2} \left[\frac{Q_1(x)}{\sqrt{N}} + \frac{Q_2(x)}{N} + \cdots \right]. \quad (2.45)$$

The Q_k are polynomials derivable from the Tschebyschew-Hermite polynomials. The first two are given as [62]

$$Q_1(x) = \frac{\hat{\kappa}_3}{6}(x^2 - 1)$$

$$= e^{x^2/2} \frac{\hat{\kappa}_3}{6} \frac{d^2}{dx^2} e^{-x^2/2}, \quad (2.46)$$

$$Q_2(x) = \frac{\hat{\kappa}_3^2}{72} x^5 + \frac{1}{24}\left(\hat{\kappa}_4 - \frac{10\hat{\kappa}_3^2}{3}\right) x^3 + \frac{1}{8}\left(\frac{5\hat{\kappa}_3^2}{3} - \hat{\kappa}_4\right) x$$

$$= -e^{x^2/2} \left(\frac{\hat{\kappa}_3^2}{72} \frac{d^5}{dx^5} + \frac{\hat{\kappa}_4}{24} \frac{d^3}{dx^3} \right) e^{-x^2/2}, \quad (2.47)$$

where $\hat{\kappa}_3$ and $\hat{\kappa}_4$ are the skewness and kurtosis of the distributions of the individual random variables as they were defined in (1.26) and (1.27). For a symmetric distribution, $\hat{\kappa}_3 = 0$, we can see that the leading-order correction is given by the kurtosis of the individual distributions divided by the number of random variables, N, that have been summed up.

2.1.7 Extreme Value Distributions

To arrive at the central limit theorem we asked ourselves what the distribution of the sum of N iid random variables x_i is. The central limit theorem gave the answer that for the centralized and normalized sum $\hat{S}_N$ the limiting distribution is Gaussian, if the distribution $p(x_i)$ of the iid random variables has non-diverging first and second moments.

Extreme value theory asks a related question. Let $M_N = \max(x_1, \ldots, x_N)$ be the maximum value of a sample of N iid random variables. What are the possible non-degenerate[3] limiting distributions of appropriately centralized and normalized maxima M_N? Knowledge about the expected extrema in a series of random experiments (or in a stochastic process) is of utmost importance in all areas where risk has to be assessed. That can be the risk of a heavy storm destroying buildings (insurance) or the risk of financial loss for a portfolio of investments (financial risk management). Here the question is not about the typical behavior of some random variable but about the extreme value it assumes for a given sample size.

To understand why the maxima need to be centralized and normalized let us consider the following. Let $F(x)$ be the cumulative distribution of the iid random variables

$$F(x) = \int^x dx' p(x')$$

and define $x_F = \sup(x \in \mathbb{R} : F(x) < 1)$. Depending on the support of $p(x)$, x_F can be finite or infinite. The cumulative probability distribution for the maximum, M_N, can then be written as (cf. (2.30))

$$F_N(M_N) = \text{Prob}(x_1 \leq M_N, \ldots, x_N \leq M_N) = F^N(M_N), \quad (2.48)$$

the last part of the equation holding due to the iid assumption. Now suppose that $x_F < \infty$. Then, if $N \to \infty$, $F^N(x) \to 0$ for all $x < x_F$ and $F^N(x) = 1$ for $x \geq x_F$, so that the limiting probability density of the maximum would have a point support at $x = x_F$. Similarly, for $x_F = \infty$ we have $F^N(x) \to 0$ for all $x \in \mathbb{R}$ so that we also do not obtain a non-degenerate probability distribution for the maximum. Therefore it only makes sense to search for a limiting distribution for properly centralized and normalized variables.

Let us exemplify this point with a specific example. Assume that we have N independent, exponentially distributed random variables x_i, that is,

$$p(x_i) = \exp(-x_i) \quad \text{and} \quad F(x_i) = \int_0^{x_i} dx'_i p(x'_i) = 1 - \exp(-x_i). \quad (2.49)$$

So we have the case $x_F = \infty$ for the exponential distribution. We therefore have

$$F_N(M_N) = \left[1 - \exp(-M_N)\right]^N.$$

The probability density for the maximum of a set of N random variables distributed according to the exponential distribution is therefore

$$p_N(M_N) = N\left[1 - \exp(-M_N)\right]^{N-1} \exp(-M_N).$$

---

[3]A non-degenerate distribution is the opposite of a degenerate distribution. A distribution of a random variable is called degenerate, if it is a delta function, i.e., it is 1 for one value of the random variable and 0 otherwise. The corresponding cumulative distribution is the Heaviside step function.

2.1 Some Basics of Probability Theory

The maximum of this distribution occurs at $b_N = \ln N$ and diverges with increasing sample size N. This value of M_N occurs for one of the x_i with probability $p(x_i = b_N) = 1/N$; so in N trials it will occur with probability 1. With $N \to \infty$ the observed maximum therefore increases to infinity, i.e., the cumulative probability density for the extrema approaches a degenerate limiting distribution.

We now centralize the maximum variable by looking at $u = M_N - \ln N$. The cumulative distribution of this variable is

$$F_N(u) = \left[1 - e^{-(u+\ln N)}\right]^N = \left(1 - \frac{e^{-u}}{N}\right)^N \xrightarrow{N \to \infty} \exp(-e^{-u}). \tag{2.50}$$

This is a well-behaved, non-degenerate limiting distribution. A second example leading to the same limiting distribution for the maxima is obtained for normally distributed independent random variables. Remember that with $\langle x \rangle = 0$ and $\sigma = 1$ we have from (1.28)

$$p_G(x) = \frac{1}{\sqrt{2\pi}} \exp\left[-\frac{x^2}{2}\right] \quad \text{and} \quad F(x) = \frac{1}{2}\left[1 + \text{erf}\left(\frac{x}{\sqrt{2}}\right)\right],$$

where

$$\text{erf}(x) = \frac{2}{\sqrt{\pi}} \int_0^x e^{-z^2} dz \tag{2.51}$$

is the error function [1].

For the Gaussian distribution one needs to centralize and normalize the assumed maxima [44]

$$u = \frac{M_N - b_N}{a_N}, \tag{2.52}$$

using for example (these expressions are not unique [44])

$$b_N = \sqrt{2\ln(N)} - \frac{\ln[\ln(N)] + \ln(4\pi)}{2\sqrt{2\ln(N)}}$$
$$a_N = \left[2\ln(N)\right]^{-\frac{1}{2}}. \tag{2.53}$$

Again, this leads to the double exponential form of the limiting distribution for the maxima.

As a matter of fact, for every probability distribution of iid random variables decaying at least exponentially for $x \to \infty$, the limiting cumulative distribution for the extrema is given by the so-called (standard) Gumbel distribution

$$F(u) = \exp(-e^{-u}). \tag{2.54}$$

Generally, there is a large degree of universality in the extreme value distributions leading to only three classes of possible limiting distributions.

Theorem 2.3 (Fisher-Tippett Theorem) *Let $(x_k)_{k=1,\ldots,N}$ be a sequence of iid random variables and $M_N = \max(x_1, \ldots, x_N)$. If there exist constants $a_N > 0$, $b_N \in \mathbb{R}$ and a non-degenerate distribution function F such that*

$$\lim_{N \to \infty} \text{Prob}\left(\frac{M_N - b_N}{a_N} \leq u\right) = F(u),$$

then F is of the type of one of the following three distribution functions:

$$\text{Gumbel:} \quad F(u) = \exp(-e^{-u}) \quad u \in \mathbb{R}. \tag{2.55}$$

$$\text{Fréchet:} \quad F_\alpha(u) = \begin{cases} 0 & u \leq 0, \\ \exp(-u^{-\alpha}) & u > 0. \end{cases} \tag{2.56}$$

$$\text{Weibull:} \quad F_\alpha(u) = \begin{cases} \exp(-(-u)^\alpha) & u \leq 0, \\ 1 & u > 0. \end{cases} \tag{2.57}$$

For Fréchet and Weibull, one has $\alpha > 0$.

These are the cumulative distribution functions of the maxima, from which the corresponding probability densities can be obtained by differentiation. Obviously, these distribution functions can be transformed into one another. When we assume that the extrema u of a variable x are distributed according to the Gumbel distribution then the extrema of the variable y defined by $x = \alpha \ln y$ for $y > 0$ are distributed according to the Fréchet distribution, and those of z defined by $x = -\alpha \ln(-z)$ for $z < 0$ are distributed according to the Weibull distribution. One can also combine the three extreme value distributions into one single form, the so-called 'von Mises representation' of the extreme value distributions

$$F_\gamma(u) = \exp\left[-(1 + \gamma u)^{-1/\gamma}\right] \quad \text{for } 1 + \gamma u > 0. \tag{2.58}$$

For $\gamma > 0$ one has the Fréchet distribution, for $\gamma < 0$ the Weibull distribution, and for $\gamma \to 0$ one obtains the Gumbel distribution.

The domain of attraction of the three distributions can also be specified but for this discussion we refer to the mathematical literature on extreme value distributions [34, 44]. Here let us only point out that the change of variables discussed above, leading from the Gumbel to the Fréchet distribution, indicates that distributions decaying asymptotically for $|x| \to \infty$ like a power law will belong to the domain of attraction of the Fréchet distribution. The Lévy stable distributions which we will discuss in Chap. 4 belong to this class.

2.2 Stochastic Processes and Their Evolution Equations

In the first section of this chapter we considered the mathematical framework to describe measurements of observables which have to be treated in a probabilistic

2.2 Stochastic Processes and Their Evolution Equations

manner, i.e., where we have incomplete knowledge about the state of a system. Now we want to go on to time-dependent phenomena and information unfolding with time. This leads to the concept of a stochastic process [104]:

Definition 2.6 A stochastic process is a family $(X_t)_{t \in I}$ of random variables.

The index set can either be a discrete set, for instance, $I = \mathbb{N}$, which would lead to a *discrete time stochastic process*, or it can be continuous, for instance, $I = \mathbb{R}$, which would lead to a *continuous time stochastic process*.

We studied a discrete time stochastic process in the introductory chapter, with $(X_n)_{n \in \mathbb{N}}$ being the coordinate ($X_n \in \mathbb{Z}$) of a random walker on the set of whole numbers. We can write a continuous time stochastic process as a function,

$$X: \mathbb{R} \times \Omega \to \mathbb{R}^d, \qquad (t, \omega) \mapsto X(t, \omega).$$

An example to which we will come back later would be the solution of the *Langevin equation*

$$\dot{X}(t, \omega) = \eta(t, \omega),$$

with η modeling the stochastic forces on the position X of a Brownian particle. They are usually taken to be a *Gaussian white noise*, i.e., random variables with a Gaussian probability distribution which is fully specified by the following two moments:

$$\langle \eta(t, \omega) \rangle = 0, \qquad \langle \eta(t, \omega) \eta(t', \omega) \rangle \propto \delta(t - t'). \qquad (2.59)$$

The angular brackets indicate the expectation value over the probability space Ω. We have used here the notation most often found in physics books and added the explicit dependence on ω which denotes different realizations of the time dependence of the stochastic force, and therefore of the path $X(t, \omega)$ of the Brownian particle. To write out the averages explicitly, we have

$$\langle \eta(t, \omega) \rangle = \int_{\Omega} \eta(t, \omega) d\mu(\omega) = 0. \qquad (2.60)$$

For every fixed n (discrete time) or t (continuous time) X_n or X_t, respectively, is a random variable so that we can use the concepts and properties discussed in the first part of this chapter. We can, for instance, define the distribution function of X at time t,

$$p(x, t) = \int_{\Omega} \delta(x - X_t(\omega)) d\mu(\omega), \qquad (2.61)$$

and thereby generate a time-dependent probability density on $\mathbb{R}^d$: $p(x, t) dx$ is the probability of finding a value of the random variable in $[x, x + dx]$ at time t. We can generalize this definition to the *n-point distribution functions*,

$$p_n(x_1, t_1; \ldots; x_n, t_n) = \int_{\Omega} \delta(x_1 - X_{t_1}(\omega)) \cdots \delta(x_n - X_{t_n}(\omega)) d\mu(\omega), \qquad (2.62)$$

which is the probability density for the stochastic process to pass through $[x_1, x_1 + dx_1]$ at time t_1 and $[x_2, x_2 + dx_2]$ at time t_2 and ... and $[x_n, x_n + dx_n]$ at time t_n. With the help of these n-point distribution functions we can calculate time dependent moments and correlation functions of the stochastic process under study:

$$\langle X(t_1) \cdots X(t_n) \rangle = \int_{\mathbb{R}^n} dx_1 \cdots dx_n x_1 \cdots x_n p_n(x_1, t_1; \ldots; x_n, t_n). \tag{2.63}$$

In a similar way as we did for the correlations between two random variables, we can define a time-dependent covariance matrix for two stochastic processes:

$$\text{Cov}[X_i(t_1), X_j(t_2)] = \sigma_{ij}^2(t_1, t_2)$$
$$= \langle X_i(t_1) X_j(t_2) \rangle - \langle X_i(t_1) \rangle \langle X_j(t_2) \rangle. \tag{2.64}$$

The diagonal elements of this matrix are called *autocorrelation functions* and those off-diagonal are the *cross-correlation functions*.

The hierarchy of distribution functions p_n (2.62) characterizes a stochastic process completely and fulfills the following relations:

- non-negativity

$$p_n \geq 0,$$

- symmetry

$$p_n(\ldots; x_k, t_k; \ldots; x_l, t_l; \ldots) = p_n(\ldots; x_l, t_l; \ldots; x_k, t_k; \ldots).$$

This property is intelligible from the meaning of p_n: $p_n(\ldots; x_k, t_k; \ldots; x_l, t_l; \ldots)$ is the probability density for the stochastic process to be in $[x_k, x_k + dx_k]$ at time t_k and $[x_l, x_l + dx_l]$ at time t_l. So the order of the 'events' k and l should not matter;

- completeness

$$\int p_n(x_1, t_1; \ldots; x_n, t_n) dx_n = p_{n-1}(x_1, t_1; \ldots; x_{n-1}, t_{n-1}). \tag{2.65}$$

Due to the symmetry property, the integration can be carried out for any x_k with $k = 1, \ldots, n$;

- probability measure

$$\int p_1(x, t) dx = 1.$$

Every hierarchy of functions which fulfills the above conditions, on the other hand, uniquely defines a stochastic process.

A stochastic process is called *stationary* iff for all n we have

$$p_n(x_1, t_1 + \Delta t; x_2, t_2 + \Delta t; \ldots; x_n, t_n + \Delta t) = p_n(x_1, t_1; x_2, t_2; \ldots; x_n, t_n), \tag{2.66}$$

where Δt is an arbitrary but fixed time shift.

2.2 Stochastic Processes and Their Evolution Equations

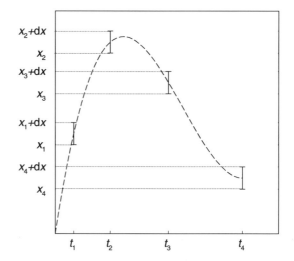

Fig. 2.3 Definition of a subensemble of all continuous paths by requiring the process to pass through a prescribed set of intervals at time points $t_1, \ldots, t_k$

The conditional probabilities we defined in the first part of this chapter can now be used to yield informations about subensembles of the stochastic process. Let us consider the following typical question (illustrated in Fig. 2.3). If we assume that a stochastic process yields a result in $[x_1, x_1 + dx_1]$ at time t_1, in $[x_2, x_2 + dx_2]$ at time $t_2, \ldots$, in $[x_k, x_k + dx_k]$ at time t_k, what is the probability of finding a value in $[x_{k+1}, x_{k+1} + dx_{k+1}]$ at time $t_{k+1}, \ldots, [x_{k+l}, x_{k+l} + dx_{k+l}]$ at time t_{k+l}?

We define the conditional probability for the events at $k+1$ to $k+l$ given the events 1 to k to be

$$p_{k+l}(x_1, t_1; \ldots; x_{k+l}, t_{k+l}) \\ = p_k(x_1, t_1; \ldots; x_k, t_k) \times p_{l|k}(x_{k+1}, t_{k+1}; \ldots; x_{k+l}, t_{k+l} | x_1, t_1; \ldots; x_k, t_k). \quad (2.67)$$

Specifically, we have

$$p_2(x_1, t_1; x_2, t_2) = p_{1|1}(x_2, t_2 | x_1, t_1) p_1(x_1, t_1) \quad (2.68)$$

and due to the completeness of p_2

$$\int dx_2 \, p_{1|1}(x_2, t_2 | x_1, t_1) = 1. \quad (2.69)$$

This normalization of the conditional probability means that a transition from x_1 to some x_2 has occurred with certainty.

2.2.1 Martingale Processes

In this section we want to discuss a special class of stochastic processes that will be of importance in the financial applications. Actually, it is also in the area of betting and gambling that the properties of this class of processes are best presented. Let us start with the concept of a *fair game*, as you would expect if you bet on the outcome of a coin toss. Whatever the history of the game, you would like the probability for head or tail in the next throw to be equal, $p_{\text{head}} = p_{\text{tail}} = 1/2$, making the expected gain on a doubling of the bet on the next throw exactly zero, whatever you bet on.

Definition 2.7 A sequence $(X_n)_{n\in\mathbb{N}}$ is called absolutely fair when for all $n = 1, 2, \ldots$ we have

$$E[X_1] = 0 \quad \text{and} \quad E[X_{n+1}|X_1, \ldots, X_n] = 0. \tag{2.70}$$

If we now define another sequence of random variables $(Y_n)_{n\in\mathbb{N}}$ by

$$Y_n = E[Y_1] + X_1 + \cdots + X_n,$$

we can easily see that we have

$$E[Y_{n+1}|Y_1, \ldots, Y_n] = E[Y_{n+1}|X_1, \ldots, X_n] = Y_n.$$

Definition 2.8 A sequence $(Y_n)_{n\in\mathbb{N}}$ is a martingale iff

$$E[Y_{n+1}|Y_1, \ldots, Y_n] = Y_n. \tag{2.71}$$

It is a submartingale iff

$$E[Y_{n+1}|Y_1, \ldots, Y_n] \geq Y_n \tag{2.72}$$

and a supermartingale iff

$$E[Y_{n+1}|Y_1, \ldots, Y_n] \leq Y_n. \tag{2.73}$$

This means that for a martingale the best estimator for the next value, given all the information obtained through the past values, is the actual value. If the process is a martingale, its changes (increment process) are a fair game, meaning that the expectation value of the increments is zero.

We have also already seen such a behavior in our discussion of the random walk on a line in the first chapter. If the walker jumps one unit distance to the left or to the right with equal probability, then the expectation value of the position change with one jump is zero. The position process of the walker is therefore a martingale, being the sum of fair-game increments.

This process has another appealing property which is called the Markov (A.A. Markov (1856–1922)) property. *Markov processes* and their description will concern us for the rest of this chapter.

2.2.2 Markov Processes

The defining property of Markov processes is that they have no memory. When we return to the definition of the conditional probabilities for a set of events given the past history of the process in (2.67) we get an idea of the simplification that this property entails.

Definition 2.9 For a Markov process, we have for all n and all $t_1 < t_2 < \cdots < t_n$

$$p_{1|n-1}(x_n, t_n | x_1, t_1; \ldots; x_{n-1}, t_{n-1}) = p_{1|1}(x_n, t_n | x_{n-1}, t_{n-1}). \tag{2.74}$$

One therefore only has to know the actual state of the system (x_{n-1}, t_{n-1}) in order to calculate the probability for the occurrence of (x_n, t_n). In this sense, the process has no memory. If we apply this iteratively to the n-point function (cf. (2.67)), we get

$$p_n(x_1, t_1; \ldots; x_n, t_n) = \prod_{l=2}^{n} p_{1|1}(x_l, t_l | x_{l-1}, t_{l-1}) p_1(x_1, t_1). \tag{2.75}$$

In the case of a Markov process, therefore, the knowledge of two functions, $p_{1|1}$ and p_1, suffices to describe the process completely and we can appreciate the enormous simplification compared to the infinite hierarchy of p_n needed in the general case.

Before we go on to elaborate on the consequences of this property, let us remind ourselves of another example of a Markov process we are all very familiar with, namely deterministic motion as described by Newton's or Hamilton's equations. If we know the positions and momenta of a system of N particles at $t = t_0$ and we know their interactions as a function of the positions and momenta, we can predict the complete future trajectory of the system for all $t > t_0$:

$$\begin{aligned} d\boldsymbol{r}_i(t) &= \frac{\boldsymbol{p}_i(t)}{m_i} dt, \\ d\boldsymbol{p}_i(t) &= \boldsymbol{F}_i(\{\boldsymbol{r}_i(t)\}) dt. \end{aligned} \tag{2.76}$$

Deterministic motion, therefore, could be described as a random process which at each time has a singular distribution

$$p(\boldsymbol{r}, \boldsymbol{p}) = \prod_i \delta(\boldsymbol{r} - \boldsymbol{r}_i(t)) \delta(\boldsymbol{p} - \boldsymbol{p}_i(t)).$$

Let us now specialize the result for the n-point function to the case $n = 3$:

$$p_3(x_1, t_1; x_2, t_2; x_3, t_3) = p_{1|1}(x_3, t_3 | x_2, t_2) p_{1|1}(x_2, t_2 | x_1, t_1) p_1(x_1, t_1).$$

Integrating over x_2 and using the completeness property of the hierarchy of the p_n we get

$$p_2(x_1, t_1; x_3, t_3) = p_1(x_1, t_1) \int dx_2 \, p_{1|1}(x_3, t_3 | x_2, t_2) p_{1|1}(x_2, t_2 | x_1, t_1).$$

With the definition of the conditional probability in (2.68), we arrive at the following result:

$$p_{1|1}(x_3, t_3|x_1, t_1)$$
$$= \int dx_2 \, p_{1|1}(x_3, t_3|x_2, t_2) p_{1|1}(x_2, t_2|x_1, t_1) \quad \text{for } t_3 \geq t_2 \geq t_1. \quad (2.77)$$

This is the *Chapman-Kolmogorov equation*. It is a consistency equation for the conditional probabilities of a Markov process and the starting point for deriving the equations of motion for Markov processes. We furthermore note the following result:

Theorem 2.4 *Two positive, normalized functions p_1 and $p_{1|1}$ which fulfill*

- the Chapman-Kolmogorov equation,
- $p_1(x_2, t_2) = \int dx_1 \, p_{1|1}(x_2, t_2|x_1, t_1) p_1(x_1, t_1)$,

completely and uniquely define a Markov process.

In the following sections we will approximate the Chapman-Kolmogorov equation further, focusing on special situations.

Stationary Markov Processes and the Master Equation

For a stationary Markov process (cf. (2.66)) we can write for the defining functions

$$p_1(x, t) = p_1(x, t + \Delta t) \stackrel{\Delta t = -t}{=} p_1(x, 0) =: p_{\text{eq}}(x),$$
$$p_{1|1}(x_2, t_2|x_1, t_1) = p_{1|1}(x_2, t_2 + \Delta t|x_1, t_1 + \Delta t) \stackrel{\Delta t = -t_1}{=} p_t(x_2|x_1), \quad (2.78)$$

where p_t denotes a *transition probability* within the time interval $t \, (= t_2 - t_1)$ from state x_1 to state x_2. Using the Chapman-Kolmogorov equation for p_t, we get

$$p_{t+t'}(x_3|x_1) = \int dx_2 \, p_{t'}(x_3|x_2) p_t(x_2|x_1). \quad (2.79)$$

When we consider a discrete probability space for x_i it is easily seen that this is a matrix multiplication, the p_t being matrices transforming one discrete state into another.

We now want to derive the differential form of the Chapman-Kolmogorov equation for stationary Markov processes. For this we consider the case of small time intervals t' and write the Taylor expansion of the transition probability up to first order in the following way [101]:

$$p_{t'}(x_3|x_2) = \left(1 - w_{\text{tot}}(x_2)t'\right)\delta(x_3 - x_2) + t'w(x_3|x_2). \quad (2.80)$$

2.2 Stochastic Processes and Their Evolution Equations

This equation defines $w(x_3|x_2)$ as the *transition rate* (transition probability per unit time) from x_2 to x_3. $(1 - w_{\text{tot}}(x_2)t')$ is the probability to remain in state x_2 up to time t'. From the normalization condition for $p_{t'}(x_3|x_2)$ (cf. (2.69)) we obtain

$$w_{\text{tot}}(x_2) = \int dx_3 \, w(x_3|x_2). \tag{2.81}$$

Using (2.80) in the Chapman-Kolmogorov equation results in

$$p_{t+t'}(x_3|x_1) = \left(1 - w_{\text{tot}}(x_3)t'\right)p_t(x_3|x_1) + t'\int dx_2 \, w(x_3|x_2)p_t(x_2|x_1),$$

and so

$$\frac{p_{t+t'}(x_3|x_1) - p_t(x_3|x_1)}{t'}$$

$$= \int dx_2 \, w(x_3|x_2)p_t(x_2|x_1) - \int dx_2 \, w(x_2|x_3)p_t(x_3|x_1), \tag{2.82}$$

in which we have used the definition of w_{tot}. In the limit $t' \to 0$ we arrive at the *master equation*, which is the differential version of the Chapman-Kolmogorov equation

$$\frac{\partial}{\partial t}p_t(x_3|x_1) = \int dx_2 \, w(x_3|x_2)p_t(x_2|x_1) - \int dx_2 \, w(x_2|x_3)p_t(x_3|x_1). \tag{2.83}$$

It is an integro-differential equation for the transition probabilities of a stationary Markov process.

When we do not assume stationarity and choose a $p_1(x_1, t) \neq p_{\text{eq}}(x)$ but keep the assumption of time-homogeneity of the transition probabilities, i.e., they only depend on time differences, we can multiply this equation by $p_1(x_1, t)$ and integrate over x_1 to get a master equation for the probability density itself:

$$\frac{\partial}{\partial t}p_1(x_3, t) = \int dx_2 \, w(x_3|x_2)p_1(x_2, t) - \int dx_2 \, w(x_2|x_3)p_1(x_3, t). \tag{2.84}$$

To simplify the notation in the following, where we will only work with the functions p_1 and w to describe the Markov process, let us make the changes $p_1 \to p$, $x_3 \to x$ and $x_2 \to x'$

$$\frac{\partial}{\partial t}p(x, t) = \int dx' \, w(x|x')p(x', t) - \int dx' \, w(x'|x)p(x, t). \tag{2.85}$$

On a discrete probability space we would write this equation in the following way:

$$\frac{\partial}{\partial t}p(n, t) = \sum_{n'}\left(w(n|n')p(n', t) - w(n'|n)p(n, t)\right). \tag{2.86}$$

We used this equation in the random walker problem in Chap. 1 from the rate-equation perspective. That was a special example of a (time-homogeneous) Markov process in discrete time and space where a master equation can be used to define the time evolution of the probability distribution.

The master equation has an intuitive interpretation. Let us take (2.85) as an example. The variation of $p(x,t)$ with time is caused by the balance of an inward flux (first term) and an outward flux (second term). The first term describes the increase of the probability of x via the flux from all states x' ($x' \neq x$ and $x' = x$) to x, while the second term represents the decrease of $p(x,t)$ through transitions from x to x'.

So far we have derived and discussed the Chapman-Kolmogorov equation and also the master equation as a consistency equation for the transition probabilities (rates) of a stationary (or time-homogeneous) Markov process. We can, however, reinterpret these equations, turning them into 'equations of motion' in a Markov-process description of a given problem. This reinterpretation proceeds through the following steps:

- We are considering a dynamic problem (physical, biological, financial, ...) that we can model as a stochastic process.
- We can neglect memory effects and the problem is stationary (or time-homogeneous).
- From independent phenomenological or theoretical knowledge about the problem, we can derive the functional form of the transition rates $w(x'|x)$ between all states x, x' of the system.

From this starting point we can use (2.85) as an equation of motion (evolution equation) for the probability density of the stochastic process used to model our problem.

Let us now set the stage for a further approximation of the master equation and make the following assumptions:

- We write $w(x|x') = w(x|x-r) =: w(x-r;r)$, with the 'jump distance' $r := x - x'$. Similarly, $w(x'|x) = w(x-r|x) =: w(x;-r)$.
- There are only small jumps, i.e., $w(x-r;r)$ as a function of r is a sharply peaked function around $r = 0$

$$\exists \delta > 0 \quad \text{with } w(x-r;r) \approx 0, \quad |r| > \delta.$$

- $w(x-r;r)$ is a slowly varying function of its first argument

$$\exists \delta' > 0 \quad \text{with } w(x-r;r) \approx w(x;r), \quad |r| < \delta'.$$

- The last also holds true for $p(x,t)$.
- w and p are sufficiently smooth functions of both arguments.

We rewrite (2.85) as

$$\frac{\partial}{\partial t} p(x,t) = \int_{-\infty}^{+\infty} w(x-r;r) p(x-r,t) \mathrm{d}r - p(x,t) \int_{-\infty}^{+\infty} w(x;-r) \mathrm{d}r. \quad (2.87)$$

2.2 Stochastic Processes and Their Evolution Equations

We can now perform a Taylor expansion in $x - r$ around $r = 0$ in the first integral on the right-hand side:

$$\frac{\partial}{\partial t} p(x,t) = p(x,t) \int_{-\infty}^{+\infty} w(x;r) dr - p(x,t) \int_{-\infty}^{+\infty} w(x;-r) dr$$

$$- \int_{-\infty}^{+\infty} r \frac{\partial}{\partial x} [w(x;r) p(x,t)] dr + \frac{1}{2} \int_{-\infty}^{+\infty} r^2 \frac{\partial^2}{\partial x^2} [w(x;r) p(x,t)] dr$$

$$\mp \cdots,$$

yielding

$$\frac{\partial}{\partial t} p(x,t) = \sum_{n=1}^{\infty} \frac{(-1)^n}{n!} \frac{\partial^n}{\partial x^n} [a_n(x) p(x,t)], \qquad (2.88)$$

$$a_n(x) = \int_{-\infty}^{+\infty} r^n w(x;r) dr. \qquad (2.89)$$

This is the *Kramers-Moyal expansion* of the master equation. As such it is only a rewriting of the master equation, changing it from an integro-differential equation into a partial differential equation of infinite order. In the next section we will encounter a useful and often used approximation to this expansion.

Fokker-Planck and Langevin Equations

In deriving the Kramers-Moyal expansion in the last section, we have not yet made use of our assumption that the transition rates $w(x,r)$ are supposed to be slowly varying functions of their first argument. For these slowly varying functions—which will give rise to slowly varying probability densities $p(x,t)$—we can make the assumption that we can truncate the Kramers-Moyal expansion at a certain order of derivatives. A truncation after the second order leads to the so-called *Fokker-Planck equation*:

$$\frac{\partial}{\partial t} p(x,t) = -\frac{\partial}{\partial x} [a_1(x) p(x,t)] + \frac{1}{2} \frac{\partial^2}{\partial x^2} [a_2(x) p(x,t)]. \qquad (2.90)$$

In this equation, a_1 is called the *drift coefficient* and a_2 the *diffusion coefficient*.

Let us remember at this point that we originally derived the master equation from the Chapman-Kolmogorov equation for the conditional probability $p(x,t|x_0,t_0)$, and so the Fokker-Planck equation for the conditional probability reads:

$$\frac{\partial}{\partial t} p(x,t|x_0,t_0) = -\frac{\partial}{\partial x} [a_1(x) p(x,t|x_0,t_0)] + \frac{1}{2} \frac{\partial^2}{\partial x^2} [a_2(x) p(x,t|x_0,t_0)]. \qquad (2.91)$$

Because it involves derivatives with respect to the final coordinates (x,t), it is also called the *forward Fokker-Planck equation*. To solve this equation we have to specify boundary conditions and especially a starting condition at t_0. To derive the corresponding *backward Fokker-Planck equation* containing derivatives with respect to the starting coordinates, let us commence with the simple observation that

$$\frac{\partial}{\partial s} p(x,t|x_0,t_0) = 0.$$

Employing the Chapman-Kolmogorov equation, we can write this as

$$0 = \frac{\partial}{\partial s} \int dz\, p(x,t|z,s) p(z,s|x_0,t_0)$$

$$= \int dz \left[p(z,s|x_0,t_0) \frac{\partial}{\partial s} p(x,t|z,s) + p(x,t|z,s) \frac{\partial}{\partial s} p(z,s|x_0,t_0) \right]$$

$$= \int dz\, p(z,s|x_0,t_0) \frac{\partial}{\partial s} p(x,t|z,s) + \int dz \left\{ p(x,t|z,s) \right.$$

$$\left. \times \left(-\frac{\partial}{\partial z}[a_1(z) p(z,s|x_0,t_0)] + \frac{1}{2}\frac{\partial^2}{\partial z^2}[a_2(z) p(z,s|x_0,t_0)] \right) \right\},$$

in which we have inserted the forward Fokker-Planck equation in the second term. When we perform partial integrations in the second integral and assume that all surface contributions vanish (this requires careful checking with respect to the different possible choices of boundary conditions one can choose for the forward equation [59, 178]), we arrive at

$$0 = \int dz\, p(z,s|x_0,t_0) \left[\frac{\partial}{\partial s} p(x,t|z,s) + a_1(z) \frac{\partial}{\partial z} p(x,t|z,s) \right.$$

$$\left. + a_2(z) \frac{1}{2} \frac{\partial^2}{\partial z^2} p(x,t|z,s) \right].$$

Since this must be valid for all $p(z,s|x_0,t_0)$, we deduce the backward Fokker-Planck equation:

$$\frac{\partial}{\partial s} p(x,t|z,s) = -a_1(z) \frac{\partial}{\partial z} p(x,t|z,s) - \frac{1}{2} a_2(z) \frac{\partial^2}{\partial z^2} p(x,t|z,s)$$

$$\equiv -\mathcal{L}^+ p(x,t|z,s), \qquad (2.92)$$

where now all derivatives are taken with respect to the initial coordinates (z,s). To solve this equation in the time interval $[t_0, T]$, we have to supply a final condition at $t = T$. The partial differential operators

$$\mathcal{L} = -\frac{\partial}{\partial z} a_1(z) + \frac{1}{2} \frac{\partial^2}{\partial z^2} a_2(z) \quad \text{and} \quad \mathcal{L}^+ = a_1(z) \frac{\partial}{\partial z} + a_2(z) \frac{1}{2} \frac{\partial^2}{\partial z^2}$$

2.2 Stochastic Processes and Their Evolution Equations

appearing in the forward and backward Fokker-Planck equations are adjoint operators on the square-integrable and twice continuously differentiable functions.

We now want to make contact between these dynamic equations for the probability density of a Markov process and the equation governing the time development of the sample path $x(t)$ of this process. For this purpose we will assume that we have $p(x, 0) = \delta(x)$ as a starting condition and will look at what the forward Fokker-Planck equation predicts for the moments $\langle x^n(\Delta t) \rangle$ of the stochastic variable for small times Δt. Here the angular brackets denote an averaging with $p(x, t)$

$$\langle x(\Delta t) \rangle = \int_0^{\Delta t} dt \langle \dot{x}(t) \rangle = \int_0^{\Delta t} dt \int_\Omega dx\, x \frac{\partial}{\partial t} p(x, t)$$

$$= -\int_0^{\Delta t} dt \int_\Omega dx\, x \frac{\partial}{\partial x} [a_1(x) p(x, t)]$$

$$+ \frac{1}{2} \int_0^{\Delta t} dt \int_\Omega dx\, x \frac{\partial^2}{\partial x^2} [a_2(x) p(x, t)]$$

$$= -\int_0^{\Delta t} dt [x a_1(x) p(x, t)]_{\partial\Omega} + \int_0^{\Delta t} dt \int_\Omega dx\, a_1(x) p(x, t)$$

$$+ \frac{1}{2} \int_0^{\Delta t} dt \left[x \frac{\partial}{\partial x} (a_2(x) p(x, t)) - a_2(x) p(x, t) \right]_{\partial\Omega}.$$

Requiring that the surface terms vanish, we get

$$\langle x(\Delta t) \rangle = \int_\Omega dx \int_0^{\Delta t} dt\, a_1(x) p(x, t)$$

$$= a_1(0)\Delta t + o(\Delta t), \qquad (2.93)$$

where we used $p(x, 0) = \delta(x)$ to obtain the last result. Similarly, one can derive

$$\langle x^2(\Delta t) \rangle = a_2(0)\Delta t + o(\Delta t) \qquad (2.94)$$

and

$$\langle x^n(\Delta t) \rangle = o(\Delta t) \quad \text{for } n \geq 3. \qquad (2.95)$$

As for the model in the introductory chapter, the Fokker-Planck equation contains one term (a_2) which generates a diffusional behavior, $x^2 \propto t$, and which is therefore referred to as the diffusion coefficient.

For the description of the sample paths we refer to the phenomenological description of stochastic motion contained in the Langevin equation. It is mostly found—in the physics literature—similar to the following form:

$$\dot{x} = v(x(t)) + b(x(t))\eta(t), \qquad (2.96)$$

where we have assumed that the coefficients may depend on the stochastic variable, but should not depend explicitly on time (stationarity). Here $v(x)$ is a deterministic term which would be just a particle velocity if we interpreted x as the position of a particle. This deterministic part is, however, augmented by a term which is meant to describe the influence of randomness on the development of the sample path $x(t)$. In the case of $x(t)$ being a particle position, we would interpret it as the effect of random collisions with the surrounding medium on the particle path. (We will return to this problem in detail in the second section of the next chapter in the discussion of Brownian motion.) One usually requires these fluctuations, or noise, to have the following properties, termed Gaussian white noise (see also (2.59)):

$$\langle \eta(t) \rangle = 0 \quad \forall t,$$

$$\langle \eta(t)\eta(t') \rangle = \delta(t - t'),$$

$$\langle \eta^n(t) \rangle = 0, \quad n \geq 3, \tag{2.97}$$

$$\langle \eta(t)v(x(t')) \rangle = \langle \eta(t) \rangle \langle v(x(t')) \rangle = 0, \quad t' \leq t,$$

$$\langle \eta(t)b(x(t')) \rangle = \langle \eta(t) \rangle \langle b(x(t')) \rangle = 0, \quad t' \leq t.$$

As we hinted earlier, one usually drops the argument ω in $\eta(t, \omega)$, by which we label the random realizations of the noise term $\eta(t)$; one also does not explicitly state the sample space Ω for these random variables, preferring to use the induced probabilistic structure on $\mathbb{R}$. If we interpret these requirements, they mean that the noise at a given time t is not correlated with the values of $x(t')$ in the past, or, put differently, $x(t')$ does not depend on the effect of the noise at later times. We therefore have

$$x(\Delta t) = \int_0^{\Delta t} dt\, v(x(t)) + \int_0^{\Delta t} dt\, b(x(t))\eta(t). \tag{2.98}$$

Averaging over the random realization of the noise, we get

$$\langle x(\Delta t) \rangle = \int_0^{\Delta t} dt\, \langle v(x(t)) \rangle + \int_0^{\Delta t} dt\, \langle b(x(t)) \rangle \langle \eta(t) \rangle$$

$$= v(0)\Delta t + o(\Delta t), \tag{2.99}$$

where we used $x(t = 0) = 0$ in the last line. Similarly, we can derive

$$\langle x^2(\Delta t) \rangle = b^2(0)\Delta t + o(\Delta t) \tag{2.100}$$

and

$$\langle x^n(\Delta t) \rangle = o(\Delta t). \tag{2.101}$$

2.2 Stochastic Processes and Their Evolution Equations

Table 2.1 Correspondence between the coefficients in the Fokker-Planck and Langevin equations

Langevin	Fokker-Planck
$v(x)$	$a_1(x)$
$b^2(x)$	$a_2(x)$

We thus found the following correspondence: If we model a diffusion process, we can use either the Fokker-Planck equation approach to describe the time evolution of its probability density or the Langevin equation approach to describe the time evolution of its sample paths (see Table 2.1).

The Langevin description as we presented it above has, however, one serious drawback—it is mathematically flawed. To see this, let us consider the special case where only random noise is present and the diffusion coefficient is set to unity

$$\dot{x}(t) = \eta(t) \Rightarrow x(t) = \int_{t_0}^{t} \eta(t')dt' \quad \text{with } x(t_0) = 0. \tag{2.102}$$

As we have just derived, the Fokker-Planck equation belonging to this process is

$$\frac{\partial}{\partial t}p(x,t) = \frac{1}{2}\frac{\partial^2}{\partial x^2}p(x,t) \tag{2.103}$$

with the solution for $p(x, t_0) = \delta(x)$ and $p(x \to \pm\infty, t) = 0$ being

$$p(x,t) = \frac{1}{\sqrt{2\pi(t-t_0)}} \exp\left[-\frac{x^2}{2(t-t_0)}\right]. \tag{2.104}$$

The process with this Gaussian probability density is called the *Wiener process*, and its properties will lie at the heart of the discussions in the next section. Let us consider sample paths with fixed $p(x, t_0) = \delta(x - x_0)$. We note the following properties:

- The Wiener process (like every diffusion process) has *continuous sample paths*. Let $k > 0$ be an arbitrary real constant and consider

$$\text{Prob}\big[|\tilde{x}| > k\big] := \text{Prob}\big[|x(t_0 + \Delta t) - x_0| > k\big]$$

$$= \int_{k}^{\infty} d\tilde{x} \frac{2}{\sqrt{2\pi \Delta t}} \exp\left[-\frac{\tilde{x}^2}{2\Delta t}\right] = \text{erfc}\left(\frac{k}{\sqrt{2\Delta t}}\right),$$

where $\text{erfc}(x)$ is the complementary error function [1]. Since $\text{erfc}(x)$ vanishes for $x \to \infty$, we have

$$\lim_{\Delta t \to 0} \text{Prob}\big[|x(t_0 + \Delta t) - x_0| > k\big] = 0 \quad \forall k > 0.$$

Thus, it is almost certain that in the limit $\Delta t \to 0$ the distance $|x(t_0 + \Delta t) - x_0|$ will never become larger than k, no matter how small k becomes. Qualitatively, this agrees with the property of a continuous trajectory. This argument can be extended to the time derivative of the path, with the following result:

- The sample paths are, with probability one, nowhere differentiable:

$$\text{Prob}\left[\left|\frac{\tilde{x}}{\Delta t}\right| > k\right] := \text{Prob}\left[\left|\frac{x(t_0 + \Delta t) - x_0}{\Delta t}\right| > k\right]$$

$$= \int_{k\Delta t}^{\infty} d\tilde{x} \frac{2}{\sqrt{2\pi \Delta t}} \exp\left[-\frac{\tilde{x}^2}{2\Delta t}\right] = \text{erfc}(k\sqrt{\Delta t/2}).$$

Since erfc(x) tends to 1 for $x \to 0$, we now have

$$\lim_{\Delta t \to 0} \text{Prob}\left[\left|\frac{x(t_0 + \Delta t) - x_0}{\Delta t}\right| > k\right] = 1 \quad \forall k > 0.$$

Thus, it is almost certain that the derivative at any point of the path is larger than k, no matter how large k becomes. That is, the derivative is almost certainly infinite.

But now we are faced with a paradox. We started from the Langevin equation $\dot{x} = \eta(t)$ and used the correspondence to the Fokker-Planck equation to derive the result that there is no such thing as $\dot{x}$ for this process!

The solution to this paradox is to treat the differential $dx(t)$ of Brownian motion with unit diffusion coefficient as a new object, the stochastic differential $dW(t)$ of the Wiener process. The next section will be concerned with studying the properties of this stochastic differential, i.e., a differential which is a random variable. Clearly, the Wiener process $W(t)$ itself is then given by

$$W(t) := \int_0^t dW(t'). \tag{2.105}$$

Since W is a sum of Gaussian distributed random variables dW, it is itself Gaussian distributed and defined through its first and second moments

$$\langle W(t) \rangle = \int_0^t \langle dW(t') \rangle = 0,$$

$$\langle W(s)W(t) \rangle = \int_0^s \int_0^t \langle dW(t')dW(t'') \rangle = \min(s, t). \tag{2.106}$$

The first of these equations is obvious from the properties of the increments dW discussed above, the second equation can be derived from the results presented in the next section.

2.3 Itô Stochastic Calculus

Fig. 2.4 Definition of the stochastic integral of a function $G(t)$ over the interval $(t_0, t_n = t)$

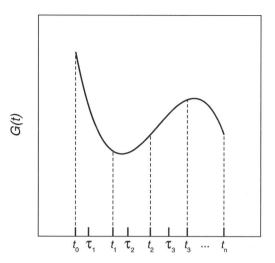

2.3 Itô Stochastic Calculus

The route to understanding the differential $dW(t)$ of the Wiener process and its meaning in the context of differential equations starts with defining how to integrate a function of time with respect to this differential.

2.3.1 Stochastic Integrals

Let us consider a function $G(t')$ over an interval $(t_0, t_n = t)$, as drawn in Fig. 2.4. In the same way as is generally done for introducing the Riemann integral, we consider a partitioning of this interval into subintervals $(t_0, t_1), (t_1, t_2), \ldots, (t_{n-1}, t_n)$ and a set of points τ_i with $t_{i-1} < \tau_i < t_i$ inside these subintervals. We then define

$$\int_{t_0}^{t_n} G(t') dW(t') = \underset{n \to \infty}{\text{ms-lim}} \sum_{i=1}^{n} G(\tau_i)\left[W(t_i) - W(t_{i-1})\right]. \tag{2.107}$$

In this equation, the difference from the definition of the Riemannian integral is the use of the mean-square (ms) limit as the type of convergence required. This mean-square limit is defined in the following way:

$$\underset{n \to \infty}{\text{ms-lim}} X_n = X \iff \lim_{n \to \infty} \langle (X_n - X)^2 \rangle = 0. \tag{2.108}$$

Convergence in the mean-square limit is a weaker requirement than the point-wise convergence used for the Riemann integral. Equation (2.107) defines a stochastic integral, but not yet uniquely, because its value depends on the choice of points τ_i at which the integrand is evaluated. To see this, let us consider the case $G(t') = W(t')$

and calculate

$$\langle S_n \rangle = \sum_{i=1}^{n} \langle W(\tau_i)[W(t_i) - W(t_{i-1})] \rangle.$$

We set $\tau_i = \alpha t_i + (1 - \alpha)t_{i-1}$, with $0 \leq \alpha \leq 1$. This gives

$$\langle S_n \rangle = \sum_{i=1}^{n} [\langle W(\alpha t_i + (1-\alpha)t_{i-1})W(t_i)\rangle - \langle W(\alpha t_i + (1-\alpha)t_{i-1})W(t_{i-1})\rangle]$$

$$= \sum_{i=1}^{n} \left(\min[\alpha t_i + (1-\alpha)t_{i-1}, t_i] - \min[\alpha t_i + (1-\alpha)t_{i-1}, t_{i-1}]\right)$$

in which we made use of a result for the covariance of the Wiener process (cf. (2.106)). So we finally arrive at

$$\langle S_n \rangle = \sum_{i=1}^{n} \alpha(t_i - t_{i-1}) = \alpha(t - t_0). \tag{2.109}$$

The two most important choices for α are

- $\alpha = 0$ (i.e. $\tau_i = t_{i-1}$), defining the *Itô stochastic integral*, and
- $\alpha = 1/2$ (i.e. $\tau_i = (t_{i-1} + t_i)/2$), defining the *Stratonovich stochastic integral*.

The Itô stochastic integral has the advantage that it is a martingale as we will show further down. This fact leads to many helpful mathematical properties and makes usage of the Itô stochastic integral ubiquitous in the literature. We will therefore solely use this definition throughout the book. For the Itô stochastic integral, we will define the meaning of a stochastic differential equation and derive a useful calculus for the manipulation of these differential equations.

A central concept for Itô stochastic integrals is that of *non-anticipating* functions.

Definition 2.10 A function $G(t)$ is called a non-anticipating function of t when $G(t)$ is statistically independent of $W(s) - W(t)$ for all s, t with $t < s$.

This is a type of causality requirement, where one does not want $G(t)$ to anticipate what the future behavior of the stochastic process $W(t)$ will be. Examples of non-anticipating functions are

- $W(t)$.
- $\int_{t_0}^{t} dt' F[W(t')]$ where F is a functional of the Wiener process.
- $\int_{t_0}^{t} dW(t') F[W(t')]$ where F is a functional of the Wiener process.
- $\int_{t_0}^{t} dt' G(t')$ if G itself is non-anticipating.
- $\int_{t_0}^{t} dW(t') G(t')$ if G itself is non-anticipating.

2.3 Itô Stochastic Calculus

For a non-anticipating functions we can now show that

$$X(t) = \int_0^t G(s) dW(s) \tag{2.110}$$

is a Martingale. In fact, for $t \geq t'$ we have

$$E[X(t)|X(t')] = E[X(t')|X(t')] + E\left[\int_{t'}^t G(s) dW(s) | X(t')\right]$$

$$= X(t') + E\left[\int_{t'}^t G(s) dW(s)\right]$$

$$= X(t'), \tag{2.111}$$

since $G(s)$ and $dW(s)$ are independent of the past, in particular of $X(t')$.

To become more acquainted with the properties of the Itô stochastic integral, we want to note the following 'mnemonic' equations:

$$[dW(t')]^2 = dt' \leftrightarrow \int_{t_0}^t [dW(t')]^2 G(t') = \int_{t_0}^t dt' G(t') \tag{2.112}$$

and

$$[dW(t')]^n = 0, \quad n \geq 3 \leftrightarrow \int_{t_0}^t [dW(t')]^n G(t') = 0 \tag{2.113}$$

for an arbitrary non-anticipating $G(t')$. One can interpret these equations to mean

$$dW(t) = O\big((dt)^{1/2}\big). \tag{2.114}$$

Let us prove the first of these relations. With $\Delta W_i^2 := [W(t_i) - W(t_{i-1})]^2$, let us define

$$I := \lim_{n \to \infty} \left\langle \left[\sum_{i=1}^n G_{i-1}(\Delta W_i^2 - \Delta t_i)\right]^2 \right\rangle$$

$$= \lim_{n \to \infty} \left[\left\langle \sum_{i=1}^n (G_{i-1})^2 (\Delta W_i^2 - \Delta t_i)^2 \right. \right.$$

$$\left. \left. + 2 \sum_{i>j} G_{i-1} G_{j-1} (\Delta W_j^2 - \Delta t_j)(\Delta W_i^2 - \Delta t_i) \right\rangle \right]$$

$$= \lim_{n \to \infty} \left[\sum_{i=1}^n \langle G_{i-1}^2 \rangle \langle (\Delta W_i^2 - \Delta t_i)^2 \rangle \right.$$

$$\left. + 2 \sum_{i>j} \langle G_{i-1} G_{j-1} (\Delta W_j^2 - \Delta t_j) \rangle \langle \Delta W_i^2 - \Delta t_i \rangle \right],$$

because G and W are non-anticipating functions. The variance of increments of the Wiener process is the variance occurring in (2.104)

$$\langle \Delta W_i^2 \rangle = \Delta t_i$$

and, due to the Gaussian nature of the Wiener process, we have

$$\langle (\Delta W_i^2 - \Delta t_i)^2 \rangle = 3\langle \Delta W_i^2 \rangle^2 - (\Delta t_i)^2 = 2(\Delta t_i)^2.$$

Using these results, we arrive at

$$I = 2 \lim_{n \to \infty} \sum_{i=1}^{n} (\Delta t_i)^2 \langle G_{i-1}^2 \rangle = 0,$$

where the last equality is valid, for instance, when G is bounded, because then

$$\sum_{i=1}^{n} (\Delta t_i)^2 \langle G_{i-1}^2 \rangle \leq \frac{(t-t_0)^2 M}{n} \quad \text{for } \langle G_{i-1}^2 \rangle \leq M$$

with M a real number ($<\infty$). We thus have derived the mean-square limit behavior as defined in (2.108):

$$\text{ms-lim}_{n \to \infty} \left(\sum_{i=1}^{n} \Delta W_i^2 G_{i-1} - \sum_{i=1}^{n} \Delta t_i G_{i-1} \right) = 0 \qquad (2.115)$$

or, equivalently,

$$\int_{t_0}^{t} G(t')[\mathrm{d}W(t')]^2 = \int_{t_0}^{t} G(t')\mathrm{d}t'.$$

The proof of the second 'mnemonic' equation uses the higher moments of the Gaussian distribution of W. Here we only want to note, for completeness, that

$$\int_{t_0}^{t} G(t')\mathrm{d}W(t')\mathrm{d}t' = \text{ms-lim}_{n \to \infty} \sum_{i=1}^{n} \Delta W_i \Delta t_i G_{i-1} = 0, \qquad (2.116)$$

and

$$\langle \mathrm{d}W(t')\mathrm{d}W(t'') \rangle$$
$$= \delta(t'-t'')\mathrm{d}t'\mathrm{d}t'' \leftrightarrow \left\langle \int_{t_0}^{t} \mathrm{d}W(t') \int_{t_0}^{t} \mathrm{d}W(t'') G(t',t'') \right\rangle = \int_{t_0}^{t} \mathrm{d}t' G(t',t').$$
$$(2.117)$$

2.3 Itô Stochastic Calculus

An important consequence that can be derived from (2.114) is the behavior of the total differential of a functional of the Wiener process [59]:

$$df[W(t), t] = \frac{\partial f}{\partial t}dt + \frac{\partial f}{\partial W}dW$$
$$+ \frac{1}{2}\frac{\partial^2 f}{\partial t^2}(dt)^2 + \frac{1}{2}\frac{\partial^2 f}{\partial W^2}(dW)^2 + \frac{1}{2}\frac{\partial^2 f}{\partial t \partial W}dt dW + \cdots$$

Making use of

$$(dt)^2 \to 0,$$
$$dt\, dW \to 0,$$
$$(dW)^2 = dt,$$

we get the result

$$df[W(t), t] = \left[\frac{\partial f}{\partial t} + \frac{1}{2}\frac{\partial^2 f}{\partial W^2}\right]dt + \frac{\partial f}{\partial W}dW(t). \quad (2.118)$$

2.3.2 Stochastic Differential Equations and the Itô Formula

This last equation already has the form of the modified Langevin equation that we wanted to give a meaning to in this section. We now can proceed to define what is meant by a stochastic differential equation.

Definition 2.11 (Itô Stochastic Differential Equation (SDE)) A stochastic process $x(t)$ obeys an Itô SDE, written as

$$dx(t) = a[x(t), t]dt + b[x(t), t]dW(t), \quad (2.119)$$

when for all t_0, t we have

$$x(t) = x(t_0) + \int_{t_0}^{t} a[x(t'), t']dt' + \int_{t_0}^{t} b[x(t'), t']dW(t'). \quad (2.120)$$

We have already defined the meaning of the stochastic integral occurring in this definition and we use it now to define the SDE (2.119) as the differential version of the stochastic integral equation (2.120), which we know how to treat.

Theorem 2.5 *For a given SDE* (2.119) *there exists a unique non-anticipating solution $x(t')$ in $[t_0, t]$ iff the following two conditions are fulfilled:*

- *Lipschitz condition*:

$$\exists k > 0: \quad |a(x,t') - a(y,t')| + |b(x,t') - b(y,t')| \leq k|x - y|,$$

for all x, y and $t' \in [t_0, t]$.
- *growth condition*:

$$\exists k > 0: \quad |a(x,t')|^2 + |b(x,t')|^2 \leq k(1 + |x|^2),$$

for all $t' \in [t_0, t]$.

The solution $x(t)$ is a Markov process.

As a final result in this chapter, which is very important in applications, we want to note the *Itô formula*. Suppose that $x(t)$ fulfills (2.119) and that f is a functional of x and t. Then one has

$$\begin{aligned}
\mathrm{d} f[x(t), t] &= \left[\frac{\partial}{\partial t} f[x(t), t] + a[x(t), t] \frac{\partial}{\partial x} f[x(t), t] + \frac{1}{2} b^2[x(t), t] \frac{\partial^2}{\partial x^2} f[x(t), t] \right] \mathrm{d}t \\
&\quad + b[x(t), t] \frac{\partial}{\partial x} f[x(t), t] \mathrm{d}W(t). \quad (2.121)
\end{aligned}$$

2.4 Summary

To conclude this chapter let us summarize with an overview of some of the types of stochastic processes we encountered and their evolution equations.

- General stochastic process (discrete time): We need to know all n-time probabilities, $p_n(x_1, t_1; \ldots; x_n, t_n)$.
- Markov process: We need to know only the one-time probabilities, $p_1(x_1, t_1)$, and the transition probabilities, $p_{1|1}(x_2, t_2|x_1, t_1)$.

  Furthermore, we have the Chapman-Kolmogorov equation

  $$p_{1|1}(x_3, t_3|x_1, t_1) = \int \mathrm{d}x_2\, p_{1|1}(x_3, t_3|x_2, t_2) p_{1|1}(x_2, t_2|x_1, t_1),$$

  with $t_1 \leq t_2 \leq t_3$ (discrete time or continuous time).
- Stationary Markov process: the master equation

  $$\partial_t p_t(x_3|x_1) = \int \mathrm{d}x_2 \left[w(x_3|x_2) p_t(x_2|x_1) - w(x_2|x_3) p_t(x_3|x_1) \right],$$

which we have written here for continuous time. For smooth jump rates w, we can apply the Kramers-Moyal expansion and derive

$$\frac{\partial}{\partial t}p(x,t) = \sum_{n=1}^{\infty} \frac{(-1)^n}{n!} \frac{\partial^n}{\partial x^n}\left[a_n(x)p(x,t)\right],$$

with

$$a_n(x) = \int_{-\infty}^{+\infty} \mathrm{d}r\, r^n w(x;r) \quad \text{(one-dimensional case)}.$$

- Diffusion process: the Fokker-Planck equation

$$\frac{\partial}{\partial t}p(x,t) = -\frac{\partial}{\partial x}\left[a_1(x)p(x,t)\right] + \frac{1}{2}\frac{\partial^2}{\partial x^2}\left[a_2(x)p(x,t)\right].$$

Equivalently, we have on the level of the individual sample paths an Itô SDE

$$\mathrm{d}x = a_1(x)\mathrm{d}t + \sqrt{a_2(x)}\mathrm{d}W(t).$$

2.5 Further Reading

In this short introduction to the mathematical background, we gave an account first of probability theory and then of the mathematics of stochastic processes.

Section 2.1

- There are many introductory textbooks on probability theory, treating discrete sample spaces as well as continuous sample spaces and the measure theoretic background of probability theory. Because of their breadth as well as depth we recommended the two volumes by Feller [51, 52] as the best source for the mathematical background of probability theory. Although the book is only partly on an introductory level, the many applications from a variety of areas that are included help to find a starting point from which to get an understanding also of the more advanced definitions and theorems. Coming back to these volumes time and again as your involvement with probabilities and stochastic processes develops, the 'non-introductory' parts will become more and more useful. For the central limit theorem and limiting distributions in general, the book by Gnedenko and Kolmogorov [62] is a classic reference.
- Statistical physics as applied probability theory is the theme of a book by Honerkamp [80]. We briefly referred to this subject in the short example on the probabilistic formulation of the canonical ensemble and more extensively when discussing the maximum entropy approach [93]. The application of the maximum entropy approach to statistical physics is also addressed in many statistical physics textbooks, e.g. in the excellent book by Reichl [176].
- Extreme value theory, and especially its applications in the context of financial risks is discussed in [44].

Section 2.2

- The mathematics of stochastic processes can also be studied using the books by Feller [51, 52]. For those interested in mathematical finance, a self-contained (but advanced) presentation can be found in the book by Bingham and Kiesel [11].
- The importance and versatility of martingale processes in probability theory can be gleaned from Williams' book [214] and, in the context of mathematical finance, again from Bingham and Kiesel [11].
- Markov processes are the most widely used level of description of stochastic phenomena. There is a vast mathematical literature on these processes, of which we only want to cite the classic books by Dynkin [40, 41].
- The applications of stochastic processes in the natural sciences are well presented in the books by van Kampen [101], Gardiner [59] and Öttinger [155]. Van Kampen critically discusses the assumptions underlying the different levels of stochastic modeling used in physics and chemistry and their inherent limits, which are sometimes forgotten in applications. He also proposes an alternative to the Kramers-Moyal expansion for a perturbational treatment of the differential version of the Chapman-Kolmogorov equation. Instead of the limit of vanishing jump distance, he proposes an expansion in the inverse volume of the system. This expansion works for self-averaging systems composed of many individual entities, where the relative size of fluctuations around a deterministic average behavior is proportional to the inverse square root of the number of entities. The book by Gardiner is an exhaustive handbook describing the tools and tricks of the trade of stochastic-method applications in the natural sciences; it also presents a well-executed introduction to the mathematics of Markov processes and especially diffusion processes. The book by Öttinger consists of two parts. The first part presents a lucid and comprehensive description of the theory of stochastic processes. This description includes Markov processes, martingales, and stochastic calculus. It serves as a background for the second part which deals with the application of stochastic processes to computer simulations of polymeric fluids. The book by Oppenheim et al. [157] is devoted to the master equation and its applications. Symmetries and ergodic properties of stochastic processes as they occur in statistical physics are discussed in Hänggi and Thomas [76].
- Diffusion processes are the theme of a book by Risken [178] which deals exclusively with the theory, solutions and applications of Fokker-Planck equations.

Section 2.3

- Itô's version of stochastic calculus is the stochastic differential equation counterpart to the Fokker-Planck (or Chapman-Kolmogorov) treatment of diffusion processes. From the many mathematical books on this topic, we want to mention the classical one by Itô and McKean [87] and also a book containing a selection of review papers on the mathematics and applications of the Itô stochastic calculus in a broad variety of disciplines [83]. A modern treatment of the mathematics of Brownian motion can be found in Karatzas and Shreve [103], and Kloeden

and Platen [106] is an extensive treatise on the mathematics and numerics of Itô stochastic differential equations. A very readable introduction to the theory of stochastic differential equations can be found in the book by Arnold [2].
- A very good discussion of stochastic calculus from the physicists point of view, comparing also e.g. the Itô and Stratonovich versions of the theory, may be found in the books by Gardiner [59] and Öttinger [155].

Chapter 3
Diffusion Processes

3.1 The Random Walk Revisited

Let us return to the 'random walk on a lattice' problem that we began to study in the introductory chapter. But now we want to be more general and consider a random walk on some d-dimensional hypercubic lattice with base vectors $(\boldsymbol{a}_1, \boldsymbol{a}_2, \ldots, \boldsymbol{a}_d) = a(\hat{\boldsymbol{e}}_1, \hat{\boldsymbol{e}}_2, \ldots, \hat{\boldsymbol{e}}_d)$, where $\hat{\boldsymbol{e}}_i$ are unit vectors and a is the lattice constant. In the following, we will choose a as our length scale and set $a = 1$. A lattice point, $\boldsymbol{r}$, can then be parameterized by its coordinate vector in this basis:

$$\boldsymbol{r} = r_1 \hat{\boldsymbol{e}}_1 + r_2 \hat{\boldsymbol{e}}_2 + \cdots + r_d \hat{\boldsymbol{e}}_d = (r_1, \ldots, r_d). \tag{3.1}$$

Furthermore, we first want to treat a finite lattice with extension $L = Na = N$ in each direction and we require *periodic boundary* conditions in each direction:

$$(r_1, \ldots, r_d) = (r_1 + N, \ldots, r_d) = \cdots = (r_1, \ldots, r_d + N). \tag{3.2}$$

The jumps of the random walker are supposed to occur in regularly spaced time intervals (so we are working in discrete time), but we want to assume that the jump distances and directions occur according to a distribution $f(\Delta \boldsymbol{r})$. With these assumptions, we are treating a stationary Markov process and can write down a discrete-time master equation for the probability of being at the lattice point $\boldsymbol{r}$ after n steps (cf. (2.82)):

$$p_n(\boldsymbol{r}) - p_{n-1}(\boldsymbol{r}) = \sum_{\boldsymbol{r}'} f(\boldsymbol{r} - \boldsymbol{r}') p_{n-1}(\boldsymbol{r}') - \sum_{\boldsymbol{r}'} f(\boldsymbol{r}' - \boldsymbol{r}) p_{n-1}(\boldsymbol{r}),$$

where we assume that $p_0(\boldsymbol{r}) = \delta_{\boldsymbol{r},\boldsymbol{0}}$. This master equation can be rewritten as

$$p_n(\boldsymbol{r}) = \sum_{\boldsymbol{r}'} f(\boldsymbol{r} - \boldsymbol{r}') p_{n-1}(\boldsymbol{r}') \tag{3.3}$$

because (compare (2.69))

$$\sum_{r'} f(r' - r) = 1. \tag{3.4}$$

This is a higher dimensional generalization of the rate equation we discussed for the one-dimensional random walker in the introductory chapter (see (1.21)).

Equation (3.3) has the form of a convolution. The tool to solve convolution equations is Fourier transformation, which we here have to apply to discrete variables. So let us define the so-called *structure function* of the random walk, which is the Fourier transform of the jump-vector distribution f. Because of the normalization of f, the structure function of the random walk is also the characteristic function (see (2.34)) of the jump-vector distribution,

$$f\left(2\pi \frac{m}{N}\right) = \sum_r f(r) \exp\left[2\pi i \frac{r \cdot m}{N}\right], \tag{3.5}$$

where $m = (m_1, m_2, \ldots, m_d)$ with $m_i = 0, 1, \ldots, N-1$ (integer). In the same way we can define the characteristic function of the probability of the random walker to be at position r at time n:

$$p_n\left(2\pi \frac{m}{N}\right) = \sum_r p_n(r) \exp\left[2\pi i \frac{r \cdot m}{N}\right]. \tag{3.6}$$

Applying the Fourier transformation to (3.3) and using the convolution theorem for Fourier transforms and the initial condition for the probability density, we get

$$p_n\left(2\pi \frac{m}{N}\right) = \left[f\left(2\pi \frac{m}{N}\right)\right]^n. \tag{3.7}$$

With a Fourier back-transformation we have formally solved our problem by expressing the lattice occupation probability at all times through the given jump-vector distribution:

$$p_n(r) = \frac{1}{N^d} \sum_{m_1=0}^{N-1} \cdots \sum_{m_d=0}^{N-1} \left[f\left(2\pi \frac{m}{N}\right)\right]^n \exp\left[-2\pi i \frac{r \cdot m}{N}\right]. \tag{3.8}$$

The prefactor $1/N^d$ is a normalization constant guaranteeing

$$\frac{1}{N^d} \sum_{m_1=0}^{N-1} \cdots \sum_{m_d=0}^{N-1} \exp\left[-2\pi i \frac{r \cdot m}{N}\right] \exp\left[2\pi i \frac{r \cdot m}{N}\right] = 1.$$

3.1 The Random Walk Revisited

To calculate the properties of this solution, it is helpful to define the *generating function* of the random walk:

$$G(r, z) := \sum_{n=0}^{\infty} p_n(r) z^n$$

$$= \frac{1}{N^d} \sum_{n=0}^{\infty} \sum_{\{m_i\}} \left[zf\left(2\pi \frac{m}{N}\right) \right]^n \exp\left[-2\pi i \frac{r \cdot m}{N}\right]$$

$$= \frac{1}{N^d} \sum_{\{m_i\}} \frac{\exp[-2\pi i r \cdot m/N]}{1 - zf(2\pi m/N)}, \qquad (3.9)$$

in which the last equation employs the properties of the geometric series and is valid for

$$\left| zf\left(2\pi \frac{m}{N}\right) \right| < 1.$$

When we consider the limit of an infinite lattice ($N \to \infty$) and denote $k = 2\pi m/N$, we have a continuum of k-space points in the interval $[0, 2\pi]^d$ or, because of the periodic boundary conditions, equivalently in $[-\pi, \pi]^d$ with a density (see Fig. 3.1)

$$\left(\frac{2\pi}{N}\right)^{-d}. \qquad (3.10)$$

The equation for the occupation probability can then be written as

$$p_n(r) = \frac{1}{(2\pi)^d} \int_{-\pi}^{+\pi} \cdots \int_{-\pi}^{+\pi} d^d k \, [f(k)]^n \exp[-i r \cdot k], \qquad (3.11)$$

and for the generating function we get:

$$G(r, z) = \frac{1}{(2\pi)^d} \int_{-\pi}^{+\pi} \cdots \int_{-\pi}^{+\pi} d^d k \, \frac{\exp[-i r \cdot k]}{1 - zf(k)}. \qquad (3.12)$$

In a completely analogous way, one can derive the corresponding formulas for a random walker in the continuum. For the probability density the equation reads

$$p_n(r) = \frac{1}{(2\pi)^d} \int_{-\infty}^{+\infty} \cdots \int_{-\infty}^{+\infty} d^d k \, [f(k)]^n \exp[-i r \cdot k], \qquad (3.13)$$

and for the generating function it is

$$G(r, z) = \frac{1}{(2\pi)^d} \int_{-\infty}^{+\infty} \cdots \int_{-\infty}^{+\infty} d^d k \, \frac{\exp[-i r \cdot k]}{1 - zf(k)}. \qquad (3.14)$$

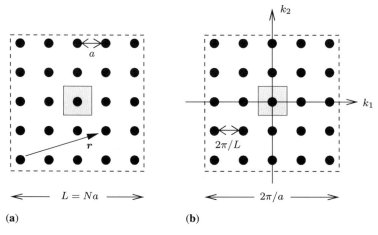

Fig. 3.1 Two-dimensional illustration of a hypercubic lattice in real (**a**) and reciprocal space (**b**). In real space the linear dimension of the lattice is $L = Na$ ($a =$ lattice constant which is chosen as the length unit, i.e., $a = 1$). The vector $\boldsymbol{r}$, defined by (3.1), denotes the distance of a lattice site from the origin (= *lower left corner*) of the coordinate system. *The shaded square* around the site in *the middle* shows the unit cell of the square lattice. *The dashed square* in (**a**) indicates the periodic boundary conditions: the lattice is periodically continued in all directions according to (3.2). Since the boundary conditions apply to any function of $\boldsymbol{r}$, we have, for instance, $\exp[ik_i r_i] = \exp[ik_i(r_i + L)]$ ($r_i, k_i = i$th component of $\boldsymbol{r}$ and $\boldsymbol{k}$). This implies $\exp[ik_i L] = 1$ so that the k-vectors of the reciprocal space are quantized: $k_i = 2\pi m_i/L$, $m_i = 0, 1, \ldots, N-1$ ($i = 1, \ldots, d; d = 2$). The integers m_i may also be chosen as $m_i = -N/2, \ldots, N/2 - 1$ due to the periodic boundary conditions. The corresponding k-vectors then lie in the first Brillouin zone (*dashed square* of linear dimension $2\pi/a$ in (**b**)). If the direct space was not a lattice, but continuous, we would have $k_i = 2\pi m_i/L$ with $m_i = 0, \pm 1, \pm 2, \ldots$. In both the discrete and continuous case the volume per lattice site in reciprocal space is $(2\pi/L)^d$ (*shaded square* in (**b**)) which justifies (3.10)

The moments of the stochastic variable $\boldsymbol{r}$ after n jumps can be calculated from the generating function as

$$\langle r^m \rangle_n = \int_{-\infty}^{+\infty} \cdots \int_{-\infty}^{+\infty} d^d r \, r^m \frac{1}{n!} \frac{d^n}{dz^n} G(\boldsymbol{r}, z)|_{z=0}, \qquad (3.15)$$

where we want to denote

$$\boldsymbol{r}^m = \begin{cases} r^{2l} & \text{for } m = 2l, \\ r^{2l} \boldsymbol{r} & \text{for } m = 2l + 1. \end{cases}$$

3.1.1 Polya Problem

Armed with this formal solution, we will proceed to discuss the famous Polya problem: *What is the probability that a random walker starting at the origin will return to this point?*

3.1 The Random Walk Revisited

First of all, let us make more explicit what *return* actually is supposed to mean: a walker starting at the origin at time 0 *returns* to that point at time n if he is at the origin at time n and has not been there for all $1 \leq j \leq n-1$. So let us define the probability $Q_n(r)$ to be at position r at time n for the first time. Its generating function is

$$F(r, z) = \sum_{n=1}^{\infty} z^n Q_n(r). \tag{3.16}$$

Since we required $p_0(r) = \delta_{r,0}$, we can write down the following relation between the p_n and Q_n:

$$p_n(r) = \sum_{j=1}^{n} p_{n-j}(0) Q_j(r). \tag{3.17}$$

After j steps the walker is at position r for the first time and for the next $n-j$ steps he proceeds through a loop, starting at r and ending at r. This is an example of what is called *renewal equation*, a versatile tool in the analysis of stochastic processes which we will use again in (3.131). According to (3.17) we can write for the generating function of p_n:

$$G(r, z) = \sum_{n=0}^{\infty} z^n p_n(r),$$

and this gives

$$G(r, z) = \delta_{r,0} + \sum_{n=1}^{\infty} z^n \sum_{j=1}^{n} p_{n-j}(0) Q_j(r)$$

$$= \delta_{r,0} + \sum_{n=1}^{\infty} \sum_{j=1}^{n} z^j Q_j(r) z^{n-j} p_{n-j}(0)$$

$$= \delta_{r,0} + \sum_{j=1}^{\infty} z^j Q_j(r) \sum_{n=j}^{\infty} z^{n-j} p_{n-j}(0)$$

$$= \delta_{r,0} + \sum_{j=1}^{\infty} z^j Q_j(r) \sum_{l=0}^{\infty} z^l p_l(0)$$

$$= \delta_{r,0} + F(r, z) G(0, z). \tag{3.18}$$

Therefore, we find that

$$F(r, z) = \begin{cases} 1 - 1/G(0, z) & \text{for } r = 0, \\ G(r, z)/G(0, z) & \text{for } r \neq 0. \end{cases} \tag{3.19}$$

Furthermore, we have by definition that

$$F(\mathbf{0}, 1) = Q_1(\mathbf{0}) + Q_2(\mathbf{0}) + \cdots,$$

so that $F(\mathbf{0}, 1)$ is the total return probability to the origin. According to (3.12), we can write

$$G(\mathbf{0}, 1) = \frac{1}{(2\pi)^d} \int_{-\pi}^{+\pi} \cdots \int_{-\pi}^{+\pi} d^d\mathbf{k} \frac{1}{1 - f(\mathbf{k})}. \qquad (3.20)$$

If this integral diverges, we have $G(\mathbf{0}, 1) = \infty$ and $F(\mathbf{0}, 1) = 1$, i.e. it is certain that the random walker returns to the origin. Otherwise, he has—according to (3.19)—a finite probability of never returning to the origin.

The singularity of the integrand occurs at $\mathbf{k} = \mathbf{0}$, since $f(\mathbf{0}) = 1$. Let us therefore consider the integral over a small hypersphere of radius k_0 around $\mathbf{k} = \mathbf{0}$. We can rewrite the integrand as

$$1 - f(\mathbf{k}) = \sum_{\mathbf{r}} f(\mathbf{r})\left(1 - e^{i\mathbf{k}\cdot\mathbf{r}}\right), \qquad (3.21)$$

because of the normalization of f, and expand the exponential

$$1 - f(\mathbf{k}) = \sum_{\mathbf{r}} f(\mathbf{r})\left(-i\mathbf{k}\cdot\mathbf{r} + \frac{1}{2}(\mathbf{k}\cdot\mathbf{r})^2 + \cdots\right).$$

To keep the exposition simple, we specialize to symmetric transition probabilities $f(\mathbf{r}) = f(-\mathbf{r})$ and $f(r\hat{\mathbf{e}}_1) = f(r\hat{\mathbf{e}}_2) = \cdots = f(r\hat{\mathbf{e}}_d)$, where the $\hat{\mathbf{e}}_i$ are unit vectors along the lattice basis directions, so that

$$\sum_{\mathbf{r}} f(\mathbf{r})\mathbf{r} = 0.$$

The leading-order term in a small-$\mathbf{k}$ expansion therefore reads

$$1 - f(\mathbf{k}) \approx \frac{1}{2}\sum_{\mathbf{r}} f(\mathbf{r}) \sum_{\alpha,\beta=1}^{d} k_\alpha k_\beta r_\alpha r_\beta = \frac{1}{2}\sum_{\alpha,\beta=1}^{d} k_\alpha k_\beta \sum_{\mathbf{r}} f(\mathbf{r}) r_\alpha r_\beta$$

$$= \frac{1}{2}\sum_{\alpha,\beta=1}^{d} k_\alpha k_\beta \langle r_\alpha^2 \rangle \delta_{\alpha\beta}$$

$$= \frac{1}{2d} k^2 \mu_2, \qquad (3.22)$$

where $\mu_2 = \langle r^2 \rangle = d\langle r_\alpha^2\rangle$. For the integral over the small hypersphere around $\mathbf{0}$, we can write

$$\int_{\text{sphere}(k_0)} d^d\mathbf{k} \frac{1}{1 - f(\mathbf{k})} = S_d \lim_{\varepsilon \to 0} \int_\varepsilon^{k_0} dk\, k^{d-1} \frac{2d}{k^2 \mu_2},$$

3.1 The Random Walk Revisited

where S_d is the surface of the unit sphere in d dimensions (see Appendix B). The radial integral is

$$I = \frac{2d}{\mu_2} \int_\varepsilon^{k_0} dk \, k^{d-3}$$

$$= \frac{2d}{\mu_2} \begin{cases} \varepsilon^{-1} - k_0^{-1} & d = 1, \\ \ln(k_0/\varepsilon) & d = 2, \\ (k_0^{d-2} - \varepsilon^{d-2})/(d-2) & d \geq 3. \end{cases} \quad (3.23)$$

Thus, we found that for $\varepsilon \to 0$ we have a diverging integral in $d = 1, 2$ and a finite integral for $d \geq 3$ and therefore

$$F(\mathbf{0}, 1) \begin{cases} = 1 & d = 1, 2 \text{ (recurrent)}, \\ < 1 & d \geq 3 \text{ (transient)}. \end{cases} \quad (3.24)$$

In one and two dimensions the random walk is recurrent, i.e., it returns to the origin with probability one, and in three and more dimensions it is transient, i.e., it has a finite probability of not returning to the origin. The exact value of $F(\mathbf{0}, 1)$ depends on the lattice type. When $d = 3$, one finds for the simple cubic lattice (SC), body-centered cubic lattice (BCC) and face-centered cubic lattice (FCC) [145]

$$F(\mathbf{0}, 1) = \begin{cases} 0.340537330 & \text{SC}, \\ 0.282229983 & \text{BCC}, \\ 0.256318237 & \text{FCC}. \end{cases} \quad (3.25)$$

Not surprisingly, the return probability decreases with increasing coordination number of the lattice, SC: 6, BCC: 8 and FCC: 12.

3.1.2 Rayleigh-Pearson Walk

We already mentioned in the introduction the problem of the two-dimensional random walker that Pearson raised in an article to Nature in 1905. Lord Rayleigh's solution to the question: 'What is the probability for the random walker to be at a distance between r and $r + dr$ from his starting position after n steps?' was

$$p(r) dr = \frac{2}{n} e^{-r^2/n} r \, dr. \quad (3.26)$$

We will derive this solution. To characterize the random walk we note that

- the steps occur at regularly spaced time points (discrete time),
- the walk is isotropic, and
- the jump distance has a distribution $p(\ell)$.

In d space dimensions we can write the jump distribution in the following way:

$$f(\boldsymbol{\ell}) = f(\ell, \theta_1, \ldots, \theta_{d-2}, \phi) = \frac{1}{S_d \ell^{d-1}} p(\ell),$$

where S_d denotes again the surface of the unit sphere in d dimensions (see Appendix B). The structure function of the walk is therefore given as

$$f(k) = \int d^d \boldsymbol{\ell} \, f(\boldsymbol{\ell}) e^{i\boldsymbol{k}\cdot\boldsymbol{\ell}}$$

$$= \frac{S_{d-1}}{S_d} \int_0^\infty d\ell \, p(\ell) \int_0^\pi d\theta (\sin\theta)^{d-2} e^{ik\ell\cos\theta}$$

$$= \frac{S_{d-1}}{S_d} \int_0^\infty d\ell \, p(\ell) \int_0^{\pi/2} d\theta \, 2(\sin\theta)^{d-2} \cos(k\ell\cos\theta)$$

$$= \Gamma(d/2) \int_0^\infty d\ell \, p(\ell) \left(\frac{2}{k\ell}\right)^{d/2-1} J_{d/2-1}(k\ell).$$

Here we have used an integral representation for the Bessel function $J_{d/2-1}$ ([67], p. 962, Eqs. (8.411–4.)), which is valid for $d > 1$. This Bessel function can be written in the following power series form (see [67], Eq. (8.440)):

$$J_{d/2-1} = \left(\frac{k\ell}{2}\right)^{d/2-1} \sum_{i=0}^\infty \frac{(-1)^i}{i!\Gamma(d/2+i)} \left(\frac{k\ell}{2}\right)^{2i}; \qquad (3.27)$$

we therefore have

$$f(k) = \int_0^\infty d\ell \, p(\ell) \sum_{i=0}^\infty \frac{(-1)^i \Gamma(d/2)}{i!\Gamma(d/2+i)} \left(\frac{k\ell}{2}\right)^{2i}.$$

Assuming that all moments of $p(\ell)$ are finite, we integrate the power series term by term. We furthermore want to focus on the large-scale properties of the walk, i.e., we require $k\ell \ll 1$ and approximate (making use of the symmetry of $p(\ell)$)

$$f(k) = \int_0^\infty d\ell \, p(\ell) \left(1 - \frac{\Gamma(d/2)}{4\Gamma(d/2+1)} k^2 \ell^2 + \cdots \right)$$

$$\approx 1 - \frac{1}{2d} k^2 \langle \ell^2 \rangle. \qquad (3.28)$$

In the same approximation we get

$$[f(k)]^n \approx 1 - \frac{n}{2d} k^2 \langle \ell^2 \rangle \approx \exp[-(n/2d) k^2 \langle \ell^2 \rangle]. \qquad (3.29)$$

3.1 The Random Walk Revisited

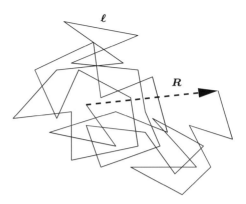

Fig. 3.2 A random coil configuration of a freely jointed chain with links ℓ and end-to-end distance $\boldsymbol{R}$

Inserting this result into (3.13) we get

$$p_n(\boldsymbol{r}) = \frac{1}{(2\pi)^d} \int d^d\boldsymbol{k}\, e^{-(n/2d)k^2\langle\ell^2\rangle} e^{-i\boldsymbol{k}\cdot\boldsymbol{r}}$$
$$= \left(\frac{d}{2\pi n\langle\ell^2\rangle}\right)^{d/2} \exp\left[-\frac{dr^2}{2n\langle\ell^2\rangle}\right]. \tag{3.30}$$

Specializing to 2 dimensions and $\langle\ell^2\rangle = 1$, we find Lord Rayleigh's result:

$$p_n(r) = 2\pi r p_n(\boldsymbol{r}) = \frac{2r}{n} e^{-r^2/n}. \tag{3.31}$$

A Polymer Model

The Rayleigh-Pearson random walk also appears in another disguise as a model in polymer physics. Here it is called the *freely jointed chain* model (see Fig. 3.2). One describes a polymer as a series of links of fixed length $\ell = \ell_0$, which are connected through random angles, and crossings of these links in space [54] are allowed. This is the Rayleigh-Pearson walk in $d = 3$ with $p(\ell) = \delta(\ell - \ell_0)$ and the number of the repeat unit or monomer when we go along the chain replacing time. For the length distribution of the end-to-end vector of a chain of N links, we can write

$$p_N(R) = 4\pi R^2 p_N(\boldsymbol{R})$$
$$= \sqrt{2/\pi}\left(\frac{N\ell_0^2}{3}\right)^{-3/2} R^2 \exp\left[-\frac{3R^2}{2N\ell_0^2}\right]. \tag{3.32}$$

From this, we find that the mean-square end-to-end distance of the polymer chain is

$$\langle R^2 \rangle = \int_0^\infty dR \sqrt{2/\pi}\left(\frac{N\ell_0^2}{3}\right)^{-3/2} R^4 \exp\left[-\frac{3R^2}{2N\ell_0^2}\right]$$
$$= N\ell_0^2. \tag{3.33}$$

This so-called Gaussian or *random coil* state of a polymer chain is, to a very good approximation, realized in long-chain polymer melts and for isolated long chains in special solvents (θ-solvents) [179]. In these situations, the excluded volume interactions between different parts of the chain are efficiently screened by the interactions with the surrounding medium (other chains in the melt case, solvent molecules for θ-solvents) so that one may neglect these excluded volume forces for the calculation of the chain size to a first approximation. Corrections to this Gaussian behavior are known for both polymer melts [216] and θ-solvents [72, 186].

Of course, a polymer chain cannot really cross itself, and in general one has to take this into account in the modeling. This leads to the concept of the self-avoiding walk (SAW). When one considers a walk on a lattice, the SAW is not allowed to visit the same lattice point more than once. In a continuum model, one would need to assign a volume to the links (monomers), for instance, treating them as tangent hard spheres. The statistical properties of the SAW differ significantly from the those of the random walk [122, 177, 192]. For instance, the end-to-end distance scales with chain length N as

$$\langle R^2 \rangle \propto N^{2\nu} \quad \text{with } \nu = \begin{cases} 0.589 & d=3, \\ 3/4 & d=2. \end{cases} \tag{3.34}$$

3.1.3 Continuous-Time Random Walk

In this chapter we have so far considered random walks with jumps occurring at fixed times with a prescribed jump-distance distribution. It is natural to also consider the time between two jumps as a random variable. Walks with a distribution of *waiting times* between the jumps are called *continuous-time random walks* (CTRWs) [145].

More precisely, a CTRW can be described as follows: Consider a random walker being at the origin at time $t = 0$. The walker waits for a time $\tau_1 = t_1 - 0$ before jumping by $\boldsymbol{\ell}_1$ away from the origin, where he waits again for some time $\tau_2 = t_2 - t_1$ before making a jump to a new position $\boldsymbol{\ell}_1 + \boldsymbol{\ell}_2$, and so on up to time t. This process, for which the jump distances $\{\boldsymbol{\ell}_1, \boldsymbol{\ell}_2, \ldots\}$ and the waiting times $\{\tau_1, \tau_2, \ldots\}$ are iid random variables, is the CTRW. In the following we will examine separable CTRWs, that is, random walks for which the joint distribution of jump distances and waiting times factorizes:

$$\Psi(\boldsymbol{\ell}, \tau) = f(\boldsymbol{\ell})\psi(\tau). \tag{3.35}$$

Here $f(\boldsymbol{\ell})$ is the jump-vector distribution introduced in Sect. 3.1 and $\psi(\tau)\mathrm{d}\tau$ is the probability for a jump to occur in the time interval $[\tau, \tau + \mathrm{d}\tau]$ when the last jump occurred at time $\tau = 0$.

In the preceding sections, the key quantity was $p_n(\boldsymbol{r})$, the probability to be at position $\boldsymbol{r}$ after the nth jump. This gave directly the probability $p(\boldsymbol{r}, t)$ to be at $\boldsymbol{r}$ at time t, since $t = n\Delta t$ and the time interval Δt between jumps was fixed. For

3.1 The Random Walk Revisited

the CTRW, however, Δt is replaced by the random waiting time τ, implying that the number of jumps (n) occurring in a given time interval $[0, t]$ also becomes a random variable. Therefore, the relation between $p_n(r)$ and $p(r, t)$ is more complicated; it involves the probability $P_n(t)$ that n jumps take the time t:

$$p(r, t) = \sum_{n=0}^{\infty} P_n(t) p_n(r). \tag{3.36}$$

Here the product $P_n(t) p_n(r)$ is the probability to be at r after n jumps and that the n jumps take exactly the time t (this holds due to the separability of the CTRW). The sum over n accounts for the fact that any number of jumps may occur in $[0, t]$.

The new quantity in (3.36), $P_n(t)$, is related to the waiting time distribution. Let us denote by $\psi_n(\tau)$ the probability density for the nth jump to occur after a waiting time τ, and by $\phi(\tau)$ the probability that no jump occurs in the interval $[0, \tau]$, i.e.,

$$\phi(\tau) = 1 - \int_0^\tau \psi(\tau') d\tau' = \int_\tau^\infty \psi(\tau') d\tau'. \tag{3.37}$$

Furthermore, let us assume that $\tau \leq t$. Then, the product $\phi(t - \tau) \psi_n(\tau)$ is the probability that n jumps take a time τ and that no further jump occurs in the remaining time $t - \tau$. Integration over all values of τ ($0 \leq \tau \leq t$) gives $P_n(t)$:

$$P_n(t) = \int_0^t \phi(t - \tau) \psi_n(\tau) d\tau. \tag{3.38}$$

By inserting this expression into (3.36) we get

$$p(r, t) = \int_0^t \phi(t - \tau) Q(r, \tau) d\tau \tag{3.39}$$

with $Q(r, \tau)$ being the probability to arrive at position r at time τ,

$$Q(r, \tau) = \sum_{n=0}^{\infty} \psi_n(\tau) p_n(r). \tag{3.40}$$

To proceed, we can write down the following recurrence relation for $\psi_n(\tau)$:

$$\psi_n(\tau) = \int_0^\tau \psi(\tau - \tau') \psi_{n-1}(\tau') d\tau', \quad \psi_0(\tau) = \delta(\tau). \tag{3.41}$$

Due to the convolution theorem for Laplace transforms this recurrence relation simplifies upon Laplace transformation ($t \to s$):

$$\psi_n(s) := \int_0^\infty d\tau e^{-s\tau} \psi_n(\tau) = \psi(s) \psi_{n-1}(s) = [\psi(s)]^n.$$

With this result Laplace transformation of (3.40) yields

$$Q(r,s) = \sum_{n=0}^{\infty} [\psi(s)]^n p_n(r), \tag{3.42}$$

which is exactly the definition of the generating function of the walk $G(r, z)$ in (3.9) with z replaced by $\psi(s)$, the Laplace transform of the waiting-time distribution:

$$Q(r,s) = G(r, \psi(s)).$$

By Laplace transformation of (3.39), we therefore get

$$p(r,s) = \phi(s) G(r, \psi(s)) = \frac{1}{s}[1 - \psi(s)] G(r, \psi(s)), \tag{3.43}$$

where we utilized an integration by parts and the normalization of $\psi(t)$ to determine the Laplace transform of $\phi(t)$ from (3.37). Now we can use the result (3.14) for the Fourier representation of the generating function G to finally get for the Fourier-Laplace transform of $p(r, t)$:

$$p(k,s) = \frac{1}{s} \frac{1 - \psi(s)}{1 - \psi(s) f(k)}, \tag{3.44}$$

where $f(k)$ is the structure function of the walk defined in (3.5).

For the Rayleigh-Pearson walk we looked at the properties for large distances, i.e., for $k\ell \ll 1$. We now extend this to derive the behavior of the CTRW for large distances and long times, i.e., we also require $s\tau \ll 1$. We therefore approximate the structure function and the Laplace transform of the waiting-time distribution in the following way:

$$f(k) \approx 1 + i k \cdot \langle \ell \rangle - \frac{1}{2d} k^2 \langle \ell^2 \rangle, \tag{3.45}$$

$$\psi(s) \approx 1 - s \langle \tau \rangle = 1 - s \tau_0, \tag{3.46}$$

where we assume

$$\langle \tau \rangle = \int_0^\infty \tau \psi(\tau) d\tau =: \tau_0 < \infty. \tag{3.47}$$

Inserting these approximations into (3.44), we get

$$p(k,s) = \frac{\tau_0}{s\tau_0 - i k \cdot \langle \ell \rangle (1 - s\tau_0) + k^2 \langle \ell^2 \rangle (1 - s\tau_0)/2d}.$$

In the same way as we did for the one-dimensional random walk with fixed jump length Δx and waiting time Δt in Chap. 1, we now perform the following limiting procedure:

$$\tau_0 \to 0, \quad \langle \ell \rangle \to \mathbf{0} \quad \text{and} \quad \frac{\langle \ell \rangle}{\tau_0} \to v, \tag{3.48}$$

$$\tau_0 \to 0, \quad \langle \ell^2 \rangle \to 0 \quad \text{and} \quad \frac{\langle \ell^2 \rangle}{2d\tau_0} \to D. \tag{3.49}$$

With this, we arrive at

$$p(\mathbf{k}, s) = \frac{1}{s - i\mathbf{k} \cdot \mathbf{v} + Dk^2}. \tag{3.50}$$

This is the Fourier-Laplace transform of a Gaussian diffusion process with conditional probability $p(\mathbf{r}, t | \mathbf{r}_0, 0)$ given by

$$p(\mathbf{r}, t | \mathbf{r}_0, 0) = (4\pi Dt)^{-d/2} \exp\left(-\frac{(\mathbf{r} - \mathbf{r}_0 - \mathbf{v}t)^2}{4Dt}\right). \tag{3.51}$$

Not surprisingly, we once more derived that, when the first two moments of a probability distribution exist (in this case of the jump-size distribution), the sum variable, i.e., the displacement on large scales, asymptotically is described by a Gaussian distribution. For the CTRW, we furthermore had to require the existence of the first moment of the waiting-time distribution.

The interesting case of what happens when either of the first two moments of the jump-size distribution or the first moment of the waiting-time distribution do not exist, will be discussed in the next chapter on Lévy distributions and Lévy flights. We will now proceed with a treatment of the classical problem that gave rise to the concept and studies of what now are called diffusion processes.

3.2 Free Brownian Motion

In this section we will discuss the archetypical stochastic process, the Brownian motion problem. As already mentioned in Chap. 1, in 1827 the Scottish botanist Brown observed a very irregular motion of pollen particles immersed in a solvent. It took about 80 years until Einstein and Smoluchowski gave a theoretical explanation of this phenomenon. With todays knowledge, we would start to model this system in the following way.

The fluid particles are much lighter than the Brownian particle (Fig. 3.3). Due to this mass difference they move much faster, exerting a myriad of (weak) collisions on the Brownian particle. The compound effect of these collisions is to impart a random force and a viscous drag on the Brownian particle. We model these two aspects as follows:

- The viscosity of the fluid leads to a viscous drag force on the Brownian particle which counteracts its motion. Since the velocity of the particle is small, the drag force can be taken to be proportional to the velocity difference between the particle and the fluid (Stokes friction). Assuming the fluid to be at rest, the drag force is thus given by

$$\mathbf{f}_{\text{drag}} = -\gamma M \mathbf{v}, \tag{3.52}$$

where γ is the friction coefficient of the Brownian particle and M its mass.

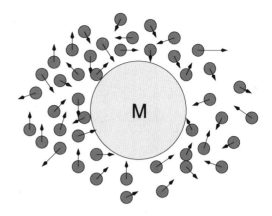

Fig. 3.3 A large Brownian particle of mass M immersed in a fluid of much smaller and lighter particles

- Because of the mass difference, the small fluid particles move much faster than the Brownian particle which we want to describe. We will approximate the collisions between the Brownian particle and the fluid particles as occurring instantaneously and imparting a random velocity change onto the Brownian particle. Furthermore, we assume that all collisions are uncorrelated. We will therefore model the velocity changes of the Brownian particle due to the stochastic collisions by a Wiener process with a strength Γ.

For the sake of later generalization we also include an external force $\boldsymbol{F}$ acting on the Brownian particle into the description. Thus, we write the equations of motion of the particle as

$$\begin{aligned} d\boldsymbol{x} &= \boldsymbol{v} dt, \\ M d\boldsymbol{v} &= \big(\boldsymbol{F}(\boldsymbol{x}) - \gamma M \boldsymbol{v}\big) dt + \Gamma d\boldsymbol{W}(t). \end{aligned} \tag{3.53}$$

These two coupled stochastic differential equations are equivalent to the following Fokker-Planck equation, which is called the Kramers equation [101, 110],

$$\frac{\partial}{\partial t} p(\boldsymbol{x}, \boldsymbol{v}, t) = -\nabla_{\boldsymbol{x}} \cdot \big[\boldsymbol{v} p(\boldsymbol{x}, \boldsymbol{v}, t)\big] + \nabla_{\boldsymbol{v}} \cdot \left[\left(\gamma \boldsymbol{v} - \frac{\boldsymbol{F}(\boldsymbol{x})}{M}\right) p(\boldsymbol{x}, \boldsymbol{v}, t)\right] \\ + \frac{1}{2}\left(\frac{\Gamma}{M}\right)^2 \Delta_{\boldsymbol{v}} p(\boldsymbol{x}, \boldsymbol{v}, t). \tag{3.54}$$

With the knowledge of Chap. 2 we can qualitatively understand how this equation comes about. The deterministic terms, $\boldsymbol{v}$ and $\boldsymbol{F}(\boldsymbol{x}) - \gamma M \boldsymbol{v}$, give rise to first order derivatives at the Fokker-Planck level, whereas the stochastic term leads to a second order derivative with an amplitude proportional to Γ^2 (see Table 2.1). Equation (3.54) was used in 1940 by Kramers to describe the kinetics of chemical reactions.

3.2 Free Brownian Motion

3.2.1 Velocity Process

Let us now specialize to the classical Brownian motion problem and choose $F = 0$ and apply a one-dimensional treatment. Without external force the velocity and position equations decouple and we can focus on the stochastic differential equation for the velocity of the Brownian particle,

$$dv = -\gamma v dt + \frac{\Gamma}{M} dW(t). \tag{3.55}$$

The equivalent Fokker-Planck equation for the probability distribution of the velocity of the Brownian particle is

$$\frac{\partial}{\partial t} p(v,t) = \frac{\partial}{\partial v}[\gamma v p(v,t)] + \frac{1}{2}\left(\frac{\Gamma}{M}\right)^2 \frac{\partial^2}{\partial v^2} p(v,t). \tag{3.56}$$

Due to (2.91), (3.56) also holds for the conditional probability $p(v,t|v_0,t_0)$. The stochastic process that is defined by (3.55) or (3.56) is called the *Ornstein-Uhlenbeck process*. Taking the average of (3.55) we get

$$\langle v(t) \rangle = \langle v_0 \rangle e^{-\gamma t}. \tag{3.57}$$

The average velocity in an ensemble of Brownian particles (realizations of the Ornstein-Uhlenbeck process) is exponentially damped with a time constant,

$$\tau_{\text{damping}} = \gamma^{-1}. \tag{3.58}$$

For a given realization of the process (realization of $W(t)$) the stochastic variable $v(t)$ will show fluctuations around the average velocity. We therefore make the following ansatz for the solution of (3.55):

$$v(t) = v_0 e^{-\gamma t}\bigl(1 + f(t)\bigr).$$

Inserting this into (3.55), we get

$$v_0 e^{-\gamma t} df(t) = \frac{\Gamma}{M} dW(t) \Rightarrow f(t) = \frac{\Gamma}{M v_0} \int_0^t e^{\gamma t'} dW(t'). \tag{3.59}$$

The equation for the fluctuations contains no drift term—as one would require and as is also implied by the choice of word. The incremental changes of the fluctuating part of the velocity are a fair game ($\langle df(t) \rangle = 0$) and therefore the equilibrium fluctuations—which are a sum of fair-game variables—are a martingale. We want to note that by subtracting the drift from the total velocity process we have generated a process (fluctuations) that is a martingale. This property is also central to the Black-Scholes theory of option pricing, and we will discuss its mathematical background in more detail in Chap. 5. The solution of (3.55) is then given by

$$v(t) = v_0 e^{-\gamma t} + \frac{\Gamma}{M} e^{-\gamma t} \int_0^t e^{\gamma t'} dW(t'). \tag{3.60}$$

With this solution we can calculate the average of the squared velocity of the Brownian particle:

$$\langle v^2(t)\rangle = \langle v_0^2\rangle e^{-2\gamma t} + 2\frac{\langle v_0\rangle \Gamma}{M} e^{-2\gamma t} \int_0^t e^{\gamma t'}\langle dW(t')\rangle$$

$$+ \frac{\Gamma^2}{M^2} e^{-2\gamma t} \int_0^t \int_0^t e^{\gamma(t'+t'')}\langle dW(t')dW(t'')\rangle$$

$$= \langle v_0^2\rangle e^{-2\gamma t} + \frac{\Gamma^2}{M^2} e^{-2\gamma t} \int_0^t e^{2\gamma t'} dt',$$

in which we made use of the properties of the Wiener process, (cf. (2.117)). Our final result is

$$\langle v^2(t)\rangle = \frac{\Gamma^2}{2\gamma M^2} + \left(\langle v_0^2\rangle - \frac{\Gamma^2}{2\gamma M^2}\right) e^{-2\gamma t}. \qquad (3.61)$$

The stationary value of the mean-square velocity is reached for $t \to \infty$

$$\langle v^2(\infty)\rangle = \frac{\Gamma^2}{2\gamma M^2}.$$

This has to be equal to the mean thermal velocity of the Brownian particle, which is thermalized by the collisions with the fluid particles. If the fluid is kept at a temperature T, we obtain the following for the mean kinetic energy of the Brownian particle using the *equipartition theorem* [176]:

$$\frac{M}{2}\langle v^2(\infty)\rangle = \frac{1}{2}k_B T \Rightarrow \Gamma^2 = 2\gamma M k_B T. \qquad (3.62)$$

We conclude that (3.55) only correctly describes the long time behavior of the Brownian particle—where initial effects have decayed and the motion of the particle is solely triggered by the contact with the solvent in thermal equilibrium—if the two parameters in (3.55) are interrelated by (3.62). This is a version of the *fluctuation–dissipation theorem* [176] relating the strength of equilibrium fluctuations (Γ) to the magnitude of the dissipation (γ).

The autocorrelation function of the velocity can also be calculated explicitly:

$$\langle v(t)v(s)\rangle = \langle v_0^2\rangle e^{-\gamma(t+s)}$$

$$+ \frac{2k_B T \gamma}{M} e^{-\gamma(t+s)} \int_0^t \int_0^s e^{\gamma(t'+t'')}\langle dW(t')dW(t'')\rangle$$

$$= \langle v_0^2\rangle e^{-\gamma(t+s)} + \frac{2k_B T \gamma}{M} e^{-\gamma(t+s)} \int_0^{\min(t,s)} e^{2\gamma t'} dt'.$$

Here we again made use of the fact that $\langle dW(t')dW(t'')\rangle = dt'\delta_{t',t''}$, so that the two stochastic integrals only contribute along the line $t' = t''$ and only up to $\min(t, s)$.

3.2 Free Brownian Motion

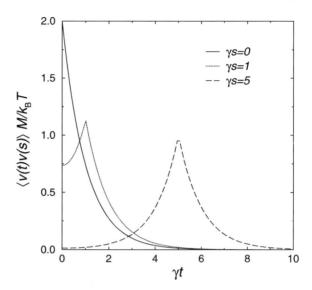

Fig. 3.4 Velocity autocorrelation function (see (3.63)) of a Brownian particle for different reference times, s, as indicated in the figure. Velocities are normalized by the thermal velocity, $v_T^2 = k_B T/M$. The starting velocity at $s = 0$ is supposed to be $v_0^2 = 2v_T^2$

So we get

$$\langle v(t)v(s)\rangle = \left(\langle v_0^2\rangle - \frac{k_B T}{M}\right) e^{-\gamma(t+s)} + \frac{k_B T}{M} e^{-\gamma|t-s|} \quad (3.63)$$

because

$$\left.\begin{array}{ll} t > s & t+s-2\min(t,s) = t-s > 0 \\ t < s & t+s-2\min(t,s) = -(t-s) > 0 \end{array}\right\} \Rightarrow t+s-2\min(t,s) = |t-s|. \quad (3.64)$$

The stationary behavior of the autocorrelation function is attained in the limit $t, s \to \infty$ with $|t - s|$ fixed. This gives

$$\langle v(t)v(s)\rangle \stackrel{t,s\to\infty}{=} \frac{k_B T}{M} e^{-\gamma|t-s|}. \quad (3.65)$$

In the stationary limit the velocity autocorrelation function thus only depends on $|t - s|$ and the equal-time ($t = s$) mean-square velocity is constant. It is given by the equilibrium value, $\langle v^2(t)\rangle = v_T^2$ with $v_T^2 = k_B T/M$ being the thermal velocity.

Equation (3.63) is plotted in Fig. 3.4 for three reference times, $\gamma s = 0$, $\gamma s = 1$ and $\gamma s = 5$, and a (non-stationary) starting velocity of $\langle v_0^2\rangle = 2v_T^2$. The decay of velocity correlations with respect to the starting condition is simple exponential (curve $\gamma s = 0$). When the reference time is equal to the damping time (curve $\gamma s = 1$), there is an asymmetric cusp around $t = s$ with a maximum value larger than one. Finally, for a reference time much larger than the damping time (curve $\gamma s = 5$), we find the stationary behavior with a cusp of height one around $t = s$.

The Velocity Distribution

We complete the analysis of the velocity of the Brownian particle by deriving its distribution function. From (3.57), (3.60) and (3.62) we see that

$$u(t) := v(t) - \langle v(t) \rangle = \sqrt{\frac{2\gamma k_B T}{M}} e^{-\gamma t} \int_0^t e^{\gamma t'} dW(t'). \tag{3.66}$$

Being the sum (integral) of Gaussian-distributed random variables $dW(t')$, $u(t)$ is a Gaussian-distributed random variable with the following first and second moments

$$\langle u(t) \rangle = 0 \quad \text{and} \quad \langle u^2(t) \rangle = \frac{k_B T}{M}\left(1 - e^{-2\gamma t}\right).$$

So the conditional probability to transition from $u(t=0) = 0$ (from $v(t=0) = v_0$) to u at time t (to $v(t)$) is

$$p(u, t|0, 0) = \frac{1}{\sqrt{2\pi \langle u^2 \rangle}} e^{-u^2/2\langle u^2 \rangle},$$

and inserting the definition of u we get

$$p(v, t|v_0, 0)$$
$$= \frac{1}{\sqrt{2\pi k_B T (1 - e^{-2\gamma t})/M}} \exp\left[-\frac{M}{2k_B T} \frac{(v - v_0 e^{-\gamma t})^2}{1 - e^{-2\gamma t}}\right]. \tag{3.67}$$

One can verify by insertion that this is the solution to the Fokker-Planck equation (3.56) with the starting condition $p(v, 0|v_0, 0) = \delta(v - v_0)$ and the boundary conditions $p(v \to \pm\infty, t|v_0, 0) = 0$. Thereby, it is also the so-called *Green's-function* solution to this partial differential equation, or, in other terms, the *Gaussian propagator* between $(v_0, 0)$ and (v, t). Since $p(v, 0) = \delta(v - v_0)$ and we have (see (2.65) and (2.68))

$$p(v, t) = \int_{-\infty}^{+\infty} p(v, t|v', 0) p(v', 0) dv',$$

Eq. (3.67) also holds for probability $p(v, t)$ of finding the velocity v at time t.

Figure 3.5 shows the behavior of the velocity distribution for three different times. The starting distribution at $t = 0$ is chosen as $p(v_0) = \delta(v_0 - 2v_T)$, as in Fig. 3.4. For $\gamma t = 0.1$ we see a rounding of the initial δ peak and already a small shift towards $v = 0$. The broadening and the shift towards $v = 0$ proceed in time until, for $\gamma t = 5$, the distribution closely approximates the normal distribution in the chosen variable.

The normal distribution is the stationary limit of $p(v, t)$. For $t \to \infty$ we find that (3.67) tends towards the Maxwell-Boltzmann velocity distribution

$$p(v, t) \xrightarrow{t \to \infty} p_{eq}(v) = \frac{1}{\sqrt{2\pi k_B T/M}} \exp\left[-\frac{Mv^2}{2k_B T}\right]. \tag{3.68}$$

3.2 Free Brownian Motion

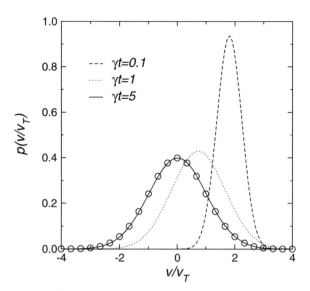

Fig. 3.5 Probability distribution for an ensemble of Brownian particles starting with $v_0 = 2v_T$ at time $t = 0$. For $t > 0$ we see a Gaussian distribution whose center position is relaxing to zero. The stationary case is a normal distribution (*circles*) with mean zero and width one in the scaled representation

If we took the starting velocity v_0 not from $\delta(v - v_0)$, but from the equilibrium distribution (i.e., $\langle v_0^2 \rangle = v_T^2$), $p_{\text{eq}}(v)$ would be preserved under the Brownian dynamics. That is,

$$p_{\text{eq}}(v) = \int_{-\infty}^{+\infty} p(v, t | v', 0) p_{\text{eq}}(v', 0) dv'. \tag{3.69}$$

This equation and (3.67) for the conditional probability show that the Ornstein-Uhlenbeck process satisfies the condition (2.78). Thus, it is a stationary, Gaussian, Markov process. According to a theorem by Doob [209], the Ornstein-Uhlenbeck process is the only stochastic process with these properties. Whenever we are to model a problem as a stationary, Gaussian, Markov process, we can therefore write down the unique solution for the time dependent-transition probabilities according to (3.67).

3.2.2 Position Process

We want to conclude the analysis of the classical Brownian motion problem by calculating the position of the Brownian particle. From

$$dx(t) = v(t) dt,$$

we get with (3.60), (3.62), and setting $x(0) = 0$,

$$x(t) = \frac{v_0}{\gamma}\left(1 - e^{-\gamma t}\right) + \sqrt{\frac{2\gamma k_B T}{M}} \int_0^t dt' e^{-\gamma t'} \int_0^{t'} dW(t'') e^{\gamma t''}. \tag{3.70}$$

The mean particle position is given by

$$\langle x(t) \rangle = \frac{v_0}{\gamma}\left(1 - e^{-\gamma t}\right), \tag{3.71}$$

behaving for small times as $\langle x(t) \rangle = v_0 t$ and approaching $x_\infty = \langle x(t \to \infty) \rangle = v_0/\gamma$, which is the asymptotic displacement reached under the influence of the friction force when the particle starts at $x = 0$ with velocity v_0.

For the second moment we have to calculate

$$\langle x^2(t) \rangle = \frac{v_0^2}{\gamma^2}\left(1 - e^{-\gamma t}\right)^2 + \frac{2\gamma k_B T}{M} \int_0^t dt' e^{-\gamma t'} \int_0^t dt'' e^{-\gamma t''}$$

$$\times \int_0^{t'} e^{\gamma \tilde{t}'} \int_0^{t''} e^{\gamma \tilde{t}''} \langle dW(\tilde{t}') dW(\tilde{t}'') \rangle$$

$$= \frac{v_0^2}{\gamma^2}\left(1 - e^{-\gamma t}\right)^2 + \frac{2\gamma k_B T}{M} \int_0^t dt' e^{-\gamma t'} \int_0^t dt'' e^{-\gamma t''}$$

$$\times \int_0^{\min(t',t'')} dt''' e^{2\gamma t'''},$$

where we used again the covariance of the Wiener process (2.117). The triple integral can be calculated with the help of (3.64) and

$$\int_0^t dt' \int_0^t dt'' e^{-\gamma |t'-t''|} = 2 \int_0^t dt' \int_0^{t'} dt'' e^{-\gamma(t'-t'')}.$$

The final result for the mean-square displacement of the Brownian particle reads

$$\langle x^2(t) \rangle = \left[\frac{v_0^2}{\gamma^2} - \frac{k_B T}{M\gamma^2}\right]\left(1 - e^{-\gamma t}\right)^2 + \frac{2k_B T}{M\gamma}\left[t - \frac{1}{\gamma}\left(1 - e^{-\gamma t}\right)\right]$$

$$= \frac{2k_B T}{M\gamma} t + \frac{v_0^2}{\gamma^2}\left(1 - e^{-\gamma t}\right)^2 - \frac{k_B T}{M\gamma^2}\left(3 - 4e^{-\gamma t} + e^{-2\gamma t}\right). \tag{3.72}$$

Asymptotically for $t \to \infty$, we recover the diffusive behavior

$$\langle x^2(t) \rangle \overset{t \to \infty}{\propto} 2Dt, \quad D = \frac{k_B T}{M\gamma}, \tag{3.73}$$

where D is the diffusion coefficient. This is Einstein's result for the diffusive motion of the Brownian particle. For short times, on the other hand, the Ornstein-Uhlenbeck description of Brownian motion captures the ballistic motion of the particle:

$$\langle x(t) \rangle = v_0 t \quad \text{and} \quad \langle x^2(t) \rangle = v_0^2 t^2. \tag{3.74}$$

From the preceding discussion we see that the position process depends on the value of v_0. Let us assume that $v_0 = v_T = \sqrt{k_B T/M}$, i.e., that the underlying velocity process is stationary. Then, the velocity autocorrelation function decays exponentially

3.2 Free Brownian Motion

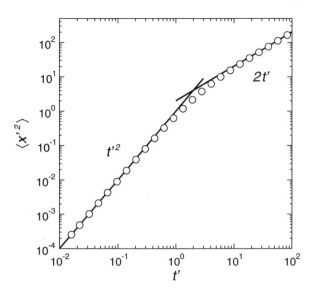

Fig. 3.6 Mean-square displacement (3.76) of a Brownian particle according to the Ornstein-Uhlenbeck description of Brownian motion (*circles*). Time and distance are measured in scaled coordinates (see text). The asymptotic behavior for $t \to 0$ (3.74) and $t \to \infty$ (3.73) is also shown (*solid lines*)

(3.65) and the autocorrelation function of the particle positions is given by

$$\langle x(t)x(s)\rangle = \int_0^t dt' \int_0^s dt'' \langle v(t')v(t'')\rangle$$

$$= \frac{k_B T}{M} \int_0^t dt' \int_0^s dt'' e^{-\gamma|t'-t''|}$$

$$= \frac{k_B T}{M}\left[\frac{2}{\gamma}\min(s,t) - \frac{1}{\gamma^2}\left(1 - e^{-\gamma s} - e^{-\gamma t} + e^{-\gamma|t-s|}\right)\right]. \quad (3.75)$$

For $t = s$ this equation gives back (3.72), if we take $v_0 = v_T$. Defining the normalized coordinates

$$x' := x/x_\infty \quad \text{and} \quad t' := \gamma t,$$

the result for the mean-square displacement can be written in the following scaled form:

$$\langle x'^2(t')\rangle = 2[t' - (1 - e^{-t'})]. \quad (3.76)$$

This function is plotted in Fig. 3.6 in a double-logarithmic fashion. At times $t' < 1$ the displacement follows a t'^2 law, showing the ballistic short time behavior of the process, and for $t' \sim 1$ we see a crossover to the linear diffusion behavior, also predicted in Einstein's treatment of Brownian motion.

The Position Distribution

Based on the results obtained in this section we want to derive the distribution function for the position of the Brownian particle. This can be done in close analogy to the velocity distribution of the Ornstein-Uhlenbeck process. Equation (3.70) shows that

$$u(t) := x(t) - \langle x(t) \rangle = x(t) - \frac{v_0}{\gamma}(1 - e^{-\gamma t})$$

is a sum (integral) of Gaussian distributed random variables and thus its distribution is also Gaussian. If we take again $v_0 = v_T$, the first and second moments are given by

$$\langle u(t) \rangle = 0$$
$$\sigma_x^2(t) := \langle u^2(t) \rangle = \frac{2v_T^2}{\gamma} t - \frac{v_T^2}{\gamma^2}(3 - 4e^{-\gamma t} + e^{-2\gamma t}). \tag{3.77}$$

These moments fully determine the conditional probability $p(x,t|0,0)$ to transition from $x_0 = 0$ at $t = 0$ to x at time t and, due to

$$p(x,t) = \int_{-\infty}^{+\infty} p(x,t|x_0,0) p(x_0,0) dx_0$$
$$= \int_{-\infty}^{+\infty} p(x,t|x_0,0) \delta(x_0) dx_0 = p(x,t|0,0),$$

also the probability $p(x,t)$ to find the Brownian particle at position x at time t. So we have

$$p(x,t) = \frac{1}{\sqrt{2\pi \sigma_x^2(t)}} \exp\left[-\frac{(x - v_T(1 - e^{-\gamma t})/\gamma)^2}{2\sigma_x^2(t)}\right] \tag{3.78}$$

$$\xrightarrow{t \text{ large}} \frac{1}{\sqrt{4\pi Dt}} \exp\left[-\frac{(x - v_T/\gamma)^2}{4Dt}\right] \tag{3.79}$$

with the diffusion coefficient $D = k_B T/M\gamma$ as in (3.73).

Let us conclude this section by two remarks. First, (3.79) shows that $p(x,t)$ vanishes for $t \to \infty$. Thus, the position process of Brownian motion is Gaussian, time homogeneous,[1] but—contrary to the velocity process— not stationary. Second, (3.79) also shows that we recover the Wiener process (see (2.104)) from it in the limit $\gamma \to \infty$, provided that $D = v_T^2/\gamma$ remains fixed. Keeping D finite implies that we have to increase the temperature in such as way that the fuelled thermal energy counterbalances the viscous slowing down of the Brownian motion caused by the increase of γ.

3.3 Caldeira-Leggett Model

Our discussion of free Brownian motion in the preceding section started from a separation of time scales. Due to the huge difference in mass, fluid particles move

---

[1] Time homogeneity means that $p(x,t|x_0,0)$ only depends on the time difference.

3.3 Caldeira-Leggett Model

much faster than the Brownian particle. Therefore, the fluid can equilibrate on the time scale where the motion of the Brownian particle is observed. This argument naturally decomposes the problem into a system, the Brownian particle, and an environment, the surrounding fluid. The environment is a reservoir with a macroscopic number of fast degrees of freedom whose microscopic dynamics is eliminated. This elimination yields a stochastic differential equation for the system (cf. (3.53)).

Although the physical reasoning leading to (3.53) is plausible, it would certainly lend further credence to our reasoning, could we justify this equation, at least for special cases, from the underlying classical equations of motion of the composite system (particle and reservoir). This programme can indeed be carried out for Brownian motion, but it involves theoretical techniques (projection operators) which will not be presented in this book (see e.g. Chap. 3 of [6] for a detailed discussion). However, in order to illustrate that a stochastic differential equation may be obtained from Newtonian dynamics we discuss in this section a simple model, the *Caldeira-Leggett model*, which is constructed such that the elimination of the degrees of freedom of the reservoir is tractable.

3.3.1 Definition of the Model

The Hamiltonian of the composite particle–reservoir system,

$$H = H_S + H_R + H_I, \tag{3.80}$$

has three terms. The Hamiltonian of the particle (the "system") of mass M and momentum p, which is subject to the potential $U(x)$ acting on its position x,

$$H_S = \frac{p^2}{2M} + U(x), \tag{3.81}$$

the Hamiltonian of the reservoir,

$$H_R = \sum_{i=1}^{N}\left(\frac{p_i^2}{2m} + \frac{1}{2}m\omega_i^2 r_i^2\right), \tag{3.82}$$

and the Hamiltonian describing the interaction between the system and the reservoir,

$$H_I = -\sum_{i=1}^{N} c_i r_i x + \frac{1}{2}\left[\sum_{i=1}^{N} \frac{c_i^2}{m\omega_i^2}\right] x^2. \tag{3.83}$$

The reservoir consists of large number (N) of independent oscillators of mass m and frequency ω_i with momenta p_i and positions r_i. The interaction Hamiltonian is bilinear in the coordinates r_i and x and introduces the coupling coefficients c_i.

Insertion of H_S, H_R, and H_I into (3.80) yields

$$H = \frac{p^2}{2M} + U(x) + \sum_{i=1}^{N}\left\{\frac{p_i^2}{2m} + \frac{1}{2}m\omega_i^2\left[r_i - \frac{c_i}{m\omega_i^2}x\right]^2\right\}. \quad (3.84)$$

This Hamiltonian has been used to discuss dissipation in system-plus-reservoir models for a long time. In the recent literature, particularly in the field of research on quantum dissipative systems, it is called *Caldeira-Leggett model* (see e.g. [212]). Here we treat this model in the framework of classical mechanics. Our presentation closely follows the discussion in [220].

3.3.2 Velocity Process and Generalized Langevin Equation

From the Hamiltonian (3.84) we obtain the equations of motion for the system and the reservoir. For the system we find

$$\dot{x} = \frac{\partial H}{\partial p} = \frac{p}{M},$$

$$\dot{p} = -\frac{\partial H}{\partial x} = -\frac{dU(x)}{dx} + \sum_{i=1}^{N} c_i\left(r_i - \frac{c_i}{m\omega_i^2}x\right), \quad (3.85)$$

and for the reservoir

$$\dot{r}_i = \frac{\partial H}{\partial p_i} = \frac{p_i}{m},$$

$$\dot{p}_i = -\frac{\partial H}{\partial r_i} = -m\omega_i^2 r_i + c_i x,$$

or equivalently,

$$\ddot{r}_i + \omega_i^2 r_i = \frac{c_i}{m}x(t) =: F_i(t). \quad (3.86)$$

Equation (3.85) shows that the dynamics of the system depends on the degrees of freedom of the reservoir only through $r_i(t)$. Therefore, the first step to eliminate the reservoir is to determine $r_i(t)$ from (3.86).

Equation (3.86) is an inhomogeneous differential equation which is solved by the sum of the general solution of the homogeneous equation,

$$\ddot{r}_i + \omega_i^2 r_i = 0, \quad (3.87)$$

and a particular solution of the inhomogeneous equation. The solution of (3.87) is given by

$$r_i(t) = A_i \cos\omega_i t + B_i \sin\omega_i t \quad (A_i, B_i = \text{constants}). \quad (3.88)$$

3.3 Caldeira-Leggett Model

To find a particular solution of (3.86) we use the method of the variation of constants. That is, we make the ansatz $r_i(t) = A_i(t)\exp(i\omega_i t)$ and determine $A_i(t)$ so that (3.86) is satisfied. Insertion into (3.86) gives for $A_i(t)$

$$\dot{g}_i(t) + 2i\omega_i g_i(t) = F_i(t)\exp(-i\omega_i t) \quad \text{with } g_i(t) = \dot{A}_i. \tag{3.89}$$

Setting $g_i(t) = C_i(t)\exp(-2i\omega_i t)$ we find from (3.89)

$$C_i(t) = \int_0^t ds\, F_i(s)\exp(i\omega_i s),$$

and so

$$A_i(t) = \int_0^t ds\, g_i(s)$$
$$= \frac{1}{2i\omega_i}\left[\int_0^t ds\, F_i(s)\,e^{-i\omega_i s} - e^{-i\omega_i t}\int_0^t ds\, F_i(s)\,e^{-i\omega_i(t-s)}\right],$$

where we used an integration by parts in order to obtain the second line. Putting these results together a particular solution of (3.86) reads

$$r_i(t) = \int_0^t ds\, F_i(s)\frac{\sin\omega_i(t-s)}{\omega_i} = \frac{c_i}{m}\int_0^t ds\, x(s)\frac{\sin\omega_i(t-s)}{\omega_i}. \tag{3.90}$$

Now we add to (3.90) the homogeneous solution (3.88) and fix the constants A_i and B_i by the initial conditions for $r_i(t)$ and $p_i(t)$. This gives

$$A_i = r_i(0) \quad \text{and} \quad B_i = \frac{p_i(0)}{m\omega_i},$$

where we also used the formula (see [67], p. 1130, Eq. (12.211))

$$\frac{d}{dt}\int_{\psi(t)}^{\varphi(t)} dx\, f(x,t) = f\big(\varphi(t),t\big)\frac{d\varphi(t)}{dt} - f\big(\psi(t),t\big)\frac{d\psi(t)}{dt} + \int_{\psi(t)}^{\varphi(t)} dx\, \frac{\partial}{\partial t} f(x,t).$$

Thus, we finally obtain

$$r_i(t) = r_i(0)\cos\omega_i t + \frac{p_i(0)}{m\omega_i}\sin\omega_i t + \frac{c_i}{m}\int_0^t ds\, x(s)\frac{\sin\omega_i(t-s)}{\omega_i},$$

or after integration by parts

$$r_i(t) - \frac{c_i}{m\omega_i^2}x(t) = \left[r_i(0) - \frac{c_i}{m\omega_i^2}x(0)\right]\cos\omega_i t + \frac{p_i(0)}{m\omega_i}\sin\omega_i t$$
$$- \frac{c_i}{m\omega_i^2}\int_0^t ds\, \dot{x}(s)\cos\omega_i(t-s). \tag{3.91}$$

The left-hand side of (3.91) is exactly the term which appears in (3.85). Therefore, we find for the velocity v of the particle (system) the following equation of motion

$$M\dot{v} = -\frac{\mathrm{d}U(x)}{\mathrm{d}x} - M\int_0^t \mathrm{d}s\gamma(t-s)v(s) + \eta(t), \qquad (3.92)$$

where the 'friction coefficient' is given by

$$\gamma(t) = \frac{1}{M}\sum_{i=1}^N \left(\frac{c_i^2}{m\omega_i^2}\right)\cos\omega_i t \qquad (3.93)$$

and the 'random force' by

$$\eta(t) = \sum_{i=1}^N c_i\left\{\left[r_i(0) - \frac{c_i}{m\omega_i^2}x(0)\right]\cos\omega_i t + \frac{p_i(0)}{m\omega_i}\sin\omega_i t\right\}. \qquad (3.94)$$

This result is interesting. We see that the elimination of the degrees of freedom of the reservoir transforms the deterministic equation of motion (3.85) for the system into a Langevin-like equation. However, this transformation is formal because no approximation has been made up to now. Equation (3.92) is still deterministic: $\gamma(t)$ and $\eta(t)$ are defined in terms of the parameters of the model and the initial positions and momenta of the oscillators; they are known functions of time.

Nevertheless, it is tempting to develop of stochastic interpretation of (3.92), with dissipation due to $\gamma(t)$ and random fluctuations due to $\eta(t)$. It is legitimate to interpret the second term of (3.92) as a friction force because it is linear in the velocity. The friction coefficient, however, is not constant as in (3.53). It depends on time through the dynamics of the reservoir (see (3.93)). This time dependence gives rise to the convolution integral of $\gamma(t-s)$ and $v(s)$ in (3.92). The motion of the particle is thus influenced by its previous history. Langevin equations with such a *memory effect* are non-Markovian and commonly referred to as *generalized Langevin equations* [6, 220].

The third term in (3.92) may be interpreted as a random force if we assume that $M \gg m$. As for the free Brownian particle, we expect this mass difference to lead to a separation of time scales between the (fast) reservoir and the (slow) particle. It is then reasonable to presume that the reservoir can 'equilibrate' for any given configuration of the particle; it should be characterized by its average properties only. If we assume that the reservoir is a heat bath, this implies that averages over all realizations of the random process of the reservoir can be calculated as

$$\langle(\cdot)\rangle = \frac{1}{Z}\int_{-\infty}^{+\infty}\prod_{j=1}^N \frac{\mathrm{d}r_j\mathrm{d}p_j}{h}(\cdot)\exp\left[-\frac{(H_\mathrm{R}+H_\mathrm{I})}{k_\mathrm{B}T}\right] \qquad (3.95)$$

3.3 Caldeira-Leggett Model

with the Gaussian Hamiltonian (see (3.82)–(3.84))

$$H_R + H_I = \sum_{i=1}^{N}\left\{\frac{p_i^2}{2m} + \frac{1}{2}m\omega_i^2\left[r_i - \frac{c_i}{m\omega_i^2}x\right]^2\right\}. \tag{3.96}$$

Using this prescription we find for the average of the random force from (3.94)

$$\langle \eta(t) \rangle = \sum_{i=1}^{N} c_i \left\{ \left\langle \left[r_i(0) - \frac{c_i}{m\omega_i^2} x(0) \right] \right\rangle \cos\omega_i t + \frac{\langle p_i(0)\rangle}{m\omega_i} \sin\omega_i t \right\}$$
$$= 0, \tag{3.97}$$

because

$$\langle p_i(0) \rangle = 0 \quad \text{and} \quad \left\langle \left[r_i(0) - \frac{c_i}{m\omega_i^2} x(0) \right] \right\rangle = 0$$

due to (3.95) and (3.96). Similarly, we can calculated the correlation function $\langle \eta(t)\eta(t')\rangle$ of the random force. Equation (3.94) shows that this correlation function contains crossterms between different oscillators and between positions and momenta. All of these crossterms vanish on average so that we have

$$\langle \eta(t)\eta(t')\rangle = \sum_{i=1}^{N} c_i^2 \left\langle \left[r_i(0) - \frac{c_i}{m\omega_i^2} x(0) \right]^2 \right\rangle \cos\omega_i t \cos\omega_i t'$$
$$+ \sum_{i=1}^{N} \left(\frac{c_i}{m\omega_i}\right)^2 \langle p_i^2(0)\rangle \sin\omega_i t \sin\omega_i t'.$$

With the equipartition theorem [176]

$$\frac{\langle p_i^2(0)\rangle}{2m} = \frac{1}{2}k_B T \quad \text{and} \quad \frac{1}{2}m\omega_i^2 \left\langle \left[r_i(0) - \frac{c_i}{m\omega_i^2} x(0) \right]^2 \right\rangle = \frac{1}{2}k_B T,$$

and

$$\cos\omega_i(t - t') = \cos\omega_i t \cos\omega_i t' + \sin\omega_i t \sin\omega_i t'$$

we obtain

$$\langle \eta(t)\eta(t')\rangle = k_B T \sum_{i=1}^{N} \left(\frac{c_i^2}{m\omega_i^2}\right) \cos\omega_i(t-t') = k_B T M \gamma(t-t'), \tag{3.98}$$

where (3.93) was used to obtain the last equality.

Equation (3.98) relates the strength of the equilibrium fluctuations of the random force to the dissipation; it is thus a version of the *fluctuation–dissipation theorem* [176]. Since $\eta(t)$ is a linear superposition of Gaussian random variables (according to (3.94)–(3.96)), it is itself a Gaussian random variable and its distribution is fully specified by the first two moments, (3.97) and (3.98). Equations (3.97) and (3.98) show that $\eta(t)$ is invariant with respect to time translation and thus a stationary process (cf. (2.66)). A stationary process with the properties,

$$\langle \eta(t) \rangle = 0 \quad \text{and} \quad \langle \eta(t) \eta(t') \rangle = k_B T M \gamma(t-t'),$$

is called *colored noise* [101]. The noise would become white (and the process thus Markovian), if $\gamma(t)$ were δ-correlated (see (2.97)). This can be achieved by the following assumption. Since the number (N) of oscillators is very large, many different ω_i should contribute to the sum of cosines in (3.93). So we may expect destructive interference to occur for finite t. Under these circumstance, only the $t=0$ term should survive, and we may write

$$\gamma(t) = \frac{1}{M} \sum_{i=1}^{N} \left(\frac{c_i^2}{m\omega_i^2} \right) \cos \omega_i t \rightsquigarrow \frac{1}{M} \left[\sum_{i=1}^{N} \frac{c_i^2}{m\omega_i^2} \right] \delta(t) =: \gamma \delta(t). \qquad (3.99)$$

This gives back (3.53) upon insertion into (3.92).

For the specific example of the Caldeira-Leggett model it is thus possible to motivate the stochastic differential equation for a 'heavy' particle and the fluctuation-dissipation theorem by eliminating the fast reservoir degrees of freedom from the classical equations of motion for the full system, particle and reservoir (heat bath). This supports the qualitative arguments presented in the previous section in order to obtain (3.53).

3.4 On the Maximal Excursion of Brownian Motion

In Chap. 2.1.7 we discussed the limiting distributions for the maxima of samples of iid random variables. Here we want to extend this discussion by asking: "What is the distribution of the maximal excursion of a Brownian motion $x(t)$ in the interval $[0, t_{\max}]$?". Let us define

$$M(t) = \sup_{0 \le t \le t_{\max}} x(t),$$

the maximal value a sample path $x(t)$ assumes in the interval $[0, t_{\max}]$. To determine the distribution of $M(t)$ we introduce the concept of stopping time, T, which is of general importance in the theory of stochastic processes. In our case this time can be defined in the following way:

$$T_a := \inf(0 \le t \le t_{\max} : x(t) = a), \qquad (3.100)$$

i.e., we assume that we stop the process when the level a is reached for the first time. In general, one would define a stopping time for a stochastic process by the first occurrence of some arbitrary predetermined event. In order for this prescription to lead to a consistent definition, however, one has to require:

3.4 On the Maximal Excursion of Brownian Motion

- that the decision of whether the process has to be stopped at $t = T$ can be made on the history of the motion $x(s); 0 \leq s \leq t$ alone, and
- that the stopping time T is finite with probability one.

Under these conditions the following theorem on the distribution of $M(t)$ holds:

Theorem 3.1 (Reflection Principle) *Given a standard Brownian motion $x(t)$, for every $a \geq 0$ we have*

$$\text{Prob}(M(t) \geq a) = 2\,\text{Prob}(x(t) \geq a) = \frac{2}{\sqrt{2\pi t}} \int_a^\infty \exp\left(-\frac{x^2}{2t}\right) dx. \quad (3.101)$$

Let us derive this relation. Clearly

$$\text{Prob}(x(t) \geq a) = \text{Prob}(x(t) \geq a, M(t) \geq a) + \text{Prob}(x(t) \geq a, M(t) < a)$$
$$= \text{Prob}(x(t) \geq a, M(t) \geq a) \quad \text{since } M(t) \geq x(t).$$

Furthermore

$$\text{Prob}(x(t) \geq a, M(t) \geq a) = \text{Prob}(x(t) \geq a | M(t) \geq a)\,\text{Prob}(M(t) \geq a)$$
$$= \text{Prob}(x(t) \geq a | T_a \leq t)\,\text{Prob}(M(t) \geq a).$$

Since $x(t)$ is a Brownian motion, also $x(T_a + s) - x(T_a) = x(T_a + s) - a$ is a Brownian motion with increments that are independent of all $x(t), t \leq T_a$. We therefore have

$$\text{Prob}(x(t) \geq a | T_a \leq t) = \text{Prob}(x(T_a + (t - T_a)) - a \geq 0 | T_a \leq t) = \frac{1}{2},$$

since $\text{Prob}(x(t) \geq 0) = \frac{1}{2}$ for all t for a Brownian motion without drift. Applying this identity we obtain

$$\text{Prob}(x(t) \geq a) = \frac{1}{2}\text{Prob}(M(t) \geq a) \quad (3.102)$$

as claimed by the reflection principle. The probability density that the maximum M occurs at time t is therefore given by

$$p(M, t) = \frac{2}{\sqrt{2\pi t}} \exp\left(-\frac{M^2}{2t}\right). \quad (3.103)$$

It is easily seen that this result can be of importance, e.g., for the assessment of risks in financial markets (see Chap. 5). When we ask with what probability a stock index drops below a threshold value in a given time interval, and if the time dependence of this variable can be modeled as Brownian motion, then the reflection principle allows us to evaluate this risk.

Fig. 3.7 External potential $U(x)$ as a function of some general reaction coordinate x. The system starts in the metastable minimum around x_A and crosses the barrier to the equilibrium state at x_C by a thermally activated stochastic process

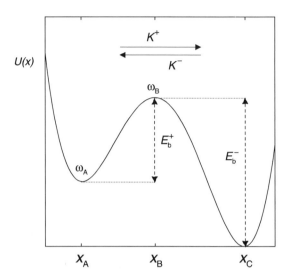

3.5 Brownian Motion in a Potential: Kramers Problem

The problem that is often referred to as the *Kramers problem* is illustrated in Fig. 3.7. Suppose we have a system that can be either in an equilibrium state C of lowest energy (or free energy) or in a *metastable state* A, which is a local minimum in energy as a function of some coordinate x, but not the absolute minimum in energy. An example of this situation is the description of a chemical reaction given by Kramers [110]. State A corresponds to the unreacted products (educts) and state C to the reacted products. For this reason, it is common to call x a (generalized) reaction coordinate. In order for the reaction to occur, the system has to cross a barrier at x_B with energy differences E_b^+ compared to the metastable state A and E_b^- compared to the stable state C. Since $E_b^- > E_b^+$ we expect the rate K^+ with which the system crosses the barrier from A to C to be larger than the rate K^- for the reverse process.

Kramers modeled this problem using the one-dimensional version of (3.53)

$$\begin{aligned} \mathrm{d}x &= v \mathrm{d}t, \\ M \mathrm{d}v &= \bigl(F(x) - \gamma M v\bigr)\mathrm{d}t + \sqrt{2M\gamma k_B T}\,\mathrm{d}W(t), \end{aligned} \quad (3.104)$$

where $F = -\mathrm{d}U/\mathrm{d}x$ is the force exerted on the Brownian particle in the external potential U. In (3.104) we have inserted the result for the strength of the stochastic forces (3.62). The time development of the reaction coordinate is therefore modeled as the Brownian motion of a particle in an external potential. Employing a Markov process description of this problem entails the following assumptions:

- the friction effects contain no memory of the form (cf. (3.92))

$$\int_0^t \gamma(t-s)v(s)\mathrm{d}s,$$

- the stochastic forces can be approximated as a Wiener process.

3.5 Brownian Motion in a Potential: Kramers Problem

When these two terms, such as in the free Brownian motion problem, model the interaction of the reaction coordinate with all other degrees of freedom of the system which are not explicitly taken into account, these other degrees of freedom have to be fast compared to the typical time scales of the reaction coordinate. Contrary to the free Brownian motion problem, the Kramers problem has more than one inherent time scale:

- the thermalization time τ_T within one of the minima: This is the time after which an ensemble of systems started, for instance, at x_A has assumed the Maxwell-Boltzmann equilibrium distribution corresponding to an infinite barrier at x_B;
- the *escape time* τ_e: This is the mean time a Brownian particle will need to go from A to C. There is of course a symmetrically defined time for going from C to A.

When τ_{bath} is a typical time scale for the fast degrees of freedom, we require the following inequalities:

$$\tau_{\text{bath}} \ll \tau_T \ll \tau_e. \tag{3.105}$$

The second inequality is the basis for calling the barrier crossing thermally activated. On a time scale τ_T, the reaction coordinate has reached thermal equilibrium with the bath variables, and it is the thermal fluctuations of these variables that create the stochastic forces which finally drive the system over the barrier.

To be consistent with the last inequality of the time scales, we have to assume the following inequality of energy scales:

$$k_B T \ll E_b^+. \tag{3.106}$$

The mean (thermal) energy of the Brownian particle has to be much smaller than the barrier height. If it were larger than the barrier height, the Brownian particle would diffuse more or less freely from A to C and back, and if it were of comparable height the time scales for equilibration and escape would not be clearly separated.

There is one more set of competing time scales contained in (3.104). The time scale for the coupling of the reaction coordinate to the bath coordinates is set by the friction coefficient γ. Without the coupling to the bath, the particle would perform a Newtonian motion in the external potential U. When the particle has a total energy smaller than the barrier, it performs oscillations around either A or C with typical frequencies $\omega_A = \sqrt{U''(x_A)/M}$ and $\omega_C = \sqrt{U''(x_C)/M}$, where we use U'' to denote the second derivative of the potential with respect to the reaction coordinate. When the particle has a total energy larger than the barrier, however, there is a time scale for the exchange between kinetic and potential energy during the barrier crossing. This time scale is given by $\omega_B = \sqrt{|U''(x_B)|/M}$. We can now distinguish two different regimes:

- strong friction, $\gamma/\omega_B \gg 1$: spatially diffusive or Smoluchowski regime,
- weak friction, $\gamma/\omega_B \ll 1$: energy-diffusive regime.

Kramers [110] presented solutions to the barrier-crossing problem in these two limiting regimes. We will focus in the following on the case of strong friction. We saw in Sect. 3.2 that $1/\gamma$ is the time scale over which the Brownian particle loses information about the starting velocity. After this time all inertia effects have died out. So one can approximate (3.104) by

$$dv = 0, \quad dx = v dt = \frac{F(x)}{\gamma M} dt + \sqrt{\frac{2k_B T}{\gamma M}} dW(t). \tag{3.107}$$

The corresponding Fokker-Planck equation is often called *Smoluchowski equation* because Smoluchowski treated free Brownian motion with this approach. For the transition probability it is given by (cf. Table 2.1)

$$\frac{\partial}{\partial t} p(x, t | x_0, 0)$$
$$= -\frac{\partial}{\partial x} \left[\frac{F(x)}{\gamma M} p(x, t | x_0, 0) \right] + \frac{1}{2} \frac{\partial^2}{\partial x^2} \left[\frac{2k_B T}{\gamma M} p(x, t | x_0, 0) \right]. \tag{3.108}$$

For this model we want to find the mean time that the Brownian particle needs to reach C for the first time, starting from A. This is an example of the more general class of *first passage time* problems for stochastic processes, which we will treat for the one-dimensional Fokker-Planck equation in the following.

3.5.1 First Passage Time for One-dimensional Fokker-Planck Equations

Appealing back to Chap. 2 we consider the following general form of a stationary one-dimensional Fokker-Planck equation for the conditional probability:

$$\frac{\partial}{\partial t} p(x', t | x, 0) = -\frac{\partial}{\partial x'} [a_1(x') p(x', t | x, 0)] + \frac{1}{2} \frac{\partial^2}{\partial x'^2} [a_2(x') p(x', t | x, 0)]. \tag{3.109}$$

For the first passage time problem we assume that the Brownian particle starts at $x \in (-\infty, x_C)$ at $t = 0$. Then, the probability that the particle is still in that interval at time t can be written as

$$G(x, t) = \int_{-\infty}^{x_C} p(x', t | x, 0) dx' = \int_{-\infty}^{x_C} p(x', 0 | x, -t) dx'. \tag{3.110}$$

In the last step we exploited the fact that time may be shifted by an arbitrary Δt, chosen here as $\Delta t = -t$.

The function $G(x, t)$ has a further significance: It is also the probability that the first passage time (τ_{xC}) of the Brownian particle to go from x to x_C is larger than t:

$$G(x, t) = \text{Prob}(\tau_{xC} \geq t) = \int_t^\infty p_{\text{fpt}}(\tau_{xC}) d\tau_{xC}, \tag{3.111}$$

3.5 Brownian Motion in a Potential: Kramers Problem

where p_{fpt} is the probability density of first passage times from x to x_C. This probability density is the key quantity of interest here. Through (3.110) and (3.111) we see that p_{fpt} is given in terms of the initial state of the Brownian particle. Therefore, we need the backward Fokker-Planck equation (2.92)

$$\frac{\partial}{\partial t}p(x',t'|x,t) = -a_1(x)\frac{\partial}{\partial x}p(x',t'|x,t) - \frac{1}{2}a_2(x)\frac{\partial^2}{\partial x^2}p(x',t'|x,t). \quad (3.112)$$

For the first passage time problem $p(x',t'|x,t)$ is subjected to the following two boundary conditions:

- absorbing boundary at $x = x_C$:

$$p(x',t'|x_C, 0) = 0. \quad (3.113)$$

If the particle is at x_C, it never returns to x_A. This implies that $E_b^- \gg E_b^+$;
- reflecting boundary for $x \to -\infty$:

$$\frac{\partial}{\partial x}p(x',t'|x,0) \xrightarrow{x \to -\infty} 0. \quad (3.114)$$

For $x \to -\infty$ there is an impenetrable barrier (created by the potential) which makes the flux ($\propto \partial p/\partial x$) vanish and reflects the particle back to the interval $(-\infty, x_C)$.

Deriving now $G(x,t)$ with respect to time and inserting (3.112) into (3.110) we obtain

$$\frac{\partial}{\partial t}G(x,t) = a_1(x)\frac{\partial}{\partial x}G(x,t) + \frac{1}{2}a_2(x)\frac{\partial^2}{\partial x^2}G(x,t). \quad (3.115)$$

Due to (3.113)–(3.110) G obeys the following boundary conditions:

$$G(x,0) = \begin{cases} 1 & x \leq x_C, \\ 0 & x > x_C, \end{cases} \quad (3.116)$$

$$G(x_C, t) = 0, \quad (3.117)$$

$$\frac{\partial}{\partial x}G(x,t) \xrightarrow{x \to -\infty} 0. \quad (3.118)$$

Since G is the complement of the cumulative probability distribution of first passage times, we can write for every function $f(\tau_{xC})$

$$\langle f(\tau_{xC})\rangle = -\int_0^\infty f(t)\frac{\partial G(x,t)}{\partial t}dt. \quad (3.119)$$

In particular, for a fixed starting point x we find for the mean first passage time from x to x_C:

$$\bar{\tau}(x) \equiv \langle \tau_{xC}\rangle = -\int_0^\infty t\frac{\partial}{\partial t}G(x,t)dt = \int_0^\infty G(x,t)dt, \quad (3.120)$$

in which we have made use of the fact that $\lim_{t\to\infty} t^n G(x,t) = 0$. Similarly, for the higher moments of the first passage time distribution, we get:

$$\overline{\tau^n}(x) \equiv \langle \tau^n_{xC} \rangle = -\int_0^\infty t^n \frac{\partial}{\partial t} G(x,t) \, dt = n \int_0^\infty t^{n-1} G(x,t) \, dt. \quad (3.121)$$

Alternatively, we can derive partial differential equations for the moments of the first passage time distribution from (3.115). Integrating that equation over time from zero to infinity yields for all $x \leq x_C$

$$\int_0^\infty dt \, \frac{\partial}{\partial t} G(x,t) = a_1(x) \frac{\partial}{\partial x} \overline{\tau}(x) + \frac{1}{2} a_2(x) \frac{\partial^2}{\partial x^2} \overline{\tau}(x)$$

or

$$-1 = a_1(x) \frac{\partial}{\partial x} \overline{\tau}(x) + \frac{1}{2} a_2(x) \frac{\partial^2}{\partial x^2} \overline{\tau}(x). \quad (3.122)$$

Similarly, for the higher moments,

$$-n\overline{\tau^{n-1}}(x) = a_1(x) \frac{\partial}{\partial x} \overline{\tau^n}(x) + \frac{1}{2} a_2(x) \frac{\partial^2}{\partial x^2} \overline{\tau^n}(x). \quad (3.123)$$

Equation (3.122) has to be solved with the boundary conditions

$$\overline{\tau}(x_C) = 0, \qquad \frac{\partial}{\partial x} \overline{\tau}(x) \xrightarrow{x\to -\infty} 0. \quad (3.124)$$

For $a_2(x) \neq 0$ the solution for the homogeneous part of (3.122) can be written as

$$\overline{\tau}'_h(x) = \overline{\tau}_0 \exp\left[-\int_{-\infty}^x dx' \frac{2a_1(x')}{a_2(x')}\right],$$

where $\overline{\tau}'_h$ is the first derivative of $\overline{\tau}_h$ with respect to x. We get a particular solution for (3.122) by the method of the variation of the constants. Choosing $\overline{\tau}_0 = \overline{\tau}_0(x)$ and inserting $\overline{\tau}'_h(x)$ into (3.122) we find

$$\overline{\tau}'_0 \exp\left[-\int_{-\infty}^x dx' \frac{2a_1(x')}{a_2(x')}\right] = -\frac{2}{a_2(x)}.$$

One particular solution is

$$\overline{\tau}_0(x) = -2 \int_{-\infty}^x dx' \frac{1}{a_2(x')} \exp\left[\int_{-\infty}^{x'} dx'' \frac{2a_1(x'')}{a_2(x'')}\right].$$

The complete solution for the first derivative of the mean first passage time can therefore be written as

$$\overline{\tau}'(x) = \overline{\tau}_0 \exp\left[-\int_{-\infty}^x dx' \frac{2a_1(x')}{a_2(x')}\right]$$

3.5 Brownian Motion in a Potential: Kramers Problem

$$-2\int_{-\infty}^{x} dx' \frac{1}{a_2(x')} \exp\left[\int_{-\infty}^{x'} dx'' \frac{2a_1(x'')}{a_2(x'')}\right]$$
$$\times \exp\left[-\int_{-\infty}^{x} dx' \frac{2a_1(x')}{a_2(x')}\right]. \qquad (3.125)$$

As $\bar{\tau}'(x)$ has to vanish for $x \to -\infty$, only the particular solution remains. With the abbreviation

$$\phi(x) = \exp\left[\int_{-\infty}^{x} dx' \frac{2a_1(x')}{a_2(x')}\right]$$

and using $\bar{\tau}(x_C) = 0$, we can write the final result as

$$\bar{\tau}(x) = 2\int_{x}^{x_C} dx' \phi^{-1}(x') \int_{-\infty}^{x'} dx'' \frac{\phi(x'')}{a_2(x'')}. \qquad (3.126)$$

3.5.2 Kramers Result

Returning to the Kramers problem, we can now insert the special forms of a_1 and a_2 into this solution, i.e.,

$$a_1(x) = -\frac{U'(x)}{M\gamma} \quad \text{and} \quad a_2(x) = \frac{2k_B T}{M\gamma},$$

to get

$$\langle \tau_{x_C} \rangle = \frac{M\gamma}{k_B T} \int_{x}^{x_C} dx' \exp\left[\frac{U(x')}{k_B T}\right] \int_{-\infty}^{x'} dx'' \exp\left[-\frac{U(x'')}{k_B T}\right]. \qquad (3.127)$$

Since we required $k_B T \ll U(x_B)$, the exponential $\exp[U(x')/k_B T]$ is sharply peaked around x_B and for the inner integral we only need to consider its behavior for $x' \approx x_B$. Because of the inverse exponential in this integral, however, it in turn is a slowly varying function of the upper limit x' of the integration, which we therefore approximate by x_B,

$$\langle \tau_{x_C} \rangle \approx \frac{M\gamma}{k_B T} \int_{x}^{x_C} dx' \exp\left[\frac{U(x')}{k_B T}\right] \int_{-\infty}^{x_B} dx'' \exp\left[-\frac{U(x'')}{k_B T}\right].$$

In the first integral we expand the external potential around $x = x_B$ and in the second around $x = x_A$:

$$U(x') = U(x_B) - \frac{1}{2} M\omega_B^2 (x' - x_B)^2,$$

$$U(x'') = U(x_A) + \frac{1}{2} M\omega_A^2 (x'' - x_A)^2,$$

where $M\omega_A^2 = U''(x_A)$ and $M\omega_B^2 = -U''(x_B)$. To be consistent with the expansions up to the second order, we replace the upper limit of integration in both integrals by $+\infty$. For $x \approx x_A$ we furthermore replace the lower limit in the first integral by $-\infty$ and arrive at

$$\langle \tau_{xC} \rangle \approx \frac{M\gamma}{k_B T} \int_{-\infty}^{+\infty} dx' \exp\left[\frac{U(x_B) - \frac{1}{2}M\omega_B^2(x'-x_B)^2}{k_B T}\right]$$
$$\times \int_{-\infty}^{+\infty} dx'' \exp\left[-\frac{U(x_A) + \frac{1}{2}M\omega_A^2(x''-x_A)^2}{k_B T}\right].$$

By solving the Gaussian integrals we obtain the following as an approximate solution for the escape time from the metastable minimum A to the stable minimum C in the case of strong friction ($\omega_B/\gamma \ll 1$):

$$\langle \tau_{AC} \rangle \approx \frac{2\pi\gamma}{\omega_B \omega_A} \exp\left[\frac{E_b^+}{k_B T}\right]; \tag{3.128}$$

for the transition rate from A to C, we obtain

$$K^+ = \langle \tau_{AC} \rangle^{-1} \approx \frac{\omega_B \omega_A}{2\pi\gamma} \exp\left[-\frac{E_b^+}{k_B T}\right]. \tag{3.129}$$

This is Kramers' celebrated result for the reaction rate. The time scale is set by a thermally activated process with activation energy E_b^+, and ω_A is the attempt frequency with which the particle tries to overcome the barrier.

Mean first passage times and other extreme-value problems are important in many applications of stochastic processes, and we have just discussed one very prominent example. The activated behavior $K \propto \exp[-E_{act}/k_B T]$ is found for many transport processes and is usually termed *Arrhenius behavior*.

In Sect. 3.7 we will discuss an application that will take us from a probabilistic formulation of a statistical physics model, through the setup of its stochastic evolution equation and approximations thereof and finally again to the formulation of a transition-rate problem.

3.6 A First Passage Problem for Unbounded Diffusion

Consider the following problem:[2] A molecule B in a solution is considered fixed at the origin. Another molecule A starts somewhere on the spherical shell with radius r_0 around B and performs a free three-dimensional diffusion modeled by a three-dimensional Wiener process. When it hits a spherical shell with radius b ($\leq r_0$)

---

[2]This is an exercise in van Kampen's book [101], Chap. XII.

3.6 A First Passage Problem for Unbounded Diffusion

around B, the two molecules react, i.e., molecule A is absorbed. What is the probability that A reacts with B and does not escape to infinity?

Obviously, the time when absorption occurs is the first passage time of A to go from the spherical shell with radius r_0 to the one with radius b. However, in contrast to the Kramers problem (Sect. 3.5), the probability to reach b is not unity, regardless of how long the diffusion process takes. This is a consequence of the fact that molecule A can escape to infinity in an unbounded system. This was not possible in the Kramers problem due to the reflecting boundary condition at $x \to -\infty$. The free diffusion problem is thus non-stationary (cf. Sect. 3.2.2), which also renders the analysis of Sect. 3.5.1 inapplicable.

Therefore, we have to develop a different approach. Let us denote the probability density for the first passage time between r_0 and b by $f(r_0, t)$. Then, the probability of absorption is given by

$$\Pi_b = \int_0^\infty f(r_0, t) dt < 1. \tag{3.130}$$

This probability gives the fraction of all molecules A that are absorbed, whereas a fraction $1 - \Pi_b$ escapes to infinity. To calculate the absorption probability we therefore have to find a solution for $f(r_0, t)$. This problem can be related to the solution of the free diffusion problem using a renewal equation like in the solution of the Polya problem in Sect. 3.1.1.

Let us define $q(r, t|r_0, 0)$ as the conditional probability to be on the shell with radius r at time t starting from the shell with radius r_0 at time 0 without ever having reached the shell with radius b. Then one can write for the conditional probability to be on the shell with radius r at time t starting from r_0 at $t = 0$

$$p(r, t|r_0, 0) = q(r, t|r_0, 0) + \int_0^t f(r_0, t') p(r, t - t'|b, 0) dt'. \tag{3.131}$$

Here $p(r, t|r_0, 0)$ is the solution of the unconstrained diffusion problem. Equation (3.131) has the following interpretation: $p(r, t|r_0, 0)$ is the sum of two disjunct events. Either A goes directly from r_0 to r with probability q or it first reaches b at some $t' \leq t$ and then goes to r in the remaining time with a probability given by the second term of (3.131). Obviously, $q(b, t|b, 0) = 0$; so we have

$$p(b, t|r_0, 0) = \int_0^t f(r_0, t') p(b, t - t'|b, 0) dt'. \tag{3.132}$$

This equation agrees with (3.17) of Sect. 3.1.1. Let us now introduce the Laplace transform

$$\hat{f}(r_0, s) = \int_0^\infty f(r_0, t) e^{-st} dt.$$

Due to the convolution theorem Laplace transformation of (3.132) yields

$$\hat{f}(r_0, s) = \frac{\hat{p}(b, s|r_0, 0)}{\hat{p}(b, s|b, 0)},$$

and from this we get

$$\Pi_b = \int_0^\infty f(r_0, t) dt = \hat{f}(r_0, 0) = \frac{\hat{p}(b, 0|r_0, 0)}{\hat{p}(b, 0|b, 0)}. \tag{3.133}$$

Therefore we have reduced the problem to finding the Laplace transform of the solution of the free diffusion problem. This problem is defined by

$$\frac{\partial}{\partial t} p(r, t|r_0, 0) = D \Delta p(r, t|r_0, 0), \tag{3.134}$$

where D is the diffusion coefficient of the molecule A. Equation (3.134) must be solved with the starting condition

$$p(r, 0|r_0, 0) = \frac{1}{4\pi r_0^2} \delta(r - r_0). \tag{3.135}$$

The prefactor $1/4\pi r_0^2$ stems from the normalization condition (see (2.69)) of the conditional probability. Since

$$\int_0^\infty \frac{\partial}{\partial t} p(r, t|r_0, 0) e^{-st} dt = -p(r, 0|r_0, 0) + s\hat{p}(r, s|r_0, 0),$$

Laplace transformation of (3.134) gives

$$-\frac{1}{4\pi r_0^2} \delta(r - r_0) + s\hat{p}(r, s|r_0, 0) = D \Delta \hat{p}(r, s|r_0, 0). \tag{3.136}$$

We now introduce the Fourier transforms

$$p(r, t|r_0, 0) = \frac{1}{(2\pi)^3} \int p(k, t|r_0, 0) e^{-i k \cdot r} d^3 k$$

$$p(k, t|r_0, 0) = \int p(r, t|r_0, 0) e^{i k \cdot r} d^3 r.$$

With

$$\int p(r, 0|r_0, 0) e^{i k \cdot r} d^3 r = \frac{1}{4\pi r_0^2} \int_0^{2\pi} d\varphi \int_0^\pi d\theta \sin\theta \int_0^\infty dr r^2 e^{ikr \cos\theta} \delta(r - r_0)$$

$$= \frac{\sin kr_0}{kr_0}$$

Fourier transformation of (3.136) yields

$$-\frac{\sin kr_0}{kr_0} + s\hat{p}(k, s|r_0, 0) = -Dk^2 \hat{p}(k, s|r_0, 0),$$

or
$$\hat{p}(k,s|r_0,0) = \frac{1}{s+Dk^2} \frac{\sin kr_0}{kr_0}. \tag{3.137}$$

Fourier back-transformation of this equation gives

$$\begin{aligned}\hat{p}(r,s|r_0,0) &= \frac{1}{(2\pi)^3}\int d^3k\, e^{-i\mathbf{k}\cdot\mathbf{r}}\frac{1}{s+Dk^2}\frac{\sin kr_0}{kr_0}\\ &\stackrel{x=kr_0}{=} \frac{1}{2\pi^2 rD}\int_0^\infty dx \sin x\, \frac{\sin[(r/r_0)x]}{(sr_0^2/D)+x^2}.\end{aligned}$$

Using
$$\sin x \sin ax = \frac{1}{2}\bigl[\cos(1-a)x - \cos(1+a)x\bigr]$$

and ([67], p. 445, Eqs. (3.723–2.))

$$\int_0^\infty \frac{\cos ax}{\beta^2+x^2}dx = \frac{\pi}{2\beta}e^{-a\beta} \quad (a\ge 0, \mathrm{Re}\,\beta>0)$$

we find

$$\hat{p}(r,s|r_0,0) = \frac{1}{8\pi rD}\sqrt{\frac{D}{sr_0^2}}\bigl[e^{-(1-r/r_0)\sqrt{sr_0^2/D}} - e^{-(1+r/r_0)\sqrt{sr_0^2/D}}\bigr]. \tag{3.138}$$

We only need the value at $s=0$ of this equation. More precisely,

$$\hat{p}(r,0|r_0,0) = 4\pi r^2 \hat{p}(r,0|r_0,0) = 4\pi r^2 \lim_{s\to 0}\hat{p}(r,s|r_0,0) = \frac{r^2}{Dr_0}.$$

Inserting this result into (3.133) we obtain

$$\Pi_b = \frac{b}{r_0}. \tag{3.139}$$

The probability to be absorbed at b starting from r_0 ($\ge b$) is therefore given purely geometrically by the ratio of the radii of both spherical shells.

3.7 Kinetic Ising Models and Monte Carlo Simulations

The *Ising model* [86] is a work horse model of statistical physics describing uniaxial magnets, i.e., systems with a preferential axis for the local magnetic moments. The Hamiltonian is given as

$$H(s) = -\sum_{(i,j)} J_{ij} s_i s_j - h\sum_i s_i, \tag{3.140}$$

where the spins $s_i = \pm 1$ represent the magnetic moments fixed in space on some lattice structure, (i, j) denotes the set of interacting spin pairs, J_{ij} is the strength of interaction between spins i and j, and h is an external magnetic field (the magnetic moment per spin, μ_B, is absorbed into the definition of the external field).

3.7.1 Probabilistic Structure

The sample space for this model is discrete, i.e, the set of all microscopic spin configurations $\Omega = \{s\}$. When we have a total of N spins, there are 2^N different microstates. We will consider this spin system in contact with a heat bath of temperature T so that the probability for a microstate is given by the canonical distribution as

$$\mu(s) = \frac{1}{Z} e^{-\beta H(s)}, \quad (3.141)$$

with $\beta = 1/k_B T$ and the partition function

$$Z = \sum_s e^{-\beta H(s)}. \quad (3.142)$$

Denoting by $\mathcal{P}(\Omega)$ the power set of Ω, which can serve as the σ-algebra of subsets of Ω for the discrete case, we can identify a probability space $(\Omega, \mathcal{P}(\Omega), \mu)$ and proceed to calculate the statistical properties of our physical system.

3.7.2 Monte Carlo Kinetics

As simple as this model appears to be, however, analytical treatments only exist for certain choices of dimensionality of the underlying lattice and interaction parameters J_{ij}. Numerical evaluation of the statistical properties for the general cases is made using so-called *Monte Carlo simulations* [10, 114]. Let us assume, for instance, that we want to determine the average magnetization $\langle M(T, h) \rangle$ at a temperature T and magnetic field h,

$$\langle M(T, h) \rangle = \sum_s \left(\sum_{i=1}^N s_i \right) \mu(s). \quad (3.143)$$

Exact enumeration of (3.143) is only possible for small numbers of spins N. In a Monte Carlo evaluation it will be very inefficient to randomly generate states s and use (3.143) to calculate the average magnetization because, many of these randomly generated states will have a high energy and therefore negligible weight in the statistical average. There is, however, a way to generate the states according to their

3.7 Kinetic Ising Models and Monte Carlo Simulations

statistical weight, which then reduces the calculation of the average magnetization to a simple arithmetic average over the thus generated set of states:

$$\langle M(T,h) \rangle_{\text{imp.samp.}} = \frac{1}{K} \sum_{k=1}^{K} \left(\sum_{i=1}^{N} s_i^k \right), \tag{3.144}$$

where $s^k = (s_1^k, \ldots, s_N^k)$ is the kth generated state. This so-called *importance sampling* technique starts from postulating a stochastic evolution equation for the state of the system, given by the following master equation in discrete time (cf. (2.82)):

$$p(s, n+1) = p(s, n) + \sum_{s'} w(s|s') p(s', n)$$

$$- \sum_{s'} w(s'|s) p(s, n). \tag{3.145}$$

With this we model the time evolution of our physical system as a Markov chain on the discrete probability space we have defined above. The basis of this approach is the following theorem [52]:

Theorem 3.2 *A finite, irreducible, aperiodic Markov chain possesses a unique invariant (stationary) probability distribution, $p_{\text{eq}}(x_k) > 0$, for all x_k. Irreducible means that every state can be reached from every other state.*

The conditions of the theorem imply requirements on the transition probabilities w in (3.145). When we derived the master equation as the differential version of the Chapman-Kolmogorov equation for stationary Markov processes in Chap. 2, we already discussed its reinterpretation as an evolution equation when the transition probabilities are given by some independent consideration. Here, we have to require that the unique stationary solution of (3.145) has to be given through the canonical probability distribution that defines the measure on our probability space:

$$p_{\text{eq}}(s) = \mu(s). \tag{3.146}$$

One way to ensure this is to require *detailed balance* behavior of the transition probabilities,

$$w(s|s') p_{\text{eq}}(s') = w(s'|s) p_{\text{eq}}(s). \tag{3.147}$$

This requirement, meaning that in equilibrium the probability flow from s' to s equals that from s to s', only determines the ratio of the transition probabilities,

$$\frac{w(s'|s)}{w(s|s')} = \frac{p_{\text{eq}}(s')}{p_{\text{eq}}(s)} = \exp[-\beta(H(s') - H(s))]. \tag{3.148}$$

Two popular choices are the Glauber [60] and the Metropolis [142] transition probabilities:

$$w(s'|s) = w_0(s'|s)\frac{1}{2}\left[1 + \tanh\left(-\frac{1}{2}\beta\Delta H\right)\right] \quad \text{(Glauber)}, \quad (3.149)$$

$$w(s'|s) = w_0(s'|s)\min(1, \exp[-\beta\Delta H]) \quad \text{(Metropolis)}, \quad (3.150)$$

where $\Delta H = H(s') - H(s)$, and $w_0(s'|s)$ is the probability to suggest s' as the next state, starting from s. Because of (3.148) we have to require the symmetry

$$w_0(s'|s) = w_0(s|s'). \quad (3.151)$$

In a computer simulation of (3.145) we would choose an arbitrary starting configuration s_0 and then use some algorithm to generate the next configurations s_1, s_2 and so on. One such algorithm is the single spin-flip kinetic Ising model, in which we select one spin at random (i.e., $w_0(s'|s) = 1/N$) and try to reverse it. For this case we have to write

$$w_G(s'|s) = \frac{1}{2N}\left[1 + \tanh\left(-\frac{1}{2}\beta\Delta H\right)\right] \quad \text{(Glauber)}, \quad (3.152)$$

$$w_M(s'|s) = \frac{1}{N}\min(1, \exp[-\beta\Delta H]) \quad \text{(Metropolis)}, \quad (3.153)$$

with the energy difference given by

$$\Delta H = H(s_1, \ldots, -s_i, \ldots, s_N) - H(s_1, \ldots, s_i, \ldots, s_N).$$

We can therefore write the master equation in the form

$$p(s_1, \ldots, s_i, \ldots, s_n, n+1)$$
$$= p(s_1, \ldots, s_i, \ldots, s_n, n) + \sum_i w(-s_i \to s_i)p(s_1, \ldots, -s_i, \ldots, s_n, n)$$
$$- \sum_i w(s_i \to -s_i)p(s_1, \ldots, s_i, \ldots, s_n, n). \quad (3.154)$$

A Monte Carlo simulation of this model would then proceed as follows:
- generate a starting configuration s_0,
- select a spin, s_i, at random,
- calculate the energy change upon spin reversal ΔH,
- calculate the probability w for this spin-flip to happen, using the chosen form of transition probability,
- generate a uniformly distributed random number, $0 < r < 1$,
- if $w > r$, flip the spin, otherwise retain the old configuration.

Theorem 3.2 guarantees that for long times the microstates we generate in this way will be distributed according to the equilibrium canonical distribution. Technically, we have to study suitable correlation functions to first judge when the correlations with the starting state have decayed and then decide to use the subsequent time series as representative of the stationary state.

3.7.3 Mean-Field Kinetic Ising Model

We now want to look at a special version of this model, which is the mean-field kinetic Ising model. In the mean-field approximation we assume that each spin is interacting with all $N-1$ other spins. To get a finite energy per spin, we require

$$J_{ij} = \frac{J}{N-1}. \tag{3.155}$$

With $J > 0$ we furthermore have a ferromagnetic Ising model, where the spins try to align parallel to each other and to the external field. In this case, the Hamiltonian (3.140) can be rewritten

$$H(m) = -\frac{1}{2} \frac{JN^2 m^2}{(N-1)} - hNm, \tag{3.156}$$

where $m = M/N$ is the magnetization per spin and in which we have omitted an irrelevant constant shift of $JN/2(N-1)$. Since each spin reversal generates a magnetization change of ± 2, the magnetization per spin can take on the $N+1$ values

$$m \in \Omega_m = \left\{ -1, -1 + \frac{2}{N}, \ldots, 1 - \frac{2}{N}, 1 \right\}.$$

For the mean-field Ising model we can work with the much simpler state space Ω_m instead of Ω. The probability measure $\mu(s)$ on $\mathcal{P}(\Omega)$ gives rise to a probability measure $\mu(m)$ on $\mathcal{P}(\Omega_m)$. Let s_m signify a spin configuration with a given magnetization m:

$$\begin{aligned}
\mu(m) &= \sum_{s_m} \mu(s) \\
&= \frac{1}{Z} e^{-\beta H(m)} \sum_{s_m} 1 \\
&= \frac{1}{Z} e^{-\beta H(m)} \frac{N!}{(N(1+m)/2)!(N(1-m)/2)!} \\
&= \frac{1}{Z} e^{-\beta H(m) + \ln N! - \ln(N(1+m)/2)! - \ln(N(1-m)/2)!} \\
&= \frac{1}{Z} e^{-\beta N f(m)}. \tag{3.157}
\end{aligned}$$

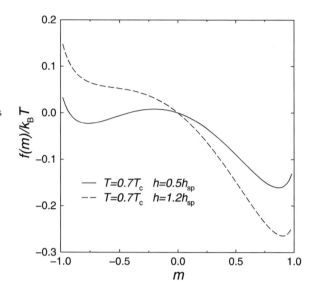

Fig. 3.8 Generalized free energy f in units of $k_B T$ for two different situations: $T = 0.7 T_c$ and $h = 0.5 h_{sp}$ (*solid curve*) and $T = 0.7 T_c$ and $h = 1.2 h_{sp}$ (*dashed curve*), where T_c is the critical temperature and h_{sp} is the spinodal field

The binomial coefficient gives the number of possibilities to have $(N(1+m)/2)$ up spins and $(N(1-m)/2)$ down spins out of a total of N spins, resulting in an overall magnetization $M = Nm$. The function f is the *generalized (Gibbs) free energy* per spin. For large N we can use Stirling's approximation for the factorials and set $N-1$ equal to N and get (again omitting a constant shift)

$$f(m) = -\frac{1}{2} J m^2 - hm + \frac{k_B T}{2} \big[(1+m)\ln(1+m) + (1-m)\ln(1-m)\big].$$
(3.158)

The partition function now is given as

$$Z = \sum_{m=-1}^{1} e^{-\beta N f(m)}.$$
(3.159)

The statistical, i.e., thermodynamic properties of our model system can therefore be obtained working with $(\Omega_m, \mathcal{P}(\Omega_m), \mu(m))$. The most probable state of the system is given by the absolute minimum of the generalized free energy f. In the thermodynamic limit, the thermodynamic free energy per spin is given exactly by the value $f(m_{\min})$,

$$-\frac{k_B T}{N} \ln Z \xrightarrow{N \to \infty} f(m_{\min}).$$

In the thermodynamic limit, this model has a phase transition for $h = 0$ from a paramagnetic phase with $m_{\min} = 0$ to a ferromagnetic phase with $m_{\min} = \pm m_{sp}$ at a temperature $T_c = J/k_B$. For small $h \neq 0$ and temperatures $T < T_c$ (see Fig. 3.8), the relative minimum with a magnetization parallel to the external field is the unique

3.7 Kinetic Ising Models and Monte Carlo Simulations

equilibrium state and the other one is a metastable state. For h larger than the so-called *spinodal field*, h_{sp}, the function f has only a single minimum, corresponding to a magnetization parallel to h.

However, not only are the probabilistic structure and calculation of the thermodynamic properties much simpler for the mean-field model, but also the kinetic equation (3.154) can be simplified. For the single spin-flip mean-field kinetic Ising model, the energy change can be written (for large N)

$$\Delta H(s_i \to -s_i) = \frac{2J}{N-1} s_i \sum_{\substack{j=1 \\ j \neq i}}^{N} s_j + 2h s_i$$

$$= \frac{2J}{N-1} s_i (Nm - s_i) + 2h s_i.$$

For large N and noting that $\delta m = -2s_i/N$, we get

$$\Delta H(m, \delta m) = -JNm\delta m - hN\delta m - \frac{2J}{N}. \tag{3.160}$$

The transition probabilities in the master equation are therefore also only functions of m and δm, and we have

$$w(s_i| - s_i) \to w(m - \delta m, \delta m)$$
$$w(-s_i|s_i) \to w(m, -\delta m)$$

if m is supposed to be the magnetization of the state $(s_1, \ldots, s_i, \ldots, s_N)$ and δm the magnetization difference between this state and the one with s_i reversed. Summing the master equation over all configurations s with a fixed m, we get

$$p(m, n+1) - p(m, n)$$
$$= \sum_{s_m} \sum_{i=1}^{N} w(m - \delta m, \delta m) p(s_1, \ldots, -s_i, \ldots, s_N, n)$$
$$- \sum_{i=1}^{N} w(m, -\delta m) p(m, n),$$

in which we have used $p(m, n) = \sum_{s_m} p(s, n)$. Performing the combinatorics [164] for magnetization changes $\delta m = 2/N$ and $\delta m = -2/N$ leading to or leaving from the magnetization m, one arrives at

$$p(m, n+1) - p(m, n)$$
$$= \frac{N}{2}\left(1 + m + \frac{2}{N}\right) w(m + 2/N, -2/N) p(m + 2/N, n)$$

$$+ \frac{N}{2}\left(1 - m + \frac{2}{N}\right) w(m - 2/N, 2/N) p(m - 2/N, n)$$

$$- \frac{N}{2}(1 + m) w(m, -2/N) p(m, n)$$

$$- \frac{N}{2}(1 - m) w(m, 2/N) p(m, n). \tag{3.161}$$

This is the master equation for a one-dimensional one-step process, van Kampen discussed in detail in [101].

Let us introduce a time scale δt for a single spin-flip and consider the limit $\delta t \to 0$, $N \delta t = \tau = \text{const}$. Dividing (3.161) by δt we derive in this limit

$$\frac{\partial}{\partial t} p(m, t) = \frac{N}{2}\left(1 + m + \frac{2}{N}\right) \tilde{w}(m + 2/N, -2/N) p(m + 2/N, n)$$

$$+ \frac{N}{2}\left(1 - m + \frac{2}{N}\right) \tilde{w}(m - 2/N, 2/N) p(m - 2/N, n)$$

$$- \frac{N}{2}(1 + m) \tilde{w}(m, -2/N) p(m, n)$$

$$- \frac{N}{2}(1 - m) \tilde{w}(m, 2/N) p(m, n), \tag{3.162}$$

where the transition *rates* are now given as (with $T_c = J/k_B$):

$$\tilde{w}_G(m, \delta m) = \frac{1}{2\tau}\left[1 + \tanh\left(\frac{T_c}{2T} N m \delta m + \frac{h}{2k_B T} N \delta m + \frac{T_c}{NT}\right)\right], \tag{3.163}$$

$$\tilde{w}_M(m, \delta m) = \frac{1}{\tau} \min\left(1, \exp\left[\frac{T_c}{T} N m \delta m + \frac{h}{k_B T} N \delta m + \frac{2T_c}{NT}\right]\right). \tag{3.164}$$

The master equation (3.162) can be elegantly rewritten [101] using the following definitions:

- rate of jumps to the right: $r(m) = (1 - m) \tilde{w}(m, 2/N)$,
- rate of jumps to the left: $l(m) = (1 + m) \tilde{w}(m, -2/N)$,
- translation by $2/N$: $\mathbb{T} = \exp(2/N \frac{\partial}{\partial m})$,
- translation by $-2/N$: $\mathbb{T}^{-1} = \exp(-2/N \frac{\partial}{\partial m})$.

The master equation becomes

$$\frac{\partial}{\partial t} p(m, t) = \frac{N}{2}(\mathbb{T} - 1)[l(m) p(m, t)] + \frac{N}{2}(\mathbb{T}^{-1} - 1)[r(m) p(m, t)]. \tag{3.165}$$

Considering that the translation operator is defined through the Taylor series of the exponential function, we see that (3.165) is a version of the Kramers-Moyal expan-

3.7 Kinetic Ising Models and Monte Carlo Simulations

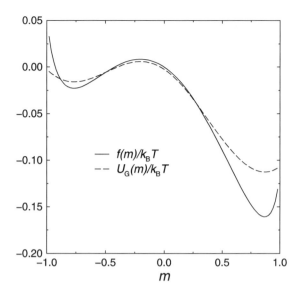

Fig. 3.9 Generalized free energy f (*solid curve*) and Glauber drift potential (*dashed curve*) in units of $k_B T$ for $T = 0.7 T_c$ and $h = 0.5 h_{sp}$

sion (see Chap. 2) of the master equation. The expansion parameter is the magnetization change $\delta m = \pm 2/N$ upon a single spin-flip. The expansion contains an infinite series of terms. For large N we will now use the Fokker-Planck approximation to this equation, truncating the expansion after the second order:

$$\tau \frac{\partial}{\partial t} p(m,t) = \frac{\partial}{\partial m}\left[U'(m) p(m,t)\right] + \frac{1}{2}\frac{\partial^2}{\partial m^2}\left[D(m) p(m,t)\right]. \qquad (3.166)$$

This equation describes a diffusion process in an external potential with the derivative

$$U'(m) = \tau\left[l(m) - r(m)\right] \qquad (3.167)$$

and a position dependent diffusion coefficient

$$D(m) = \frac{2\tau}{N}\left[l(m) + r(m)\right]. \qquad (3.168)$$

For the Glauber transition rates and in the limit $N \to \infty$, the drift potential for the Glauber rate is given as

$$\beta U_G(m) = \frac{1}{2}m^2 - \frac{T}{T_c}\ln\left[\cosh\left(\frac{k_B T_c m + h}{k_B T}\right)\right]. \qquad (3.169)$$

Figure 3.9 shows a comparison between the Glauber drift potential and the generalized free energy which defines the stationary distribution. They agree in the positions but not in the relative height of the extrema. From the Fokker-Planck equation we also see that the diffusion coefficient vanishes in the thermodynamic limit. For the mean-field system in the thermodynamic limit, the metastable state therefore has an infinite lifetime. For large, but finite N, we can consider a system which, as a starting condition, has a magnetization somewhere in the vicinity of the metastable magnetization, for instance, by preparing it in a negative external field. At time $t = 0$ we then instantaneously switch the external field to the positive value $h = 0.5h_{sp}$, for which the drift potential is shown in Fig. 3.9. We can then again ask the question at what rate the system will traverse the barrier between the metastable and the stable state in the drift potential. Thus we are led back to the mean first passage time problem we solved in Sect. 3.5 for Fokker-Planck equations of the form (3.166).

For the case of short-range interactions no description with a single reaction coordinate, m, is possible, and the lifetime of the metastable state is finite, even in the thermodynamic limit. The decay of the metastable state then proceeds by the spontaneous nucleation of droplets of the stable phase, which display a growth instability once they exceed a threshold size [141].

The problems we have discussed so far in this chapter were classical applications of the theory of Brownian motion, i.e., diffusion processes, in physics. We will now go on to discuss an application of the theory of Brownian motion in an area which is usually treated as completely disjunct from the theory of stochastic processes.

3.8 Quantum Mechanics as a Diffusion Process

This section presents an alternative derivation of quantum mechanics suggested by E. Nelson [149]. To this end, we first derive what may be called the *hydrodynamics of Brownian motion*. This framework then enables us to generate a conservative, i.e., energy-conserving, diffusion process. One special case of these *conservative diffusion* processes turns out to be equivalent to non-relativistic quantum mechanics.

3.8.1 Hydrodynamics of Brownian Motion

In the following we consider Brownian particles described by the following Itô stochastic differential equation:

$$dx(t) = b_+(x,t)dt + \sigma dW(t). \tag{3.170}$$

When we discussed the definition of this Itô SDE in Sect. 2.3, we had to require $W(t)$ and $x(t)$ to be non-anticipating functions; $dW(t)$ is independent of all $x(s)$ with $s < t$. In the context of the Ornstein-Uhlenbeck process, we also discussed the irreversible character of that process. This well studied example of Brownian motion often leads to the belief that irreversibility is a general property of stochastic processes. How could this be reconciled with the conservative dynamics we use to

3.8 Quantum Mechanics as a Diffusion Process

describe quantum particles? In order to answer this question, we have to look closer at what happens when we reverse the flow of time in an Itô SDE.

For this, let us recall Itô's formula, (cf. (2.121))

$$df(\mathbf{x},t) = \frac{\partial}{\partial t} f(\mathbf{x},t) dt + \nabla f(\mathbf{x},t) \cdot d\mathbf{x} + \frac{1}{2} \Delta f(\mathbf{x},t)(d\mathbf{x})^2,$$

where the last term is only evaluated up to linear order in dt. Using (3.170) we get

$$df(\mathbf{x},t) = \left(\frac{\partial}{\partial t} f(\mathbf{x},t) + \mathbf{b}_+(\mathbf{x},t) \cdot \nabla f(\mathbf{x},t) + \frac{1}{2}\sigma^2 \Delta f(\mathbf{x},t) \right) dt$$
$$+ \sigma \nabla f(\mathbf{x},t) \cdot d\mathbf{W}(t). \qquad (3.171)$$

We want to proceed to a kinematic treatment of Brownian motion, trying to parallel the usual treatment of Newtonian motion. We have, however, already stated that the Brownian path is not differentiable with respect to time, so there is no simple equivalent to the particle velocity. We therefore define the *mean forward derivative* of a function of the position variable as

$$D_+ f(\mathbf{x},t) = \lim_{\Delta t \to 0^+} \mathrm{E}\left[\frac{f(\mathbf{x}(t+\Delta t), t+\Delta t) - f(\mathbf{x}(t),t)}{\Delta t} \right]$$
$$= \frac{\partial}{\partial t} f(\mathbf{x},t) + \mathbf{b}_+(\mathbf{x},t) \cdot \nabla f(\mathbf{x},t) + \frac{1}{2}\sigma^2 \Delta f(\mathbf{x},t), \qquad (3.172)$$

where 0^+ signifies approaching zero from above. Combining this with (3.170), we see that the mean forward derivative of the particle position itself is given as

$$D_+ \mathbf{x}(t) = \mathbf{b}_+(\mathbf{x},t). \qquad (3.173)$$

On the level of probability distributions, this diffusion process is described by the following Fokker-Planck equation (cf. Table 2.1):

$$\frac{\partial}{\partial t} p(\mathbf{x},t) = -\nabla \cdot \left[\mathbf{b}_+(\mathbf{x},t) p(\mathbf{x},t) \right] + \frac{\sigma^2}{2} \Delta p(\mathbf{x},t). \qquad (3.174)$$

Since the position coordinate is symmetric under time reversal, we have $\mathbf{x}(t) = \mathbf{x}(-t)$ and we can write

$$d\mathbf{x}(t) = \mathbf{b}_-(\mathbf{x},t) dt + \sigma d\mathbf{W}_*(t), \qquad (3.175)$$

where $d\mathbf{W}_*(t)$ is independent of all $\mathbf{x}(s)$ with $s > t$. For the variance of this time reversed Wiener process, however, we still have

$$\mathrm{E}\left[(d\mathbf{W}_*(t))^2 \right] = |dt| = -dt \qquad (3.176)$$

for the reversed time flow. Using again the Itô formula but now (3.175) as the basic time evolution, we get

$$df(x,t) = \left(\frac{\partial}{\partial t} f(x,t) + b_-(x,t) \cdot \nabla f(x,t) - \frac{1}{2}\sigma^2 \Delta f(x,t)\right) dt$$
$$+ \sigma \nabla f(x,t) \cdot dW_*(t). \qquad (3.177)$$

From this we find for the *mean backward derivative* of a function $f(x(t), t)$

$$D_- f(x,t) = \lim_{\Delta t \to 0^+} E\left[\frac{f(x(t), t) - f(x(t - \Delta t), t - \Delta t)}{\Delta t}\right]$$
$$= \frac{\partial}{\partial t} f(x,t) + b_-(x,t) \cdot \nabla f(x,t) - \frac{1}{2}\sigma^2 \Delta f(x,t). \qquad (3.178)$$

Due to the definition in (3.175), we have

$$D_- x(t) = b_-(x,t). \qquad (3.179)$$

Performing the time reversal on (3.174) results in

$$\frac{\partial}{\partial t} p(x,t) = -\nabla \cdot \left[b_-(x,t) p(x,t)\right] - \frac{\sigma^2}{2} \Delta p(x,t). \qquad (3.180)$$

Combining the two definitions for the mean forward and mean backward derivative, we can extend the Newtonian definition of velocity to define the *mean velocity* of the Brownian particle:

$$v(x,t) = \frac{1}{2}\bigl(b_+(x,t) + b_-(x,t)\bigr) = \frac{1}{2}(D_+ + D_-) x(t). \qquad (3.181)$$

For Newtonian motion, the mean forward and backward derivatives agree and we just have

$$v(x,t) = D_+ x(t) = D_- x(t) = b_+(x,t) = b_-(x,t).$$

The relevance of the mean velocity is elucidated by adding (3.180) to (3.174) to get the *continuity equation*:

$$\frac{\partial}{\partial t} p(x,t) + \nabla \cdot \left[v(x,t) p(x,t)\right] = 0. \qquad (3.182)$$

The mean velocity describes the probability current and thus is the first ingredient to a hydrodynamic description of Brownian motion. It is only different from zero when there is a drift vector present in the Itô SDE.

Whenever the mean forward and mean backward derivative are different, there is another velocity field one can define:

$$u(x,t) = \frac{1}{2}\bigl(b_+(x,t) - b_-(x,t)\bigr) = \frac{1}{2}(D_+ - D_-) x(t). \qquad (3.183)$$

3.8 Quantum Mechanics as a Diffusion Process

The vector field $u(x, t)$ is usually called the *osmotic velocity*. Subtracting (3.180) from (3.174), we get:

$$\nabla \cdot \left[u(x,t) p(x,t) - \frac{1}{2}\sigma^2 \nabla p(x,t) \right] = 0. \tag{3.184}$$

From this we have for u

$$u(x, t) = \frac{\sigma^2}{2} \nabla \ln p(x, t). \tag{3.185}$$

An integration constant in form of a rotation, $\nabla \times c(x, t)$, can be shown to vanish when one derives (3.185) on a different route [149]. Starting from (3.185) we can derive the first of two hydrodynamic equations for the description of the Brownian motion process by taking the partial derivative with respect to time:

$$\frac{\partial}{\partial t} u = \frac{\sigma^2}{2} \nabla \frac{\partial}{\partial t} \ln p = \frac{\sigma^2}{2} \nabla \left(\frac{1}{p} \frac{\partial p}{\partial t} \right) = -\frac{\sigma^2}{2} \nabla \left(\frac{1}{p} \nabla \cdot (vp) \right)$$

$$= -\frac{\sigma^2}{2} \nabla (\nabla \cdot v) - \frac{\sigma^2}{2} \nabla \left(\frac{1}{p} v \cdot \nabla p \right),$$

in which we have used the continuity equation (3.182). Inserting the result for the osmotic velocity (3.185) again, we arrive at

$$\frac{\partial}{\partial t} u = -\frac{\sigma^2}{2} \nabla (\nabla \cdot v) - \nabla (v \cdot u). \tag{3.186}$$

To proceed towards a dynamic description of Brownian motion and to derive the hydrodynamic equation for the flow velocity v, we also have to generalize the concept of acceleration and define the *mean acceleration* as

$$a(x, t) = \frac{1}{2}(D_+ D_- + D_- D_+) x(t)$$

$$= \frac{1}{2}(D_+ b_-(x, t) + D_- b_+(x, t)). \tag{3.187}$$

To convince ourselves that this is a sensible generalization of the concept of acceleration, let us first reconsider the Ornstein-Uhlenbeck process in (3.53). For this process, the paths are differentiable [because $dx = v dt$] and we have $D_+ x = D_- x = v$. So one derives

$$a(x, t) = \frac{1}{2}(D_+ D_- + D_- D_+) x(t)$$

$$= \frac{1}{2}(D_+ + D_-) v(x, t)$$

$$= \frac{1}{m} F, \tag{3.188}$$

which is the stochastic generalization of Newton's second law for the Ornstein-Uhlenbeck process. Applying the results for the mean forward (3.172) and the mean backward derivative (3.178), we get

$$a = \frac{1}{2}\frac{\partial}{\partial t}(b_+ + b_-) + \frac{1}{2}b_+ \cdot \nabla b_- + \frac{1}{2}b_- \cdot \nabla b_+ - \frac{\sigma^2}{4}\Delta(b_+ - b_-).$$

Inserting the definitions of u (3.183) and v (3.181), we finally have

$$\frac{\partial}{\partial t}v = a - v \cdot \nabla v + u \cdot \nabla u + \frac{\sigma^2}{2}\Delta u. \tag{3.189}$$

With (3.186) and (3.189) we have a complete general description of the hydrodynamics of Brownian motion.

3.8.2 Conservative Diffusion Processes

To specialize to a conservative Markovian diffusion process, we have to proceed in the same way as Newton did in postulating his second law, i.e., by making the force on the particle a dynamic variable. If the Brownian particle moves in the external potential $U(x)$, we require the stochastic acceleration in general to be given by

$$a(x,t) = -\frac{1}{M}\nabla U(x). \tag{3.190}$$

The hydrodynamics of conservative diffusion processes is therefore given by two velocity fields satisfying the following partial differential equations:

$$\begin{aligned}\frac{\partial}{\partial t}u &= -\frac{\sigma^2}{2}\nabla(\nabla \cdot v) - \nabla(v \cdot u), \\ \frac{\partial}{\partial t}v &= -\frac{1}{M}\nabla U - v \cdot \nabla v + u \cdot \nabla u + \frac{\sigma^2}{2}\Delta u.\end{aligned} \tag{3.191}$$

We have already concluded from the result for u that it can be written as the gradient of the logarithm of the probability density (see (3.185)). Let us therefore define a scalar field $R(x,t)$ by requiring

$$p(x,t) = \exp[2R(x,t)]$$

so that we can write

$$u(x,t) = \sigma^2 \nabla R(x,t). \tag{3.192}$$

In the same way as we did for u, we now require that v contains no closed flow lines and define another scalar field $S(x,t)$ by

$$v(x,t) = \frac{1}{M}\nabla S(x,t). \tag{3.193}$$

3.8 Quantum Mechanics as a Diffusion Process

In classical mechanics, this equation relates the velocity of a particle to the *action* $S(x,t)$. Inserting the two definitions (3.192) and (3.193) into (3.191), we get

$$\sigma^2 \nabla \frac{\partial}{\partial t} R = -\frac{\sigma^2}{2M} \nabla \Delta S - \frac{\sigma^2}{M} \nabla (\nabla R \cdot \nabla S),$$

$$\frac{1}{M} \nabla \frac{\partial}{\partial t} S = -\frac{1}{M} \nabla U - \frac{1}{M^2} \nabla S \cdot \nabla \nabla S + \sigma^4 \nabla R \cdot \nabla \nabla R + \frac{\sigma^4}{2} \Delta \nabla R.$$

Now we make use of the vector identity

$$g \cdot \nabla g = \frac{1}{2} \nabla g^2 - g \times (\nabla \times g)$$

in the second equation. For $g = \nabla R$ or $g = \nabla S$, the rotation parts vanish and we get

$$\sigma^2 \nabla \left(\frac{\partial}{\partial t} R + \frac{1}{2M} \Delta S + \frac{1}{M} \nabla R \cdot \nabla S \right) = 0,$$
$$\frac{1}{M} \nabla \left(\frac{\partial}{\partial t} S + U + \frac{1}{2M} (\nabla S)^2 - \frac{M\sigma^4}{2} \left[(\nabla R)^2 + \Delta R \right] \right) = 0.$$
(3.194)

Requiring

$$\frac{\partial}{\partial t} R + \frac{1}{2M} \Delta S + \frac{1}{M} \nabla R \cdot \nabla S = 0,$$
$$\frac{\partial}{\partial t} S + U + \frac{1}{2M} (\nabla S)^2 - \frac{M\sigma^4}{2} \left[(\nabla R)^2 + \Delta R \right] = 0,$$
(3.195)

we can determine R and S up to a position-independent phase $\phi(t)$ which can be absorbed in the action S (note that R is fixed through the normalization of the probability p).

The equations in (3.195) are the stochastic mechanic counterpart to the *Hamilton-Jacobi equations* in classical mechanics, which constitute the hydrodynamic formulation of Newtonian mechanics. They differ from the classical equations by the additional terms depending on the strength σ of the stochastic forces. For $\sigma = 0$, the classical Hamilton-Jacobi equations are exactly recovered from (3.195). The first of these two equations is nothing but the conservation of probability (3.182) in another form.

3.8.3 Hypothesis of Universal Brownian Motion

As the final step in the derivation of non-relativistic quantum mechanics from the theory of stochastic processes, we invoke what Nelson termed the *hypothesis of universal Brownian motion* [149]. The physical reasoning behind this hypothesis is very simple and convincing. When we write down a Hamiltonian for a classical mechanical system, we always (mostly implicitly) assume that we can treat that system as isolated from its surroundings in the following sense. The complete Hamiltonian of a system plus its surroundings has the same form as in the Caldeira-Leggett model (cf. (3.80)):

$$H = H_S + H_R + H_I,$$

where H_S is assumed to contain the effect of the surroundings on the system that can be described through the introduction of external potentials. In classical mechanics the interaction term (H_I) is assumed to be negligible and since the variables in the system and the surroundings then separate we only treat the simple Hamiltonian

$$H = H_S.$$

This assumption works empirically very well for macroscopic bodies, the domain of applicability of classical mechanics. For quantum phenomena, this assumption of an isolated system breaks down; an argument which is also the basis of Heisenberg's derivation of the uncertainty relation. We are, however, not able to treat all the intricate interactions between the quantum particle and its surroundings explicitly. This is the same situation we encountered in the discussion of the motion of a Brownian particle immersed in a fluid. We therefore postulate a stochastic description of the motion of a quantum particle in the form of a conservative diffusion process of the type we discussed in Sect. 3.8.2. We know that such a process reduces to Newtonian mechanics in the limit $\sigma \to 0$, so we require the strength of the fluctuation term to be inversely proportional to the mass of the particle (or system):

$$\sigma^2 \propto \frac{1}{M}.$$

The proportionality constant in this equation necessarily has the physical units of an action and we write it as

$$\sigma^2 = \frac{\hbar}{M}. \tag{3.196}$$

With this identification we can write (3.195) as

$$\frac{\partial}{\partial t}R + \frac{1}{2M}\Delta S + \frac{1}{M}\nabla R \cdot \nabla S = 0,$$
$$\frac{\partial}{\partial t}S + U + \frac{1}{2M}(\nabla S)^2 - \frac{\hbar^2}{2M}\left[(\nabla R)^2 + \Delta R\right] = 0. \tag{3.197}$$

These equations are known as the *Madelung fluid* [121]. Madelung derived them in 1926, starting from the Schrödinger equation. In fact, one can show that the fact that

$$\psi(x,t) = \exp\left(R(x,t) + \frac{i}{\hbar}S(x,t)\right) \tag{3.198}$$

3.8 Quantum Mechanics as a Diffusion Process

satisfies the Schrödinger equation

$$i\hbar \frac{\partial}{\partial t}\psi = \left(-\frac{\hbar^2}{2M}\Delta + U\right)\psi \qquad (3.199)$$

is equivalent to R and S satisfying (3.197). Inserting (3.198) into the Schrödinger equation we get

$$\left(i\hbar \frac{\partial}{\partial t}R - \frac{\partial}{\partial t}S\right)\psi$$
$$= U\psi - \left(\frac{\hbar^2}{2M}\Delta R + \frac{i\hbar}{2M}\Delta S\right)\psi - \frac{\hbar^2}{2M}\left(\nabla R + \frac{i}{\hbar}\nabla S\right)^2 \psi.$$

Separating real and imaginary parts and—for $\psi \neq 0$—dividing by ψ, we arrive at the defining equations of the Madelung fluid (3.197). In the stochastic mechanics context, the Schrödinger equation therefore appears as an auxiliary equation, linearizing the complicated coupled nonlinear partial differential equations which define the hydrodynamics of a conservative diffusion process. When we choose the strength of the fluctuations in our starting Itô SDE to be given as $\sigma^2 = \hbar/M$, the conservative Brownian motion description and the Schrödinger equation are equivalent. To each solution of the Schrödinger equation there correspond trajectories of quantum particles defined by (3.170) in the same way, as to each solution of the classical Hamilton-Jacobi equations there correspond particle trajectories defined through Newton's second law or Hamilton's equations.

Which type of physical interactions are summarily treated by the quantum-fluctuation term $\sigma^2 = \hbar/M$ is not yet clear [150]. The clarification of this point would need a derivation of (3.170) from a deterministic treatment of the complete Hamiltonian for system, surroundings and all interactions.

Interpretation

First, we want to note that defining the function ψ as in (3.198) we have

$$p(\mathbf{x}, t) = |\psi(\mathbf{x}, t)|^2 = \exp(2R(\mathbf{x}, t)), \qquad (3.200)$$

so that Born's probability interpretation of the wave function is a natural outcome of the stochastic mechanics approach and not an independent postulate for the interpretation of the Schrödinger equation. $|\psi(\mathbf{x}, t)|^2 d^3 \mathbf{x}$, by construction, is the probability to find the quantum particle in a volume $d^3 \mathbf{x}$ around $\mathbf{x}$. We anticipated this in Chap. 2 when we presented $(\mathbb{R}^3, \mathcal{B}, |\psi(\mathbf{x}, t)|^2 d^3 \mathbf{x})$ as an example for a probability space occurring in quantum mechanics.

Expectation values of quantum-mechanical observables $O(\mathbf{x})$ which are functions of the position operator are defined as

$$\langle \psi | \hat{O} | \psi \rangle = \int d^3 \mathbf{x}\, O(\mathbf{x}) |\psi(\mathbf{x}, t)|^2 = \int d^3 \mathbf{x}\, O(\mathbf{x}) p(\mathbf{x}, t). \qquad (3.201)$$

This is the standard definition of the expectation value of a stochastic variable, which in this case is a function of the underlying stochastic process $x(t)$:

$$\mathrm{E}[O(x)] = \langle \psi | \hat{O} | \psi \rangle. \tag{3.202}$$

For the particle momentum and functions thereof, the situation is different. However, one can show that the expectation value of the momentum operator $\hat{p}$ in the quantum mechanical state ψ is given as

$$\frac{1}{M} \langle \psi | \hat{p} | \psi \rangle = \mathrm{E}[v] = \mathrm{E}[b_+] = \mathrm{E}[b_-], \tag{3.203}$$

because the expectation value of the osmotic velocity is zero

$$\mathrm{E}[u] = 0. \tag{3.204}$$

Nevertheless, the osmotic velocity is of physical and measurable significance, since its second moment contributes to the kinetic energy of the particle [150]. In fact, for the conservative diffusion process we have derived, the conservation of energy can be shown to hold in the following form:

$$\left(\frac{\partial}{\partial t} + v \cdot \nabla \right) E_{\text{tot}}(x, t) = 0, \tag{3.205}$$

with

$$E_{\text{tot}}(x, t) = \frac{M}{2} \left(v^2(x, t) + u^2(x, t) \right) + U(x, t). \tag{3.206}$$

For all quantum-mechanical observables defined by self-adjoined operators on the Hilbert space of our system, we can define the corresponding stochastic functions on the probability space. The reverse is not true. The microscopic and path-oriented picture of quantum processes that is given in the stochastic mechanics treatment allows for the definition of more complicated observables, such as mean first passage times of the type we discussed for the Kramers problem.

As a final comment at this point, let us mention that the motion of a quantum particle in an external electromagnetic field is treated by the minimal coupling familiar from quantum mechanics. We identify U with the electrostatic field Φ and replace (3.193) by

$$v(x, t) = \frac{1}{M} \left(\nabla S(x, t) - \frac{e}{c} A(x, t) \right), \tag{3.207}$$

where A is the vector potential for the magnetic field.

3.8.4 Tunnel Effect

One of the most intriguing effects in quantum mechanics is the tunneling effect, i.e., the ability of a quantum particle with energy E to penetrate a barrier of energy

3.8 Quantum Mechanics as a Diffusion Process

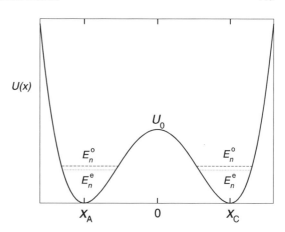

Fig. 3.10 A symmetric double well potential with a low lying energy states E_n^e and E_n^o belonging to even and odd solutions of the Schrödinger equation, respectively

$U_0 > E$ with a certain probability. Let us consider a quantum particle in a double well potential like the one shown in Fig. 3.10. Due to the symmetry of the potential the wave function of a quantum particle moving in this potential has to be either even or odd. Let us denote the energies corresponding to the even solutions as E_n^e and those belonging to the odd solutions as $E_n^o = E_n^e + \Delta E_n$. One has $\Delta E_n > 0$ because all odd solutions have an additional node at $x = 0$. Let the wave functions ψ_e and ψ_o be solutions of the one-dimensional Schrödinger equation

$$i\hbar \frac{\partial}{\partial t} \psi(x,t) = \left(-\frac{\hbar^2}{2M} \frac{\partial^2}{\partial x^2} + U(x) \right) \psi(x,t) \tag{3.208}$$

with energies E_n^e and E_n^o. In quantum mechanics the energy difference ΔE_n is called the tunnel splitting of the two states and is related to the possibility for the particle in the even state to tunnel from one well of the potential in Fig. 3.10 to the other. However, the wave functions ψ_e and ψ_o are stationary, so there is no net probability current from one well to the other in these states. To relate the energy splitting ΔE_n to the tunnel frequency one therefore resorts to the following approximate construction.

Suppose that $\psi_R(x)$ is a wave function localized on the positive real axis and $\psi_L(x) = \psi_R(-x)$ its symmetric counterpart on the negative real axis. When we assume that $\psi_R(0) = 0$ and $\psi_R'(0) = 0$, which is a viable approximation for $U_0 \gg E$, we can combine them to yield the odd and even states:

$$\psi_e(x) = \frac{1}{\sqrt{2}} [\psi_R(x) + \psi_L(x)], \qquad \psi_o(x) = \frac{1}{\sqrt{2}} [\psi_R(x) - \psi_L(x)]. \tag{3.209}$$

In the following we will therefore assume that we are treating low lying energy states. Turning (3.209) around, we obtain

$$\psi_R(x) = \frac{1}{\sqrt{2}} [\psi_e(x) + \psi_o(x)], \qquad \psi_L(x) = \frac{1}{\sqrt{2}} [\psi_e(x) - \psi_o(x)]. \tag{3.210}$$

Suppose we now start at time $t = 0$ with a particle located on the positive axis, i.e., described by $\psi(x, 0) = \psi_R(x)$. Its time evolution is given by

$$\psi(x, t) = \frac{1}{\sqrt{2}} e^{-iE_n^e t/\hbar} \left[\psi_e(x) + e^{-i\Delta E_n t/\hbar} \psi_o(x) \right], \quad (3.211)$$

i.e., $\psi(x, t)$ is not a stationary solution of the Schrödinger equation. From this we see that for $e^{-i\Delta E_n t/\hbar} = -1$ the particle is located on the negative real axis. So we can read off the tunneling times between the wells,

$$\tau_n = \frac{\pi \hbar}{\Delta E_n}, \quad (3.212)$$

which are determined by the energy differences between the eigenvalues of the even and odd solutions. If we can solve the Schrödinger equation to obtain the eigenvalues (which is often only possible approximately, often one uses the so-called WKB approximation [139]), tunnel frequencies may be calculated.

However, one can also derive an explicit connection between the tunneling time and the wave function of the tunneling state by considering the probability flux from one well to the other. The probability to be in the well on the right side is

$$P_R(t) = \int_0^\infty p(x, t) dx = \int_0^\infty |\psi(x, t)|^2 dx. \quad (3.213)$$

Its time derivative is

$$\frac{dP_R(t)}{dt} = \int_0^\infty \frac{\partial p(x, t)}{\partial t} dx. \quad (3.214)$$

We evaluate (3.214) in two ways, by directly inserting the wave function from (3.211) and by first using the quantum mechanical continuity equation. The first way results in

$$\frac{dP_R(t)}{dt} = -\frac{\Delta E_n}{2\hbar} \sin\left(\frac{\Delta E_n t}{\hbar}\right). \quad (3.215)$$

The quantum mechanical continuity equation as derived from the Schrödinger equation reads

$$\frac{\partial p(x, t)}{\partial t} = \frac{\hbar}{2Mi} \frac{\partial}{\partial x} \left(\psi(x, t) \frac{\partial \psi^*(x, t)}{\partial x} - \psi^*(x, t) \frac{\partial \psi(x, t)}{\partial x} \right) \quad (3.216)$$

with the probability current

$$j(x, t) = \frac{\hbar}{M} \Im \left(\psi^*(x, t) \frac{\partial \psi(x, t)}{\partial x} \right), \quad (3.217)$$

where $\Im(\cdot)$ denotes the imaginary part. Inserting this result into (3.214) and requiring a vanishing probability current for $x \to \infty$ we obtain

$$\frac{dP_R(t)}{dt} = \frac{\hbar}{M} \Im \left(\psi^*(0, t) \frac{\partial \psi(0, t)}{\partial x} \right)$$

3.8 Quantum Mechanics as a Diffusion Process

$$= -\frac{\hbar}{2M}\psi_e(0)\psi'_o(0)\sin\left(\frac{\Delta E_n t}{\hbar}\right). \tag{3.218}$$

For the last equality we have let the prime denote derivative with respect to x and we have used $\psi_o(0) = \psi'_e(0) = 0$. Comparing (3.215) and (3.218) we obtain the tunneling time as

$$\tau_n = \frac{M}{\hbar}\frac{\pi}{\psi_e(0)\psi'_o(0)}. \tag{3.219}$$

The discussion above is completely rooted within probability theory, employing the interpretation of $|\psi(x,t)|^2$ as the probability to find a particle at position x at time t, and the existence of a continuity equation for this probability. As such, the whole argument runs identically in the stochastic mechanics interpretation of quantum mechanics. The connection becomes obvious when we rewrite (3.217) as

$$j(x,t) = p(x,t)\Im\left(\frac{\hbar}{M}\frac{\partial \ln[\psi(x,t)]}{\partial x}\right) = p(x,t)v(x,t), \tag{3.220}$$

where we wrote the probability current according to (3.182) and the introduced velocity $v(x,t)$ is consistent with (3.193) and (3.198).

However, the correspondence between the Schrödinger equation and an underlying conservative stochastic process allows us to go beyond this discussion and to define a tunneling time for the stationary solution $\psi_e(x,t)$ itself. With our prior knowledge about diffusion processes we identify a quantum mechanical Kramers problem in Fig. 3.10. Suppose we can determine the solution of the Schrödinger equation for this problem with energy E_n. The wave functions for these stationary states have the form $\psi_e(x,t) = \exp(R_n(x) - iE_n t/\hbar)$ (see (3.198)). Using the definitions (3.181), (3.183) and (3.192), (3.193) we obtain for the drift of the quantum diffusion process

$$b_+(x) = v(x) + u(x) = \frac{\hbar}{M}\frac{d}{dx}R_n(x) = \frac{\hbar}{M}\frac{\partial}{\partial x}\ln|\psi_e(x,t)|.$$

The equation of motion of the quantum particle in the eigenstate $\psi_e(x,t)$ of the double well potential is therefore given by

$$dx(t) = \frac{\hbar}{M}\left(\partial_x \ln|\psi_e(x,t)|\right)dt + \sqrt{\frac{\hbar}{M}}dW(t). \tag{3.221}$$

This Itô SDE is equivalent to the one-dimensional Fokker-Planck equation (see Table 2.1 and Sect. 2.4)

$$\frac{\partial}{\partial t}p(x,t) = -\frac{\hbar}{M}\frac{\partial}{\partial x}\left[\left(\partial_x \ln|\psi_e(x,t)|\right)p(x,t)\right] + \frac{\hbar}{2M}\frac{\partial^2}{\partial x^2}p(x,t). \tag{3.222}$$

In Sect. 3.5 we have derived the general solution for the mean first passage time problem for a one-dimensional diffusion process. Comparing to (3.109) in our discussion of the Kramers problem we can make the following identifications

$$a_1(x) = \frac{\hbar}{M}\partial_x \ln|\psi_e(x,t)| \quad \text{and} \quad a_2(x) = \frac{\hbar}{M}.$$

Inserting these identifications into the general solution for the escape time (3.126) over the barrier, the escape time in this quantum mechanical Kramers problem is given by

$$\bar{\tau}_n(x) = \frac{2M}{\hbar} \int_x^{x_c} dx' \frac{1}{p_n(x')} \int_{-\infty}^{x'} dx'' p_n(x''), \tag{3.223}$$

where $p_n(x) = |\psi_e(x,t)|^2$ denotes the stationary probability distribution in the eigenstate to energy E_n. Note that both, (3.223) and (3.223), predict the tunneling time to be inversely proportional to the diffusion coefficient $\hbar/M$ of the particle, while the other terms having dimensions of a squared length will differ in general between the two equations.

The diffusion process approach therefore yields a tunneling time also for the stationary solution, and this time behaves qualitatively like the one constructed for the non-stationary case. But we have to keep in mind that within the diffusion process approach the particle never has an energy smaller than the potential energy at the point where it is found. It does not tunnel through the barrier but is driven over it by the omnipresent quantum fluctuations.

3.8.5 Harmonic Oscillator and Quantum Fields

We now want to treat the harmonic oscillator as a further illustration of the approach to quantum mechanics developed in this section [70]. In classical mechanics, the one-dimensional harmonic oscillator is defined through the Hamiltonian

$$H(x,p) = \frac{p^2}{2M} + \frac{M\omega^2}{2}x^2. \tag{3.224}$$

The corresponding Hamiltonian equations of motion

$$\begin{aligned}\dot{x} &= \frac{p}{M}, \\ \dot{p} &= -M\omega^2 x,\end{aligned} \tag{3.225}$$

have the general solution

$$\begin{aligned}x_{cl}(t) &= x_0 \cos\omega t + \frac{p_0}{M\omega}\sin\omega t, \\ p_{cl}(t) &= p_0 \cos\omega t - M\omega x_0 \sin\omega t.\end{aligned} \tag{3.226}$$

3.8 Quantum Mechanics as a Diffusion Process

The quantum-mechanic harmonic oscillator obeys the Schrödinger equation,

$$i\hbar \frac{\partial}{\partial t}\psi = -\frac{\hbar^2}{2M}\frac{\partial^2}{\partial x^2}\psi + \frac{M\omega^2}{2}x^2\psi. \tag{3.227}$$

This equation is solved by a family of states called the *coherent states* [168], which are associated with the classical solutions (3.226):

$$\psi(x,t) = (2\pi\tilde{\sigma}^2)^{-1/4} \exp\left[-\frac{(x-x_{\text{cl}}(t))^2}{4\tilde{\sigma}^2} + \frac{i}{\hbar}x p_{\text{cl}}(t)\right.$$
$$\left. - \frac{i}{2\hbar}x_{\text{cl}}(t)p_{\text{cl}}(t) - i\frac{\omega t}{2}\right], \tag{3.228}$$

in which we have defined

$$\tilde{\sigma}^2 = \frac{\sigma^2}{2\omega} = \frac{\hbar}{2M\omega}. \tag{3.229}$$

For these states, the probability density is a Gaussian with mean x_{cl} and variance $\tilde{\sigma}^2$:

$$p(x,t) = \exp[2R(x,t)] = (2\pi\tilde{\sigma}^2)^{-1/2}\exp\left[-\frac{(x-x_{\text{cl}}(t))^2}{2\tilde{\sigma}^2}\right], \tag{3.230}$$

and the action is given by

$$S(x,t) = x p_{\text{cl}}(t) - \frac{1}{2}x_{\text{cl}}(t)p_{\text{cl}}(t) - \frac{1}{2}\hbar\omega t. \tag{3.231}$$

From this we can determine the velocity fields defining the conservative diffusion process, using (3.192) and (3.193):

$$v(x,t) = \frac{p_{\text{cl}}(t)}{M},$$
$$u(x,t) = -\omega(x - x_{\text{cl}}(t)), \tag{3.232}$$
$$b_{\pm}(x,t) = \frac{p_{\text{cl}}(t)}{M} \mp \omega(x - x_{\text{cl}}(t)).$$

In a coherent state, the quantum-mechanical harmonic oscillator therefore obeys the following stochastic differential equation:

$$dx = \left[\frac{p_{\text{cl}}(t)}{M} - \omega(x - x_{\text{cl}}(t))\right]dt + \sigma dW(t). \tag{3.233}$$

To solve this equation let us look at the difference process between the quantum-mechanical and classical paths:

$$x_0(t) = x(t) - x_{\text{cl}}(t). \tag{3.234}$$

By insertion we see that x_0 satisfies the following Itô SDE:

$$dx_0 = -\omega x_0 dt + \sigma dW(t). \tag{3.235}$$

The corresponding coherent-state wave functions are given by

$$\psi_0(x,t) = \left(2\pi\tilde{\sigma}^2\right)^{-1/4} \exp\left[-\frac{x^2}{4\tilde{\sigma}^2} - \frac{i}{2}\omega t\right] \tag{3.236}$$

with the stationary probability density

$$p_0(x) = \left(2\pi\tilde{\sigma}^2\right)^{-1/2} \exp\left[-\frac{x^2}{2\tilde{\sigma}^2}\right]. \tag{3.237}$$

This is the *ground-state process* that we obtain by choosing the classical ground state $x_{cl} = 0$ and $p_{cl} = 0$. The complete family of solutions of the quantum-mechanical harmonic oscillator can be generated by adding the ground-state process to all solutions of the classical harmonic oscillator in the form prescribed by the coherent-state wave functions.

We are well acquainted with the solution for the ground state process, namely the Ornstein-Uhlenbeck process (see Sect. 3.2); however, here it is occurring for the position coordinate and not for the velocity. The ground-state process of the quantum-mechanical harmonic oscillator is therefore a stationary, Gaussian Markov process and we can use all the results derived in Sect. 3.2.

The importance and usefulness of this result becomes more obvious when we look into the way field quantization is usually introduced in quantum mechanics, see, for instance, Messiah [140]. Let us consider a free scalar field with Hamiltonian

$$H = \frac{1}{2}\int d^3x \left[\left(\frac{\partial}{\partial t}\phi\right)^2 + (\nabla\phi)^2 + M^2\phi^2\right], \tag{3.238}$$

where we set, for simplicity, Planck's constant, $\hbar = 1$, and also the velocity of light, $c = 1$. The corresponding Lagrangian [61] is

$$\mathcal{L} = \frac{1}{2}\int d^3x \left[\left(\frac{\partial}{\partial t}\phi\right)^2 - (\nabla\phi)^2 - M^2\phi^2\right]. \tag{3.239}$$

The Lagrangian equation of motion for this scalar field therefore is

$$\frac{\partial^2}{\partial t^2}\phi - \Delta\phi + M^2\phi = 0. \tag{3.240}$$

When we study this model in a finite rectangular box of volume V and with periodic boundary conditions, we know that we can expand the field ϕ in plane wave functions $f_n(\mathbf{x}) = V^{-1/2}\exp(i\mathbf{k}_n \cdot \mathbf{x})$. They form a countable set of orthogonal and

3.8 Quantum Mechanics as a Diffusion Process

normalized basis functions, $f_n(\mathbf{x})$, which are eigenfunctions of the Laplacian operator:

$$\phi(\mathbf{x},t) = \sum_{n=-\infty}^{\infty} f_n(\mathbf{x}) q_n(t), \tag{3.241}$$

with

$$\Delta f_n(\mathbf{x}) = -k_n^2 f_n(\mathbf{x}), \quad \int_V d^3\mathbf{x}\, f_n(\mathbf{x}) f_{n'}^*(\mathbf{x}) = \delta_{nn'}. \tag{3.242}$$

Here f_n^* indicates the complex conjugate of f_n. When we insert (3.241) into (3.240), we get a harmonic-oscillator equation for the expansion coefficients, $q_n(t)$:

$$\frac{\partial^2}{\partial t^2} q_n(t) + \omega_n^2 q_n(t) = 0, \tag{3.243}$$

where the eigenfrequencies ω_n are defined as

$$\omega_n = \sqrt{k_n^2 + M^2}. \tag{3.244}$$

From this point, the transition to quantum mechanics is usually done by quantizing the expansion coefficients or Fourier amplitudes, q_n. We proceed with the stochastic quantization we have already performed for the harmonic oscillator. For the ground state process of the field amplitudes we get (see (3.235))

$$dq_n = -\omega_n q_n dt + dW_n(t). \tag{3.245}$$

For the increments of the Wiener processes, we have

$$\mathrm{E}\big[dW_n(t) dW_{n'}(t')\big] = \delta_{nn'} \delta(t - t') dt. \tag{3.246}$$

The $q_n(t)$ are independent, stationary, Gaussian Markov processes with mean and covariance function in the stationary state given by (see also (3.65))

$$\mathrm{E}\big[q_n(t)\big] = 0,$$
$$\mathrm{E}\big[q_n(t) q_{n'}(t')\big] = \delta_{nn'} \frac{1}{2\omega_n} \exp\big[-\omega_n |t - t'|\big]. \tag{3.247}$$

Therefore, the field ϕ also is a Gaussian stochastic field with the following first two moments:

$$\mathrm{E}\big[\phi(\mathbf{x},t)\big] = 0 \tag{3.248}$$

$$\mathrm{E}\big[\phi(\mathbf{x},t) \phi^*(\mathbf{x}',t')\big] = \sum_{n=-\infty}^{\infty} \frac{f_n(\mathbf{x}) f_n^*(\mathbf{x}')}{2\omega_n} \exp\big[-\omega_n |t - t'|\big]. \tag{3.249}$$

In the limit of an infinite box size we get a continuum of k-space points with density $V/(2\pi)^3 \mathrm{d}^3 k$ (cf. Fig. 3.1) and the covariance functional of the stochastic field ϕ can be written

$$\mathrm{E}[\phi(x,t)\phi^*(x',t')] = \frac{1}{(2\pi)^3} \int \frac{\mathrm{d}^3 k}{2\sqrt{k^2 + M^2}} \exp[-i k \cdot (x - x')]$$
$$\times \exp[-\sqrt{k^2 + M^2}|t - t'|]. \qquad (3.250)$$

Gaussian random fields with zero mean and covariance given by (3.250) are also called Euclidean quantum fields. In this way we have rather straightforwardly derived the stochastic field theory corresponding to the modern Euclidean formulation [61] of quantum field theory from a simple stochastic quantization of the harmonic oscillator.

We choose to stop this exposition of quantum mechanics as a conservative diffusion process at this point, although much more could be said about it. We refer the interested reader to the further reading section at the end of this chapter.

3.9 Summary

In this chapter we have discussed basically two types of models of stochastic processes: random walks and Brownian motion. For both models we only treated the Markov version, i.e., we neglected possible memory effects. The natural starting point for the description of Markovian random walks is a master equation for the conditional probability to be at position x at time t, given the walk starts at x_0 at time t_0. In the most general formulation, the jump distance and direction ℓ and the time lag τ between consecutive jumps are random variables with a common distribution, $\Psi(\ell, \tau)$. In the case of the separable continuous-time random walk, this distribution is assumed to factorize $\Psi(\ell, \tau) = f(\ell)\psi(\tau)$, and we could write down a closed-form solution for the Fourier-Laplace transform of the probability distribution of the walk.

Finally, we approximated the large-scale behavior (in space and time) of the walks by assuming that the first two moments of the jump-size distribution, $\langle \ell \rangle$ and $\langle \ell^2 \rangle$, and the first moment of the waiting-time distribution, $\langle \tau \rangle$, are finite. We required

$$\langle \tau \rangle \to 0, \quad \langle \ell \rangle \to 0 \quad \text{and} \quad \frac{\langle \ell \rangle}{\langle \tau \rangle} \to v, \qquad (3.251)$$

$$\langle \tau \rangle \to 0, \quad \langle \ell^2 \rangle \to 0 \quad \text{and} \quad \frac{\langle \ell^2 \rangle}{2d\langle \tau \rangle} \to D. \qquad (3.252)$$

This is the diffusion-process approximation of random walks. Diffusion processes are described by a Fokker-Planck equation for the conditional probability or, equivalently, an Itô stochastic differential equation for the stochastic variable itself. The classical example of a diffusion process is Brownian motion. Einstein's description of Brownian motion was mathematically formalized through the Wiener process, a stochastic process with Gaussian-distributed independent increments. This is a Markov process as well as a martingale and on the latter property rests the theory of stochastic integration. The Gaussian distribution of the Wiener increments corresponds to the two requirements we made above for going to the diffusion limit in the random-walk models. If we only specify the first two moments of a distribution and then require these to be a complete description of the stochastic process, the central limit theorem tells us that we will end up with a Gaussian distribution for the sum variable. It is therefore perhaps not surprising that the Ornstein-Uhlenbeck process is the most general form for a stationary Gaussian Markov process.

It is a common misconception that diffusion processes are always dissipative—again stemming from the famous Ornstein-Uhlenbeck treatment of Brownian motion. In the final example of this chapter, we derived the hydrodynamics of Brownian motion and identified how to turn the diffusion process into a conservative diffusion. We essentially followed Newton's derivation of classical mechanics in doing so. For a special choice of diffusion coefficient, this stochastic mechanics then turned out to be non-relativistic quantum mechanics.

Diffusion processes come about through the cumulative effect of many small (the scales are set by the moments of $f(\ell)$ and $\psi(\tau)$) random events. What happens when these scales diverge is the topic of the next chapter which addresses Lévy distributions and the corresponding Lévy processes. Both concepts, Brownian motion and Lévy processes, will find their applications in the modern area of econophysics, which is examined in the final chapter.

3.10 Further Reading

This has been a long chapter discussing rather diverse problems. Therefore, we organize our suggestions for further reading according to the different applications.

Section 3.1

- The book by Wax [209] contains a selection of outstanding original work and review articles by the early masters of the field. The articles by Chandrasekhar [26] and Kac [100] contain discussions of random walks in the context of Brownian motion. Very elegant expositions of the techniques and concepts for the analysis of random walks can be found in reviews by Montroll et al. [145, 147] and also in the book by Weiss [211].
- The concept of continuous-time random walks was introduced by Montroll and Weiss [146], and [145, 147, 211] contain excellent discussions of the theory and applications of CTRWs.
- We did not treat disorder effects on diffusion processes and especially on random walks as models for diffusion processes. These topics are nicely reviewed, for instance, in the articles by Bouchaud and Georges [17] and of Haus and Kehr [77].

Section 3.2

- The original treatments by Einstein and Smoluchowski of the position of the Brownian particle as a stochastic process can be found in [43, 191]. The more refined treatment of the velocity process of the Brownian particle was given in two manuscripts by Uhlenbeck et al. [203, 208], which are also contained in [209]. On the occasion of the 100th anniversary of Einstein's 1905 paper, Frey and Kroy published a very commendable review on Brownian motion and its applications to soft condensed matter and biological physics [55].
- Doob [39] (also in [209]) proved the fundamental result that the Ornstein-Uhlenbeck process is the only stationary, Gaussian Markov process.
- Brownian motion from a mathematical physicist's perspective can be found in [151].
- The book by Coffey, Kalmykov and Waldron [28] presents an extensive survey of the Langevin equation and its applications to physics, chemistry, and engineering. It contains a skillfully combined presentation of the historical background and basic theoretical concepts of Brownian motion, an explicit discussion of many examples which are hard to find elsewhere (Brownian motion in various potentials, rotational Brownian motion, etc.), and a good introduction to anomalous diffusion.

Section 3.5

- The question regarding the rate of activated transitions over a potential barrier, often in short called the Kramers problem, dates back to a famous work of Kramers in 1940 [110]. This problem is, however, still a field of active research and an overview of the 50 years after Kramers can be found in [75].
- A general macroscopic treatment of the Kramers problem is given in [68].
- The multidimensional situation, memory and quantum effects are discussed in [73]. Fokker-Planck and master equation treatments of bistable systems are compared in [74].
- The mean first passage time problem for Fokker-Planck equations is, of course, discussed in Risken's book [178], and the connections between mean first passage times and the lifetime of metastable states is the theme of [202].
- A review of state-of-the-art applications of the theory of activated barrier crossing in the natural sciences can be found in [53].

Section 3.7

- The Ising model is the prototypical lattice model of statistical physics and the corresponding probability structure is analytically known in one and two dimensions [137, 156].
- For the statistical physics of lattice systems in general there exists a good amount of rigorous results, which can be found, for instance, in the books by Lavis and Bell [116, 117] and Ruelle [180].
- With regard to the stochastic process side, we want to mention a book by Prum and Fort [175] for lattice processes and one by Lindenberg and West [118] for stochastic evolution equations in nonequilibrium statistical mechanics.

3.10 Further Reading

- The statistical mechanics of the kinetics of phase transitions is nicely discussed in [141].

Section 3.8

- We presented Nelson's derivation of the Schrödinger equation for a special choice of conservative Brownian motion, following the lines of thought in Nelson's original publication [149] and also in a review by Guerra [70]. A more advanced treatment, starting from the differential geometric formulation of classical mechanics, is chosen in Nelson's book [150]. This approach is very elucidating in treating classical mechanics and non-relativistic quantum mechanics in the same mathematical language. To follow this derivation, however, would have required at least a basic explanation of analysis on manifolds which would have been beyond the scope of this book. A discussion of Brownian motion on a manifold using the Kramers equation can be found in [102]. Emery's book [45] contains the mathematics of stochastic calculus in manifolds. (As an aside: being embedded, physicists seem to live *on* manifolds, whereas mathematicians live *in* manifolds.)
- If one requires quantum-mechanical motion to be a diffusion process on the configuration manifold, a many-particle state necessarily must be either symmetric (Bose-Einstein statistics) or antisymmetric (Fermi statistics) under particle exchange [150].
- The treatment of diffusion processes on manifolds furthermore allows for an inclusion of spin [33, 37, 38]. Again, the diffusion-process requirement leads to the consequence that there are only integer and half-integer spin values [150].
- The structural identity with classical mechanics has been much emphasized through the development of variational principles for the derivation of the equations of motion of a conservative diffusion process [71, 161–163, 218].
- The quantum-mechanical version of the Kramers problem within stochastic mechanics is treated in [98] and simulated in [136].
- The stochastic-mechanics interpretation of quantum mechanics can also help to resolve conceptual and interpretational problems in the context of evanescent waves and the question of the existence of superluminal velocities [85, 115].

Chapter 4
Beyond the Central Limit Theorem: Lévy Distributions

The theory of Brownian motion relies on the central limit theorem. The theorem states that a sum of N independent and identically distributed random variables,

$$S_N = \sum_{n=1}^{N} x_n, \tag{4.1}$$

obeys a Gaussian distribution in the limit $N \to \infty$, provided the first and second moments of x_n do not diverge. These restrictions are so mild that many distributions belong to the domain of attraction of the Gaussian.

However, not all! A well-known exception is the *Cauchy distribution*,

$$p(x) = \frac{a}{\pi} \frac{1}{a^2 + x^2}, \tag{4.2}$$

whose second moment is infinite. The Cauchy distribution occurs in many physical situations, for instance, in the Ornstein-Zernike theory of critical opalescence or in the lifetime broadening of spectral lines. Therefore the questions arise of whether it could also emerge as a limiting distribution for S_N, and what the limiting distribution would look like if the random variables were distributed according to (4.2).

The answer to these questions is the topic of the present chapter. The Cauchy distribution is just one example of a whole class of distributions which possess long, inverse-power-law tails:

$$p(x) \sim \frac{1}{|x|^{1+\alpha}}, \quad 0 < \alpha < 2 \quad (|x| \to \infty). \tag{4.3}$$

These 'broad' tails preclude the convergence to the Gaussian for $N \to \infty$, but *not* the existence of a limiting distribution.

The premises for, the form and the properties of these limiting distributions were worked out in the 1930s by P. Lévy, A. Khintchine and others. They are today called *Lévy* or *stable distributions*. The following sections give an introduction to the mathematics and properties of these distributions.

4.1 Back to Mathematics: Stable Distributions

Consider a set of random variables $\{\ell_n\}_{n=1,\ldots,N}$ which are independent and identically distributed according to

$$\text{Prob}(\ell < \ell_n < \ell + d\ell) = p(\ell) d\ell \quad (n = 1, \ldots, N). \tag{4.4}$$

Then one can ask the following questions:

- Is it possible to find (real) constants A_N and B_N (> 0) so that the distribution of the normalized sum,

$$\hat{S}_N = \frac{1}{B_N} \sum_{n=1}^{N} \ell_n - A_N, \tag{4.5}$$

converges to a limiting distribution if N tends to infinity, i.e.,

$$\text{Prob}(x < \hat{S}_N < x + dx) \xrightarrow{N \to \infty} L(x) dx? \tag{4.6}$$

- What are the form and the properties of all possible limiting distributions?
- When does the probability density $p(\ell)$ belong to the domain of attraction of a specific $L(x)$?

The answer to these questions requires the definition of a *stable distribution*.

Definition 4.1 A probability density is called 'stable' if it is invariant under convolution, i.e., if there are constants $a > 0$ and b such that

$$p(a_1 \ell + b_1) * p(a_2 \ell + b_2) := \int_{-\infty}^{+\infty} d\ell\, p\big(a_1(z - \ell) + b_1\big) p(a_2 \ell + b_2)$$

$$= p(az + b) \tag{4.7}$$

for all (real) constants $a_1 > 0$, b_1, $a_2 > 0$, b_2.

Example 4.1 A Gaussian distribution satisfies (4.7) and is therefore stable. To see this, consider

$$p(y) = \frac{1}{\sqrt{2\pi}} \exp\left(-\frac{y^2}{2}\right),$$

and set $y = a_1 \ell + b_1$ so that

$$p(a_1 \ell + b_1) = p(y) \frac{dy}{d\ell} = \frac{1}{\sqrt{2\pi \sigma_1^2}} \exp\left(-\frac{(\ell - \mu_1)^2}{2\sigma_1^2}\right),$$

4.1 Back to Mathematics: Stable Distributions

in which $\mu_1 = -b_1/a_1$ and $\sigma_1 = 1/a_1$. The same formula also holds for a_2 and b_2. Insertion into (4.7) yields

$$\frac{1}{2\pi\sigma_1\sigma_2}\int_{-\infty}^{+\infty} d\ell \exp\left(-\frac{(z-\ell-\mu_1)^2}{2\sigma_1^2} - \frac{(\ell-\mu_2)^2}{2\sigma_2^2}\right)$$

$$\stackrel{v=\ell-\mu_2}{=} \frac{1}{2\pi\sigma_1\sigma_2}\int_{-\infty}^{+\infty} dv \exp\left(-\frac{((z-\mu)-v)^2}{2\sigma_1^2} - \frac{v^2}{2\sigma_2^2}\right),$$

where $\mu = \mu_1 + \mu_2$. If we define

$$w := \frac{\sigma}{\sigma_1\sigma_2}v - \frac{\sigma_2}{\sigma\sigma_1}(z-\mu) \quad \text{and} \quad \sigma^2 := \sigma_1^2 + \sigma_2^2,$$

we can write

$$\frac{((z-\mu)-v)^2}{\sigma_1^2} + \frac{v^2}{\sigma_2^2} = w^2 + \frac{(z-\mu)^2}{\sigma^2}$$

so that

$$\frac{1}{2\pi\sigma_1\sigma_2}\int_{-\infty}^{+\infty} d\ell \exp\left(-\frac{(z-\ell-\mu_1)^2}{2\sigma_1^2} - \frac{(\ell-\mu_2)^2}{2\sigma_2^2}\right)$$

$$= \frac{1}{2\pi\sigma}\exp\left(-\frac{(z-\mu)^2}{2\sigma^2}\right)\int_{-\infty}^{+\infty} dw\, e^{-w^2/2}$$

$$= \frac{1}{\sqrt{2\pi(1/a)^2}}\exp\left(-\frac{(z+(b/a))^2}{2(1/a)^2}\right),$$

in which $a = 1/\sigma$ and $b = -\mu/\sigma$. A comparison with (4.7) shows that the Gaussian distribution is stable. To find other possible stable distributions, we can exploit the following property of convolution integrals.

Equation (4.7) becomes particularly simple in Fourier space, where the convolution $p(z) = f(\ell) * g(\ell)$ reduces to a product of the Fourier transforms:

$$p(k) =: \mathcal{F}[p(z)] = \int_{-\infty}^{+\infty} dz\, p(z)\exp(ikz)$$

$$= \int_{-\infty}^{+\infty} dz\, e^{ikz}\left(\int_{-\infty}^{+\infty} d\ell\, f(z-\ell)g(\ell)\right)$$

$$= f(k)g(k). \tag{4.8}$$

Let us consider some examples. Choose a Gaussian as the probability density, and take $a_1 = a_2 = 1$ and $b_1 = b_2 = 0$. Then

$$\mathcal{F}[p(\ell) * p(\ell)] = \exp\left(-\frac{k^2}{2}\right)\exp\left(-\frac{k^2}{2}\right) = \exp(-k^2), \tag{4.9}$$

so that $a = 1/\sqrt{2}$ and $b = 0$. Equation (4.9) is not specific to the Gaussian. It is also satisfied by exponentials with different arguments. For instance, by the simple exponential

$$\exp\left(-\frac{|k|}{2}\right)\exp\left(-\frac{|k|}{2}\right) = \exp(-|k|),$$

which is the Fourier transform of the Cauchy distribution

$$\mathcal{F}^{-1}[\exp(-|k|)] = \frac{1}{2\pi}\int_{-\infty}^{+\infty} dk \exp(-|k|)\exp(-ik\ell)$$

$$= \frac{1}{\pi}\frac{1}{1+\ell^2}. \tag{4.10}$$

Therefore, both the Gaussian and Cauchy distributions are, up to scale factors, invariant under convolution and thus stable. The importance of this common characteristic is due to the following theorem by Lévy and Khintchine:

Theorem 4.1 *A probability density $L(x)$ can only be a limiting distribution of the sum (4.5) of independent and identically distributed random variables if it is stable.*

The Gaussian and Cauchy distributions are therefore potential limiting distributions. However, there are many more. Lévy and Khintchine have completely specified the form of all possible stable distributions:

Theorem 4.2 (Canonical Representation) *A probability density $L_{\alpha,\beta}(x)$ is stable iff the logarithm of its characteristic function,*

$$L_{\alpha,\beta}(k) = \langle e^{ikx}\rangle = \int_{-\infty}^{+\infty} dx\, L_{\alpha,\beta}(x)\exp(ikx),$$

reads

$$\ln L_{\alpha,\beta}(k) = i\gamma k - c|k|^{\alpha}\left(1 + i\beta\frac{k}{|k|}\omega(k,\alpha)\right), \tag{4.11}$$

where $\gamma, c, \alpha,$ and β are real constants taking the values: γ arbitrary, $c \geq 0$,

$$0 < \alpha \leq 2, \quad -1 \leq \beta \leq 1, \tag{4.12}$$

and the function $\omega(k,\alpha)$ is given by

$$\omega(k,\alpha) = \begin{cases} \tan(\pi\alpha/2) & \text{for } \alpha \neq 1, \\ (2/\pi)\ln|k| & \text{for } \alpha = 1. \end{cases} \tag{4.13}$$

The constants γ and c are scale factors. Replacing $x - \gamma$ with $c^{1/\alpha}x$ shifts the origin and rescales the abscissa, but does not alter the function $L_{\alpha,\beta}(x)$ (unless $\alpha = 1, \beta \neq 0$). In contrast, α and β define the shape and the properties of $L_{\alpha,\beta}(x)$. These

4.1 Back to Mathematics: Stable Distributions

parameters are therefore used as indices to distinguish different stable distributions. The parameter α characterizes the large-x behavior of $L_{\alpha,\beta}$ and determines which moments exist:

- $0 < \alpha < 2$: Each stable distribution behaves as

$$L_{\alpha,\beta}(x) \sim \frac{1}{|x|^{1+\alpha}} \quad \text{for } x \to \pm\infty, \tag{4.14}$$

and has finite absolute moments of order δ

$$\langle |x|^\delta \rangle = \int_{-\infty}^{+\infty} dx \, |x|^\delta L_{\alpha,\beta}(x) \quad \text{if } 0 \le \delta < \alpha. \tag{4.15}$$

In particular, the latter property implies that the variance does not exist if $\alpha < 2$ and that both mean value and variance do not exist if $\alpha < 1$.
- $\alpha = 2$: $L_{\alpha,\beta}(x)$ is independent of β, since $\omega(k, 2) = 0$, and is Gaussian.

Due to these properties, α is called *characteristic exponent*. The second characteristic parameter, β, determines the asymmetry of $L_{\alpha,\beta}(x)$:

- $\beta = 0$: $L_{\alpha,\beta}(x)$ is an even function of x.
- $\beta = \pm 1$: $L_{\alpha,\beta}(x)$ exhibits a pronounced asymmetry for some choices of α. For instance, if $0 < \alpha < 1$, its support lies in the intervals $(-\infty, \gamma]$ for $\beta = 1$ and $[\gamma, \infty)$ for $\beta = -1$.

Theorem 4.2 defines the general expression for all possible stable distributions. However, it does not specify the conditions which the probability density $p(\ell)$ has to satisfy so that the distribution of the normalized sum $\hat{S}_N$ converges to a particular $L_{\alpha,\beta}(x)$ in the limit $N \to \infty$. If this is the case, one can say '$p(\ell)$ belongs to the domain of attraction of $L_{\alpha,\beta}(x)$'. This problem has been solved completely, and the answer can be summarized by the following theorem (see also Appendix A):

Theorem 4.3 *The probability density $p(\ell)$ belongs to the domain of attraction of a stable density $L_{\alpha,\beta}(x)$ with characteristic exponent α $(0 < \alpha < 2)$ if and only if*

$$p(\ell) \sim \frac{\alpha a^\alpha c_\pm}{|\ell|^{1+\alpha}} \quad \text{for } \ell \to \pm\infty, \tag{4.16}$$

where $c_+ \ge 0$, $c_- \ge 0$ and $a > 0$ are constants.

These constants are directly related to the prefactor c and the asymmetry parameter β by

$$c = \begin{cases} \frac{\pi(c_+ + c_-)}{2\alpha \Gamma(\alpha) \sin(\pi\alpha/2)} & \text{for } \alpha \ne 1, \\ \frac{\pi}{2}(c_+ + c_-) & \text{for } \alpha = 1, \end{cases} \tag{4.17}$$

$$\beta = \begin{cases} \frac{c_- - c_+}{c_+ + c_-} & \text{for } \alpha \ne 1, \\ \frac{c_+ - c_-}{c_+ + c_-} & \text{for } \alpha = 1. \end{cases} \tag{4.18}$$

Furthermore, if $p(\ell)$ belongs to the domain of attraction of a stable distribution, its absolute moments of order δ exist for $\delta < \alpha$:

$$\langle |\ell|^\delta \rangle = \int_{-\infty}^{+\infty} d\ell |\ell|^\delta p(\ell) \begin{cases} < \infty & \text{for } 0 \leq \delta < \alpha \ (\alpha \leq 2), \\ = \infty & \text{for } \delta > \alpha \ (\alpha < 2), \end{cases} \tag{4.19}$$

and the normalization constant in (4.5), which characterizes the typical (scaling) behavior of S_N, is given by

$$B_N = aN^{1/\alpha} \tag{4.20}$$

so that

$$\lim_{N \to \infty} \text{Prob}\left(x < \frac{1}{aN^{1/\alpha}} S_N - A_N < x + dx \right) = L_{\alpha,\beta}(x) dx, \tag{4.21}$$

where a is the same constant as in (4.16) and

$$\begin{aligned} 0 < \alpha < 1: &\quad A_N = 0, \\ 1 < \alpha < 2: &\quad A_N B_N = N\langle x \rangle. \end{aligned} \tag{4.22}$$

In particular, we have $B_N = \sigma N^{1/2}$ and $A_N B_N = N\langle x \rangle$ for a Gaussian.

Equations (4.16) and (4.19) echo the behavior of the limiting distribution $L_{\alpha,\beta}(x)$ for $\alpha < 2$. This reflects the difference between the Gaussian and other stable distributions. Whereas all probability densities which decay rapidly enough at large ℓ (at least as $|\ell|^{-3}$) belong to the attraction domain of the Gaussian, stable distributions with $\alpha < 2$ only attract those $p(\ell)$ which have the same asymptotic behavior for large x. This restricting condition is the reason for the prevalence of the Gaussian in nature. However, there are exceptions to this dominance, examples of which will be discussed in the following sections.

4.2 The Weierstrass Random Walk

The distinguishing property of Lévy distributions is the presence of long-range power-law tails which may lead to a divergence of even the lowest-order moments. For instance, both the first and second moments are infinite if the characteristic exponent is smaller than one.

Physically, these lower-order moments have a particular significance: They set the pertinent scales. In the theory of Brownian motion, we have seen that the time scale was determined by the first moment of the waiting-time distribution, while the second moment of the jump-length distribution defined the physically relevant length scale (see Sect. 3.1.3). The divergence of the corresponding moments for certain Lévy distributions implies the absence of underlying physical scales, provided these distributions are realizable in nature.

4.2 The Weierstrass Random Walk

This is not unreasonable. The absence of physical scales could be interpreted as 'scale invariance' which in turn invokes the notions of 'self-similarity' and 'fractals'. The intimate relation between Lévy distributions and self-similar behavior is the link that ties the mathematical properties to physical applications. In the present section, we want to illustrate this idea by discussing a random walk with Lévy-type jump-length distribution, the 'Weierstrass random walk'.

4.2.1 Definition and Solution

Consider a random walk on a one-dimensional lattice with lattice spacing a and with a jump-length distribution given by ($b > 1$, $M > 1$)

$$\begin{array}{ll} \pm a & \text{with probability } C, \\ \pm ba & \text{with probability } C/M, \\ \vdots & \\ \pm b^j a & \text{with probability } C/M^j, \end{array} \quad (4.23)$$

and so forth. The probability density for a jump of length ℓ can then be written as

$$p(\ell) = C \sum_{j=0}^{\infty} \frac{1}{M^j} \left[\delta(\ell - b^j a) + \delta(\ell + b^j a) \right] \quad (\ell \text{ real}),$$

where the constant C is determined by normalization. Therefore,

$$p(\ell) = \frac{M-1}{2M} \sum_{j=0}^{\infty} \frac{1}{M^j} \left[\delta(\ell - b^j a) + \delta(\ell + b^j a) \right]. \quad (4.24)$$

Equations (4.23) and (4.24) have an intuitive interpretation. If M is large, most jumps occur to the left- and right-neighboring sites with probability $C \simeq 1/2$. The probability of a larger jump of length ba is suppressed by a factor $1/M$, that of the next larger length $b^2 a$ by a factor $1/M^2$, and so forth. Although the distribution allows jumps of all lengths $a, ba, b^2 a, \ldots$, their probability decreases by an order of magnitude (in base M) if the length increases by an order of magnitude (in base b). On average, about M jumps of length a have to be completed before a larger jump can be observed. These M jumps form a cluster of visited sites separated by a distance a. About M such clusters, separated by a distance ba, are generated before a jump of length $b^2 a$ occurs. This continues on and on, leading to a hierarchy of clusters within clusters (see Fig. 4.1).

An important premise has to be satisfied for a random walk to develop such a self-similar infinite hierarchy of clusters: The second moment of $p(\ell)$ must be

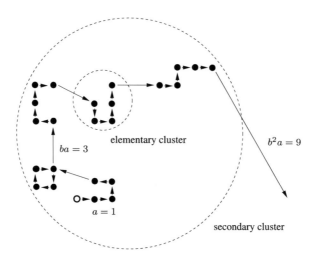

Fig. 4.1 Schematic sketch of the first steps for an idealized Weierstrass random walk with $M = 4$ and $b = 3$ so that $C = 3/8$. In this idealized walk exactly four steps have to be completed before the next larger jump takes place. The walk starts at the hollow circle and performs 4 steps of length $a\ (= 1)$. The visited points form the elementary cluster. Then, the next larger step $ba = 3$ occurs, which is followed by another cluster of step size a, and so on, until four steps of length $ba = 3$ are completed. Then, the next larger jump of size $b^2 a = 9$ takes place, and so on

infinite; otherwise, the central limit theorem imposes (regular) Brownian motion. This requires $b^2 > M$, since

$$\langle \ell^2 \rangle = \frac{(M-1)a^2}{M} \sum_{j=0}^{\infty} \left(\frac{b^2}{M}\right)^j = \infty \quad \text{if } b^2 > M. \tag{4.25}$$

If this condition holds, the probability density $p(\ell)$ tends to a stable distribution for large displacements ℓ. To show this, let us calculate the characteristic function of $p(\ell)$:

$$p(k) = \int_{-\infty}^{+\infty} d\ell\, p(\ell) \exp(ik\ell)$$

$$= \frac{M-1}{M} \sum_{j=0}^{\infty} \frac{1}{M^j} \cos(kb^j a). \tag{4.26}$$

This series was invented by K. Weierstrass in the second half of the 19th century as an example of a function which is everywhere continuous, but nowhere differentiable with respect to k when $b > M$. This is the reason why the random walk we are considering is called the 'Weierstrass random walk' [145].

Solution and Properties

Detailed analysis of the Weierstrass random walk requires a certain amount of algebra (see Appendix C). However, important qualitative features can also be inferred by observing that $p(k)$ satisfies a scaling equation. Consider

$$p(bk) = \frac{M-1}{M} \sum_{j=0}^{\infty} \frac{1}{M^j} \cos(kb^{j+1}a)$$

$$= Mp(k) - (M-1)\cos(ka),$$

or

$$p(k) = \frac{1}{M} p(bk) + \frac{M-1}{M} \cos(ka). \tag{4.27}$$

This is an inhomogeneous difference equation, the solution of which consists of the general solution of the homogeneous equation plus a particular solution of the inhomogeneous equation (as in the case of the differential equation counterpart).

To find the particular solution $p_i(k)$, let us write (4.27) in the form

$$\cos(ka) = \sum_{j=0}^{\infty} \frac{(-1)^j}{(2j)!} (ka)^{2j} = \frac{M}{M-1} p_i(k) - \frac{1}{M-1} p_i(bk). \tag{4.28}$$

This suggests that a series expansion of $p_i(k)$ should be attempted,

$$p_i(k) = \sum_{j=0}^{\infty} \frac{(-1)^j}{(2j)!} C_{2j} (ka)^{2j},$$

and that the coefficients C_{2j} should be determined by requiring consistency with (4.28), i.e.,

$$\sum_{j=0}^{\infty} \frac{(-1)^j}{(2j)!} (ka)^{2j} \stackrel{!}{=} \sum_{j=0}^{\infty} \frac{(-1)^j}{(2j)!} (ka)^{2j} \left(\frac{M}{M-1} C_{2j} - \frac{1}{M-1} C_{2j} b^{2j} \right),$$

which yields

$$C_{2j} = \frac{M-1}{M} \frac{1}{1 - b^{2j}/M},$$

so that

$$p_i(k) = \frac{M-1}{M} \sum_{j=0}^{\infty} \frac{(-1)^j}{(2j)!} \frac{(ka)^{2j}}{1 - b^{2j}/M}$$

$$= 1 + \frac{M-1}{M} \sum_{j=1}^{\infty} \frac{(-1)^j}{(2j)!} \frac{(ka)^{2j}}{1 - b^{2j}/M}. \tag{4.29}$$

This particular solution cannot be responsible for the singular behavior of the Weierstrass random walk because (4.25) requires

$$-\frac{d^2 p(k)}{dk^2}\bigg|_{k=0} = \int_{-\infty}^{+\infty} d\ell\, \ell^2 p(\ell) e^{ik\ell}\bigg|_{k=0} = \langle \ell^2 \rangle = \infty,$$

whereas (remember $b^2 > M > 1$)

$$\frac{d^2 p_i(k)}{dk^2}\bigg|_{k=0} = \frac{M-1}{M}\frac{a^2}{b^2/M - 1} \quad \text{(finite)}. \tag{4.30}$$

Therefore, the singular behavior must result from the solution of the homogeneous equation

$$p_h(k) = \frac{1}{M} p_h(bk). \tag{4.31}$$

Equation (4.31) shows that a rescaling of k by b only reduces the amplitude by $1/M$, but otherwise preserves the functional form of $p_h(k)$. This suggests that the homogeneous solution has the following properties:

- $p_h(k)$ should be a function of $\exp(\pm 2\pi i \ln|ka|/\ln b)$, since

$$p_h(bk) \propto \text{function}\left(\exp\left[\pm 2\pi i \frac{\ln|bka|}{\ln b}\right]\right)$$

$$= \text{function}\left(\exp[\pm 2\pi i] \exp\left[2\pi i \frac{\ln|ka|}{\ln b}\right]\right) \propto p_h(k).$$

Let us call this function $Q(k)$. It satisfies the relation $Q(k) = Q(kb)$, i.e., it is periodic in $\ln k$ with period $\ln b$.
- Furthermore, (4.31) can be solved by a power law. Therefore, we make the ansatz

$$p_h(k) = |ka|^\alpha Q(k), \tag{4.32}$$

where the absolute value takes into account that $p_h(k)$ is an even function of k (see (4.26)). The exponent α may be determined by insertion into (4.31). This gives

$$\alpha = \frac{\ln M}{\ln b} \tag{4.33}$$

so that $0 < \alpha < 2$ because $1 < M < b^2$ (see (4.25)).

4.2 The Weierstrass Random Walk

Explicitly, $Q(k)$ is given by (see Appendix C)

$$Q(k) = \frac{M-1}{M \ln b} \sum_{n=-\infty}^{\infty} \Gamma(\omega_n) \cos\left(\frac{\pi \omega_n}{2}\right) \exp\left(-\frac{2\pi i n \ln |ka|}{\ln b}\right)$$

$$= -c(\alpha) + \frac{M-1}{M \ln b} \sum_{\substack{n=-\infty \\ (n \neq 0)}}^{\infty} \Gamma(\omega_n) \cos\left(\frac{\pi \omega_n}{2}\right) |ka|^{-in(2\pi/\ln b)}, \quad (4.34)$$

where

$$\omega_n = -\alpha + \frac{2\pi i n}{\ln b},$$

and

$$\frac{M-1}{M \ln b} \Gamma(-\alpha) \cos\left(\frac{\pi}{2}\alpha\right) =: -c(\alpha) \quad (4.35)$$

denotes the $n=0$ term of the series $Q(k)$. For $0 < \alpha < 2$ ($\alpha \neq 1$), the left-hand side of (4.35) is negative, so $c(\alpha)$ is a positive constant. Equation (4.34) shows that the log–periodic oscillations can be interpreted as an imaginary correction to the exponent α. This is a typical signature of systems with a self-similar discrete hierarchy of clusters (i.e., discrete scale invariance), such as the Weierstrass random walk [194].

Combining (4.29) and (4.32), the full solution of (4.27) takes the following form:

$$p(k) = p_h(k) + p_i(k)$$
$$= |ka|^\alpha Q(k) + 1 + \frac{M-1}{M} \sum_{j=1}^{\infty} \frac{(-1)^j}{(2j)!} \frac{(ka)^{2j}}{1 - b^{2j}/M}. \quad (4.36)$$

Since $\alpha < 2$, the dominant small-k (large-distance) behavior of $p(k)$ comes from the homogeneous solution, as required above. It scales as $|k|^\alpha$, with superimposed log–periodic oscillations. If we keep only the constant $n=0$ term in the limit $k \to 0$, we find

$$p(k) \simeq 1 - c(\alpha)|ka|^\alpha \simeq \exp\left(-c(\alpha)|ka|^\alpha\right) \quad (k \to 0). \quad (4.37)$$

Thus, the small-k behavior of the jump-length distribution (4.26) is well approximated by that of a symmetric Lévy distribution with the characteristic exponent $\alpha = \ln M / \ln b$. The discussion in Sect. 4.1 implies that the Weierstrass random walk has the following additional properties:

- Asymptotically, the probability density of a single jump (4.24) behaves as

$$p(\ell) \sim \frac{1}{|\ell|^{1+\alpha}} \quad (\ell \to \pm\infty). \quad (4.38)$$

- Therefore, $p(\ell)$ belongs to the domain of attraction of the stable distribution $L_{\alpha,0}(x)$. In other words, being the sum of independent and identically distributed random variables obeying (4.38), the distribution of displacements of the Weierstrass random walk tends to $L_{\alpha,0}(x)$ for $N \to \infty$, i.e.,

$$\lim_{N\to\infty} \mathrm{Prob}\left(x < \frac{1}{B_N}\sum_{n=1}^{N}\ell_n < x+\mathrm{d}x\right) = L_{\alpha,0}(x)\mathrm{d}x, \quad (4.39)$$

where $B_N \propto N^{1/\alpha}$ and $A_N = 0$ also for $1 < \alpha < 2$, because the walk is symmetric, i.e., $\langle \ell_n \rangle = 0$ (see (4.20) and (4.22)).
- The normalization by $N^{1/\alpha}$ shows that a typical displacement, X_N, after N steps scales as

$$X_N^2 \sim N^{2/\alpha}. \quad (4.40)$$

Since $0 < \alpha < 2$, the displacement increases faster with the number of steps than in the case of normal diffusion. Therefore, the Weierstrass random walk exhibits *superdiffusive* behavior. We will return to this point in the next section.

Fractional Diffusion Equation

Before leaving this section, we want to derive an analogue of the diffusion equation for the Weierstrass random walk. Since the individual steps of the walk are independent random variables and identically distributed according to $p(\ell)$, the probability density for a displacement between x and $x + \mathrm{d}x$ after N steps is given by the convolution (cf. (2.41))

$$p_N(x) = p(\ell) * \cdots * p(\ell) \quad (N \text{ factors}).$$

This simplifies in Fourier space to

$$p_N(k) = [p(k)]^N \stackrel{(4.37)}{\simeq} \exp\bigl(-Nc(\alpha)|ka|^\alpha\bigr) \quad (k \to 0)$$
$$\stackrel{t=N\Delta t}{=} \exp\left(-c(\alpha)\left(\frac{a^\alpha}{\Delta t}\right)t|k|^\alpha\right), \quad (4.41)$$

in which we have introduced an elementary time scale Δt for the duration of a jump. Provided that the limit $\lim_{a,\Delta t \to 0} a^\alpha/\Delta t$ exists, one can define an 'effective' diffusion coefficient by

$$D_{\mathrm{eff}} := \lim_{a,\Delta t \to 0} c(\alpha)\frac{a^\alpha}{\Delta t}, \quad (4.42)$$

so that

$$p_N(k,t) = \exp\bigl(-D_{\mathrm{eff}}t|k|^\alpha\bigr) \quad \text{(for small } k\text{)}. \quad (4.43)$$

4.2 The Weierstrass Random Walk

This is the solution of the following equation:

$$\frac{\partial}{\partial t} p_N(k, t) = -D_{\text{eff}} |k|^\alpha p_N(k, t). \tag{4.44}$$

If α were 2, D_{eff} would be the diffusion coefficient D and (4.44) the Fourier transform of the diffusion equation

$$\frac{\partial}{\partial t} p_N(x, t) = \frac{1}{2\pi} \int_{-\infty}^{+\infty} dk e^{-ikx} \frac{\partial}{\partial t} p_N(k, t)$$

$$= -D \frac{1}{2\pi} \int_{-\infty}^{+\infty} dk e^{-ikx} k^2 p_N(k, t) = D \frac{\partial^2}{\partial x^2} p_N(x, t).$$

Similarly, one can Fourier transform (4.44) for any α and define a *fractional derivative* [79, 181]

$$\frac{\partial^\alpha}{\partial x^\alpha} := -\frac{1}{2\pi} \int_{-\infty}^{+\infty} dk e^{-ikx} |k|^\alpha, \tag{4.45}$$

which leads to the fractional diffusion equation

$$\frac{\partial}{\partial t} p_N(x, t) = D_{\text{eff}} \frac{\partial^\alpha}{\partial x^\alpha} p_N(x, t). \tag{4.46}$$

This shows that normal diffusion gives rise to ordinary partial differential equations, while Lévy-type diffusive behavior naturally leads to fractional calculus [79, 181]. Fractional calculus is therefore an important tool for studying non-ordinary diffusion and related phenomena [78, 143].

4.2.2 Superdiffusive Behavior

The Weierstrass random walk has two characteristic features:

- All steps take the same (average) time.
- The distribution of the step lengths exhibits a Lévy-type power-law decay for the largest jumps.

These properties lead to superdiffusive behavior, since a Weierstrass random walker can jump from point to point, irrespective of the distance between the two points, in the same amount of time. Therefore, one may expect that the overall displacement is dominated by the largest jumps.

Fig. 4.2 Monte Carlo simulation results of the two-dimensional Weierstrass random walk (4.47) for (**a**) $M = 30$, $b = 10$ and (**b**) $M = 80$, $b = 10$. The first set of parameters yields $\alpha \simeq 1.48$, and the second $\alpha \simeq 1.90$ (see (4.33)). In the simulations, a direction (x or y) was first drawn at random, and then a jump length was chosen according to (4.24) (see [174] for a method how to achieve this). For both simulations, $a = 1$, and 10^7 steps were simulated. Note that there are scattered clusters of small jumps separated by much larger ones for $\alpha \simeq 1.48$, whereas these large jumps almost disappear for $\alpha \simeq 1.90$. This reflects the suppression of the tails of $p(\ell)$ if $\alpha \to 2$

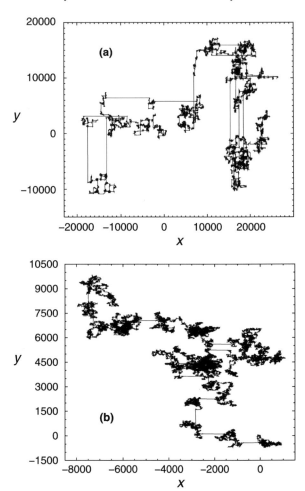

Figure 4.2 supports this idea. It shows the results of a Monte Carlo simulation for the two-dimensional Weierstrass random walk,

$$p(\ell_x, \ell_y) = \frac{M-1}{4M} \sum_{j=0}^{\infty} \frac{1}{M^j} \left[\delta(\ell_x - b^j a) + \delta(\ell_x + b^j a) \right.$$
$$\left. + \delta(\ell_y - b^j a) + \delta(\ell_y + b^j a) \right], \quad (4.47)$$

for $\alpha \simeq 1.48$ (a) and $\alpha \simeq 1.90$ (b). If $\alpha \simeq 1.48$, one can see 'islands' of small jumps that are connected by larger ones to bigger aggregates, which are in turn connected again by even larger jumps, and so on. This reflects the hierarchical structure of the Weierstrass random walk. The disparity between the largest jumps and the clusters of short ones gradually vanishes as $\alpha \to 2$ (see (b)), but would be even more pro-

4.2 The Weierstrass Random Walk

nounced for $\alpha < 1.48$. The typical displacement of the walker after N steps should therefore be determined by the largest jumps if α substantially deviated from 2.

This expectation can be corroborated by the following argument [17]. Consider a realization of the Weierstrass random walk with N steps and jump lengths $\ell_1, \ldots, \ell_N$. Let us now calculate the probability that the largest jump out of the N steps has length L. This probability is given by

$$\begin{aligned}
\text{prob. of } L(N) &= \binom{N}{1} p(L) \left(\int_0^L p(\ell) d\ell \right)^{N-1} \\
&= Np(L) \left(1 - \frac{1}{N-1} \left[(N-1) \int_L^\infty p(\ell) d\ell \right] \right)^{N-1} \\
&\stackrel{N \gg 1}{\simeq} Np(L) \exp\left(-N \int_L^\infty p(\ell) d\ell \right).
\end{aligned} \tag{4.48}$$

Our aim is to extract from (4.48) a typical value for the largest jump. Usually, we would calculate the lowest-order moment, which is non-zero, such as $\langle L \rangle$ or $\langle L^2 \rangle$. However, this is not a good choice in the present case. Depending on the value of α, these moments diverge. To circumvent this problem, one can take the most probable value L_m from the maximum of (4.48) as a characteristic jump length, i.e.,

$$\begin{aligned}
\frac{d}{dL} \text{ prob. of } L(N) \big|_{L=L_m} &= N \exp\left(-N \int_{L_m}^\infty p(\ell) d\ell \right) \\
&\quad \times \left[\frac{d}{dL} p + Np^2 \right]_{L=L_m} \stackrel{!}{=} 0.
\end{aligned} \tag{4.49}$$

Provided that the exponential prefactor is finite, which is realized by the requirement

$$N \int_{L_m}^\infty p(\ell) d\ell \simeq 1, \tag{4.50}$$

we obtain for L_m

$$L_m \sim N^{1/\alpha} \tag{4.51}$$

if the asymptotic behavior (4.38) of $p(\ell)$ is used. The interpretation of (4.50) is that there is typically one jump larger than L_m in a sequence of N steps. If N is finite, but becomes very large, this event is completely negligible and L_m represents an upper cutoff for the jump length.

Therefore, the large-distance behavior of a typical displacement, X_N, for the Weierstrass random walk may be estimated as an average of the sum $S_N = \sum_{n=1}^N \ell_n$, with ℓ large, but smaller than L_m:

$0 < \alpha < 1$:

$$X_N \sim N \int^{L_m} d\ell \, \ell p(\ell) \stackrel{(4.38)}{\sim} N L_m^{1-\alpha} \sim N^{1/\alpha}, \tag{4.52}$$

$\alpha = 1$:

$$X_N \sim N \int^{L_m} d\ell \frac{1}{\ell} \sim N \ln L_m \sim N \ln N, \qquad (4.53)$$

$1 < \alpha < 2$:

$$\begin{aligned}(X_N - \langle S_N \rangle)^2 &\sim N \int^{L_m} d\ell (\ell - \langle \ell \rangle)^2 p(\ell) \\ &\stackrel{(4.38)}{\sim} N L_m^{2-\alpha}[1 + O(\langle \ell \rangle / L_m)] \\ &\stackrel{(4.51)}{\sim} N^{2/\alpha}[1 + O(N^{-1/\alpha})] \sim N^{2/\alpha}. \qquad (4.54)\end{aligned}$$

In all cases we used the asymptotic form (4.38) for $p(\ell)$ because the large-distance behavior of X_N should be determined by the tail of the jump-length distribution. Therefore, the integrals also have a finite lower bound fixed by the smallest ℓ-value at which (4.38) becomes valid. This lower bound is a property of $p(\ell)$ (i.e., a consequence of the parameters M and b) and independent of N; therefore, the integrals are dominated by L_m. The results (4.52) to (4.54) thus represent the typical leading-order behavior of the sum S_N for large N.

The presented discussion shows that the total displacement is superdiffusive, in agreement with (4.40), and dominated by the largest jump. The typical behavior of S_N resembles a single step. This reflects again the self-similar character of the Weierstrass random walk or of any random process with the same properties: Lévy-like jump-length distribution and finite time for jumps of any length. Such processes are called *Lévy flights*.

Fractal Dimension

The term 'self-similarity' is used to characterize an object which looks the same at all length scales. If we identify a length scale with a specific magnification, the structure of a self-similar object is invariant under a change of the magnification. It exhibits 'dilation invariance'.

Another, intimately related concept is that of a *fractal*. The term 'fractal' (from the Latin *fractus* = broken) was coined by B. Mandelbrot in the 1960s to describe dilation-invariant objects which exhibit irregularities on all scales. Specific examples are clouds or coastlines of islands.

If one were to measure the volume of the clouds or the length of a coastline, the standard approach would fail. The standard approach consists in choosing a unit volume λ^d (unit length in $d = 1$, surface element in $d = 2$, and so on) and counting the number, N, of units which are necessary to cover the given volume R^d, i.e., $R^d = N\lambda^d$. Applying the same method to measure the length of a jagged coastline, for instance, yields a result that depends on λ. The length of the coastline is so irregular that it cannot be partitioned in one-dimensional line segments. It has an intrinsic *fractal dimension*, d_f, different from the Euclidean dimension, d, of the unit volume used. The fractal dimension can be estimated by counting the number of line segments inside a sphere of radius R. The scaling of N with R defines the (Hausdorff-Besicovitch) fractal dimension[1]

$$R^{d_f} \propto N. \tag{4.55}$$

The Weierstrass random walk and Lévy flights in general can also be characterized in this way by substituting X_N with R. This gives

$$R^\alpha \sim N \overset{(4.55)}{\Longrightarrow} d_f = \alpha. \tag{4.56}$$

The exponent α may therefore be interpreted as the fractal dimension of a Lévy flight [81]. The same reasoning also applies to ordinary random walks. Since $R \sim N^{1/2}$, we find $d_f = 2$. Therefore, a random walk should rather be considered as a two-dimensional than as a one-dimensional object.

4.2.3 Generalization to Higher Dimensions

The Weierstrass random walk is a one-dimensional example of a Lévy flight. Quite generally, the term 'Lévy flight' is used to denote a random walk in a continuous d-dimensional space which exhibits a Lévy distribution of the jump lengths (at least for the largest ones) and a finite (average) time between jumps. This term was coined by Mandelbrot [124] as a generalization of 'random flights', i.e., random walks in continuous space.[2] In the present section we want to derive the jump distribution $p(\ell)$ for a symmetric Lévy flight.

In complete analogy to the one-dimensional case, the jump distribution of a symmetric Lévy flight in Fourier space is given by

$$p(\mathbf{k}) = \exp(-c|\mathbf{k}|^\alpha), \tag{4.57}$$

where c is a positive constant rendering the exponent dimensionless. Using d-dimensional hyperspherical coordinates (see Appendix B), the inverse Fourier

---

[1] This is not the only definition of a fractal dimension. There are other methods. However, provided certain mathematical premises, one can establish that all definitions agree with one another [66]. Here, we take this for granted and only mention the Hausdorff dimension, which is the most intuitive one.

[2] Note that the terms 'random flight' and 'random walk' are sometimes used to distinguish a detail of the stochastic process. One speaks of a random flight if the process occurs in continuous space and consists of a sequence of independent displacements with random direction and magnitude (but finite second moment). On the other hand, if the displacements are restricted to a lattice, a random walk results. Usually, we do not make this distinction and laxly use random flights and random walks synonymously.

transform yields ($\ell = |\boldsymbol{\ell}|$)

$$p(\boldsymbol{\ell}) = \frac{1}{(2\pi)^d} \int_{-\infty}^{+\infty} d^d\boldsymbol{k}\, e^{-c|\boldsymbol{k}|^\alpha} \exp(-i\boldsymbol{k}\cdot\boldsymbol{\ell})$$

$$= \frac{S_{d-1}}{(2\pi)^d} \int_0^\infty dk\, k^{d-1} e^{-ck^\alpha} \int_0^\pi d\theta\, (\sin\theta)^{d-2} \exp(-ik\ell\cos\theta)$$

$$= \frac{S_{d-1}}{(2\pi)^d} \int_0^\infty dk\, k^{d-1} e^{-ck^\alpha} \left[\int_0^{\pi/2} d\theta\, (\sin\theta)^{d-2} \right.$$
$$\times \exp(-ik\ell\cos\theta) + \left. \int_{\pi/2}^\pi d\theta'\, (\sin\theta')^{d-2} \exp(-ik\ell\cos\theta') \right]$$

$$= \frac{S_{d-1}}{(2\pi)^d} \int_0^\infty dk\, k^{d-1} e^{-ck^\alpha} \int_0^{\pi/2} d\theta\, 2(\sin\theta)^{d-2} \cos(k\ell\cos\theta)$$

$$= \frac{1}{(2\pi)^{d/2}} \int_0^\infty dk\, k^{d-1} \left(\frac{1}{k\ell}\right)^{(d/2)-1} J_{(d/2)-1}(k\ell) e^{-ck^\alpha}, \qquad (4.58)$$

in which we substituted $\theta' = \pi - \theta$ in the penultimate step; $S_d = 2\pi^{d/2}/\Gamma(d/2)$ is the surface of a d-dimensional unit sphere and the following integral representation of the Bessel function $J_{(d/2)-1}(x)$ was used (see [67], p. 962, Eqs. (8.411–4.))

$$J_{(d/2)-1}(x)$$
$$= \frac{2}{\sqrt{\pi}\,\Gamma([d-1]/2)} \left(\frac{x}{2}\right)^{(d/2)-1} \int_0^{\pi/2} d\theta\, (\sin\theta)^{2[(d/2)-1]} \cos(x\cos\theta)$$
$$= \frac{2 S_{d-1}}{(2\pi)^{d/2}} x^{(d/2)-1} \int_0^{\pi/2} d\theta\, (\sin\theta)^{2[(d/2)-1]} \cos(x\cos\theta). \qquad (4.59)$$

This equation holds as long as $d > 1$. Furthermore, if we use (see [67], p. 979, Eqs. (8.472–3.))

$$x^\nu J_{\nu-1}(x) = \frac{d}{dx}\left[x^\nu J_\nu(x)\right] \qquad (4.60)$$

and substitute $x = k\ell$, a partial integration finally yields

$$p(\boldsymbol{\ell}) = \frac{\alpha c}{(2\pi)^{d/2}} \frac{1}{|\boldsymbol{\ell}|^{d+\alpha}} \int_0^\infty dx\, x^{d/2+\alpha-1} J_{d/2}(x) \exp\left[-c(x/\ell)^\alpha\right]. \qquad (4.61)$$

There are two cases when this expression becomes simple: $\alpha = 1$ and $\alpha = 2$. In the first case, we find the d-dimensional generalization of the Cauchy distribution (via Eq. (6.623), p. 733 of [67])

$$p(\boldsymbol{\ell}) = \frac{c\,\Gamma([d+1]/2)}{\pi^{(d+1)/2}} \frac{1}{(c^2 + \ell^2)^{(d+1)/2}}, \qquad (4.62)$$

4.2 The Weierstrass Random Walk

and in the second a Gaussian distribution (see Eqs. (6.631–4.), p. 738 of [67])

$$p(\ell) = \frac{1}{(4\pi c)^{d/2}} \exp\left(-\frac{\ell^2}{4c}\right). \tag{4.63}$$

From the general result (4.61), one can derive the asymptotic behavior for large ℓ (for $d = 2, 3$ [213]):

$$p(\ell) \sim \frac{1}{|\ell|^{d+\alpha}} \quad (|\ell| \to \infty), \tag{4.64}$$

i.e., the generalization of (4.38).

Polya's Problem Revisited

Imagine a random walk on a d-dimensional lattice. One of its important properties is whether the walk is transient or recurrent. Transience means that the walker has a finite probability of never returning to his starting point again, whereas he is certain to come back if the walk is recurrent. In Sect. 3.1.1 we found that the random walk is recurrent in $d \leq 2$, but transient otherwise. This result is Polya's theorem.

For a d-dimensional Weierstrass random walk, the superdiffusive behavior should modify this statement. One can expect that superdiffusion makes the walk transient for dimensions smaller than that of the random walk. To test this idea, we have to calculate the probability of return to the origin $F(\mathbf{0}, 1)$, which is given by (3.19) and (3.20) (note that $a = 1$ and $p(\mathbf{k})$ is called $f(\mathbf{k})$ in (3.20))

$$F(\mathbf{0}, 1) = 1 - \left(\frac{a^d}{(2\pi)^d} \int_{-\pi/a}^{+\pi/a} \cdots \int_{-\pi/a}^{+\pi/a} d^d k \frac{1}{1 - p(\mathbf{k})}\right)^{-1}. \tag{4.65}$$

If the integral diverges, $F(\mathbf{0}, 1) = 1$ and the walk is recurrent. Otherwise, the walk is transient. Since $p(\mathbf{k}) \to 1$ for $\mathbf{k} \to 0$, the divergence results from the small-k behavior. Using (4.57), we have

$$1 - p(\mathbf{k}) \simeq c|\mathbf{k}|^\alpha \quad (\mathbf{k} \to 0),$$

and so

$$F(\mathbf{0}, 1) \simeq 1 - \left(\frac{a^d S_d}{(2\pi)^d c} \int_0^{k_0} dk \frac{k^{d_{\text{eff}}-1}}{k^2}\right)^{-1}, \tag{4.66}$$

where k_0 is an upper cutoff, S_d is the surface of a d-dimensional unit sphere and

$$d_{\text{eff}} := d + 2 - \alpha. \tag{4.67}$$

If d_{eff} were equal to d, we would recover Polya's theorem for random walks. The superdiffusive behavior effectively changes the dimensions for d-dimensional Weierstrass random walks. Therefore, they are transient, not if $d > 2$, but if $d_{\text{eff}} > 2$, i.e., if

$$d > \alpha. \tag{4.68}$$

This implies that the one-dimensional walk is transient for $0 < \alpha < 1$, but recurrent for $1 < \alpha < 2$, whereas higher-dimensional walks are transient for all possible values of α.

4.3 Fractal-Time Random Walks

Let us consider the probability density $\Psi(\ell, \tau)$ that a random walker performs a jump from ℓ to $\ell + d^d\ell$ in the time interval $d\tau$ after having stayed at some site for a 'waiting time' τ. This probability defines a continuous-time random walk (CTRW) (see Sect. 3.1.3). The CTRW generalizes the examples discussed up to now, i.e., Brownian motion and the Weierstrass random walk. They can be considered as special cases for which $\Psi(\ell, \tau)$ factorizes (separable CTRW)

$$\Psi(\ell, \tau) = p(\ell)\psi(\tau). \tag{4.69}$$

Furthermore, the ordinary random walk and the Weierstrass random walk share the same waiting-time distribution $\psi(t)$ given by

$$\psi(\tau) = \delta(\tau - \tau_0), \tag{4.70}$$

i.e., all jumps take the same finite time τ_0. The main difference resides in $p(\ell)$. For the random walk we have

$$\langle \ell^2 \rangle = \int_{-\infty}^{+\infty} d^d\ell\, \ell^2 p(\ell) < \infty,$$

$$\langle \tau \rangle = \int_0^\infty d\tau\, \tau \psi(\tau) = \tau_0 < \infty,$$

whereas the Weierstrass random walk is characterized by

$$\langle \ell^2 \rangle = \infty \quad \text{for } 0 < \alpha < 2,$$

$$\langle \tau \rangle = \tau_0 < \infty.$$

The divergence of $\langle \ell^2 \rangle$ leads to the superdiffusive behavior of the Weierstrass random walk because jumps of arbitrary length all take the same time. On the other hand, a finite $\langle \ell^2 \rangle$ necessarily leads to ordinary diffusion, as long as the waiting-time distribution has a finite first moment. Therefore, the question arises of what would happen if we reversed the situation, choosing a finite $\langle \ell^2 \rangle$, but a divergent $\langle \tau \rangle$. The answer to this question is the topic of the present section.

4.3.1 A Fractal-Time Poisson Process

Consider a random walk with a waiting-time distribution $\psi(\tau)$ for each jump. A typical example is the *Poisson process*

$$\psi(\tau) = \frac{1}{\tau_0} \exp\left(-\frac{\tau}{\tau_0}\right), \tag{4.71}$$

for which jumps occur on one characteristic time scale, τ_0. Since $\tau_0 \, (= \langle \tau \rangle)$ is finite (and so is $\langle \ell^2 \rangle$ by assumption), the process gives rise to normal diffusion.

Motivated by experience with the Weierstrass random walk, we can guess how to construct a process with non-normal diffusive behavior. Two opposing factors are needed. Instead of a single characteristic time scale, there should be a whole (discrete) hierarchy of scales. For instance, we can introduce the parameter $b < 1$ and set ($t_0 = $ time unit)

$$\tau_j := \frac{t_0}{b^j} \stackrel{j \to \infty}{\longrightarrow} \infty \quad (j = 1, 2, \ldots, \infty). \tag{4.72}$$

However, not all time scales should have the same probability of occurrence. Long ones must be penalized by a suppression factor to model an opposing trend. Let us assume that the probability for the time scale τ_{j+1} is reduced by the factor $M < 1$ compared to the next smaller time τ_j. Combining these ideas, a viable generalization of the Poisson process in the spirit of a Weierstrass random walk could read

$$\psi(\tau) = \frac{1-M}{M} \sum_{j=1}^{\infty} \frac{M^j}{\tau_j} \exp\left(-\frac{\tau}{\tau_j}\right), \tag{4.73}$$

where the prefactor warrants the normalization of $\psi(\tau)$. If $\langle \tau \rangle$ is to be infinite, we must require $b < M$, since

$$\langle \tau \rangle = \int_0^\infty d\tau \, \tau \psi(\tau)$$

$$= t_0 \left(\frac{1-M}{M}\right) \sum_{j=1}^{\infty} \left(\frac{M}{b}\right)^j \begin{cases} = \infty & b < M, \\ < \infty & b > M, \end{cases}$$

which shows

$$\langle \tau \rangle = \infty \iff b < M < 1. \tag{4.74}$$

This condition should suffice to generate a 'fractal-time Poisson process' with a waiting-time distribution that adopts the typical form of stable distributions in the limit of large τ. The corresponding characteristic exponent should be $0 < \alpha < 1$ because $\langle \tau \rangle$ is divergent. This idea may be tested by an analysis which closely parallels that of the Weierstrass random walk. Therefore, we only sketch the main steps. More details can be found in [145, 188].

We introduce the Laplace transform of $\psi(\tau)$:

$$\psi(s) = \int_0^\infty d\tau e^{-s\tau} \psi(\tau)$$

$$= \frac{1-M}{M} \sum_{j=1}^\infty \frac{M^j}{1+s t_0 b^{-j}}. \tag{4.75}$$

The function $\psi(s)$ satisfies an inhomogeneous difference equation which we find by considering its behavior upon rescaling the argument, i.e.,

$$\psi(sb^{-m}) = \frac{1-M}{M} \sum_{j=1}^\infty \frac{M^j}{1+s t_0 b^{-(j+m)}}$$

$$= \frac{1}{M^m} \psi(s) - \frac{1}{M^m} \frac{1-M}{M} \sum_{j=1}^m \frac{M^j}{1+s\tau_j}$$

or

$$\psi(s) = M^m \psi(sb^{-m}) + \frac{1-M}{M} \sum_{j=1}^m \frac{M^j}{1+s\tau_j}. \tag{4.76}$$

The solution of (4.76) consists, again, of two parts: the general solution of the homogeneous equation and a particular solution of the inhomogeneous equation:

$$\psi(s) = \psi_h(s) + \psi_i(s).$$

As in the case of the Weierstrass random walk, one can show that the singular behavior responsible for $\langle \tau \rangle = \infty$ arises from the homogeneous part,

$$\psi_h(s) = M^m \psi_h(sb^{-m}). \tag{4.77}$$

Equation (4.77) is solved by the ansatz

$$\psi_h(s) = (st_0)^\alpha Q(s), \tag{4.78}$$

in which

$$Q(s) = \text{function}\left(\exp\left[\pm 2\pi i \frac{\ln(st_0)}{\ln b}\right]\right),$$

so that

$$Q(s) = Q(s/b) = Q(s/b^m) \quad (m \text{ is an integer}). \tag{4.79}$$

Therefore,

$$\psi_h(s) = M^m \psi_h(s/b^m) = \left(\frac{M}{b^\alpha}\right)^m (st_0)^\alpha Q(s/b^m) = \left(\frac{M}{b^\alpha}\right)^m \psi_h(s),$$

4.3 Fractal-Time Random Walks

which gives

$$\alpha = \frac{\ln(1/M)}{\ln(1/b)}, \qquad (4.80)$$

so $0 < \alpha < 1$, since $b < M < 1$. This range of α-values is that of stable distributions which have a divergent first moment. In order to show that $\psi(s)$ gives rise to a Lévy distribution if $s \to 0$, we use the result for the full solution of (4.76) [145, 188]:

$$\psi(s) = \psi_{\mathrm{h}}(s) + \psi_{\mathrm{i}}(s)$$
$$= (st_0)^\alpha Q(s) + 1 + \frac{1-M}{M} \sum_{j=1}^\infty \frac{(-1)^j (st_0)^j}{(b^j/M) - 1}, \qquad (4.81)$$

where

$$Q(s) = -\frac{1-M}{M \ln b} \sum_{n=-\infty}^\infty \frac{\pi M b^{\omega_n}}{\sin(\pi \omega_n)} \exp\left(-\frac{2\pi i n \ln(st_0)}{\ln b}\right) \qquad (4.82)$$

with

$$\omega_n = -\alpha + \frac{2\pi i n}{\ln b}.$$

In the limit of vanishing s, we keep only the $n = 0$ term, which is given by

$$Q(s) \xrightarrow{s \to 0} -\frac{1-M}{\ln(1/b)} \frac{\pi}{b^\alpha \sin(\pi \alpha)} =: -c(\alpha), \qquad (4.83)$$

where $c(\alpha) > 0$ for $0 < \alpha < 1$, so that

$$\psi(s) \xrightarrow{s \to 0} 1 - c(\alpha)(st_0)^\alpha \simeq \exp\bigl(-c(\alpha)(st_0)^\alpha\bigr). \qquad (4.84)$$

The leading-order term of the inhomogeneous solution scales as s^1 and can be neglected compared to s^α in the limit $s \to 0$. The large-τ behavior of the waiting-time distribution is obtained by the inverse Laplace transform of (4.84), i.e.,

$$\psi(\tau) \simeq \frac{1}{2\pi i} \int_{\delta-i\infty}^{\delta+i\infty} ds e^{s\tau} e^{-(c(\alpha)t_0^\alpha)s^\alpha} = L_{\alpha,-1}(\tau) \quad (\tau \text{ large}), \qquad (4.85)$$

and thus corresponds to an asymmetric Lévy distribution with $\beta = -1$ (see Appendix A) which decays asymptotically as (see (A.24))

$$\psi(\tau) \sim \frac{\alpha(c(\alpha)t_0^\alpha)}{\Gamma(1-\alpha)} \frac{1}{\tau^{1+\alpha}}. \qquad (4.86)$$

As in the case of the Weierstrass random walk, the characteristic exponent α can be interpreted as a fractal dimension, namely as the fractal dimension of the set of waiting-times between jumps.

The Kohlrausch Function

Equation (4.85) is interesting for many physical applications. When studying relaxation phenomena, one often finds that pertinent time-displaced correlation functions, $\Phi(t)$ ($0 \leq \Phi(t) \leq 1$, $\Phi(0) = 1$, $\Phi(t \to \infty) = 0$), do not decay as a simple exponential, but as a *stretched exponential*, i.e., as

$$\Phi(t) \propto \exp(-ct^\alpha) \quad (0 < \alpha < 1, c = \text{const.}), \tag{4.87}$$

in some time window. This function was (presumably) used first by R. Kohlrausch in 1854, as a convenient description of dielectric-polarization-decay data, and since then in many other applications. A prominent current example is the research on random systems, such as supercooled liquids and glasses, where the stretched exponential is commonly called 'Kohlrausch-Williams-Watts (KWW)' function [63, 64].

Although (4.87) is widely used as a convenient fit function, a theoretical understanding of its microscopic origin is difficult to establish. Equation (4.85) suggests that the Kohlrausch function quite naturally appears in the framework of stable distributions. By replacing s with the time t and τ with the relaxation rate γ, (4.87) is obtained, if the probability density of γ is stable, i.e.,

$$\exp(-ct^\alpha) = \int_0^\infty d\gamma \, L_{\alpha,-1}(\gamma) e^{-\gamma t}. \tag{4.88}$$

Such an interpretation has been proposed, for instance, for dielectric relaxation experiments [144, 187] and for the dynamics of supercooled liquids [57].

4.3.2 Subdiffusive Behavior

Intuitively, one would expect that a random walk with finite $\langle \ell^2 \rangle$ and a Lévy-like waiting-time distribution propagates more slowly than ordinary diffusion. The walker always risks being trapped at a certain site for a long time before he can finally advance (on average) over the distance $\langle \ell^2 \rangle$. Thus, the number of steps, N, in a fixed time interval, t, is not linear in t as in normal diffusion, but rather scales as $N \sim t^\alpha$ with $\alpha < 1$. The typical displacement, X_N, after N steps should therefore be *subdiffusive*

$$X_N^2 \sim t^\alpha \quad (0 < \alpha < 1). \tag{4.89}$$

To substantiate this idea, we use a statistical reasoning similar to that utilized in Sect. 4.2.2 [17].

Each step of the random walk is associated with a waiting time. If the walker performs N steps, there are N waiting times; therefore, the time t needed to complete all steps is given by

$$t = \sum_{n=1}^{N} \tau_n. \tag{4.90}$$

Let us now assume that the τ_n are independent and identically distributed with a probability density that behaves for large τ as ($0 < \alpha < 1$):

$$t\psi(\tau) \sim \frac{t_0^\alpha}{\tau^{1+\alpha}} \quad (\tau \to \infty). \tag{4.91}$$

Thus, t is a sum of N random variables which scales, for large N, as (see (4.20))

$$t \sim t_0 N^{1/\alpha}. \tag{4.92}$$

This immediately leads to subdiffusive behavior, since

$$X_N^2 \sim \langle \ell^2 \rangle N \sim \left(\frac{t}{t_0}\right)^\alpha \quad (0 < \alpha < 1). \tag{4.93}$$

In contrast, if α were larger than 1, the first moment of the waiting-time distribution would exist, and asymptotically normal diffusion results:

$$X_N^2 \sim \langle \ell^2 \rangle t \quad (1 < \alpha < 2). \tag{4.94}$$

For intermediate times, there are additive power-law corrections to the normal large-t behavior if $\psi(\tau)$ is given by (4.91) with $1 < \alpha < 2$ [17].

4.4 A Way to Avoid Diverging Variance: The Truncated Lévy Flight

From the point of view of practical applications, stable distributions have both appealing and non-appealing features. Certainly, a very attractive feature is the scaling property

$$p_N(x) = \frac{1}{N^{1/\alpha}} p\left(\frac{x}{N^{1/\alpha}}\right), \tag{4.95}$$

which means that the probability $p_N(x)$ for a sum of N random variables is, up to a certain scale factor, identical to that of the individual variables. This is a typical signature of fractal behavior: The whole looks like its parts.

On the other hand, stable distributions possess divergent lower-order moments. The divergence of the variance, for instance, is a very disturbing feature if a time-series analysis gives rise to a Lévy distribution, although the variance is a priori known to exist. A good example comes from an analysis of the human heart beat [167]. In the study of [167] the duration of a beat $B(i)$ was monitored over a long period of time ($i = 1, \ldots, \sim 10^5$ beats, i.e., up to 24 hours). Then, the probability of the difference between the durations of successive beats, $\Delta B = B(i+1) - B(i)$, was determined. This probability compared much better to a symmetric Lévy distribution with $\alpha = 1.7$ than to a Gaussian, implying that large differences occur more frequently than predicted by the normal distribution. Physiologically, this is not unreasonable. The heart can respond more easily to changing external influences if successive beat durations may substantially deviate from one another. However, the deviation cannot be arbitrarily large. The beat durations are bounded by a smallest and a largest value. These physiologically imposed thresholds cut off the tails of the Lévy distribution and impose a finite variance of the time series.

Therefore, a possible way to reconcile Lévy distributions with a finite variance consists in truncating the tails. Such a *truncated Lévy flight* was proposed by Mantegna and Stanley [127, 128]:

$$p(\ell) = \begin{cases} \mathcal{N} L_{\alpha,0}(\ell) & \text{for } -\ell_{\text{cut}} \leq \ell \leq \ell_{\text{cut}}, \\ 0 & \text{otherwise}, \end{cases} \quad (4.96)$$

where $\mathcal{N}$ is a normalization constant and ℓ_{cut} the cutoff parameter. Numerical simulations [127, 128] illustrate the influence of ℓ_{cut} on the statistical properties of the process. If ℓ_{cut} is large, $p(\ell)$ has a pronounced Lévy character. This leads to a drastic retardation of the convergence to a Gaussian when summing random variables distributed according to (4.96).

This numerical simulation inspired another study, in which the truncated Lévy flight was defined for large ℓ as [108]

$$p(\ell) \sim \begin{cases} c_- e^{-\lambda |\ell|} |\ell|^{-(1+\alpha)} & \ell < 0, \\ c_+ e^{-\lambda \ell} \ell^{-(1+\alpha)} & \ell > 0. \end{cases} \quad (4.97)$$

Contrary to the abrupt truncation of (4.96), the smooth exponential damping allows analytical calculation of the characteristic function. For $0 < \alpha < 2$ ($\alpha \neq 1$), the result for the sum of N random variables is (see Appendix D)

$$\ln L_{\alpha,\beta}^t(k, N) = c_0 - N\lambda^\alpha c \frac{[1 + (k/\lambda)^2]^{\alpha/2}}{\cos(\pi\alpha/2)} \cos\left(\alpha \arctan \frac{|k|}{\lambda}\right)$$
$$\times \left[1 + i\beta \frac{k}{|k|} \tan\left(\alpha \arctan \frac{|k|}{\lambda}\right)\right], \quad (4.98)$$

where c is given by (4.17), i.e.,

$$c = \frac{\pi(c_+ + c_-)}{2\alpha \Gamma(\alpha) \sin(\pi\alpha/2)} \quad (0 < \alpha < 2, \alpha \neq 1), \quad (4.99)$$

the constant c_0 accounts for the normalization (i.e., $L_{\alpha,\beta}^t(k=0, N) = 1$),

$$c_0 = \frac{N\lambda^\alpha c}{\cos(\pi\alpha/2)}, \quad (4.100)$$

and the asymmetry parameter β has the same value as in (4.18),

$$\beta = \frac{c_- - c_+}{c_+ + c_-} \quad (0 < \alpha < 2, \alpha \neq 1). \quad (4.101)$$

4.4 A Way to Avoid Diverging Variance: The Truncated Lévy Flight

This function exhibits a crossover from a Lévy to a Gaussian regime if N increases. If $N\lambda^\alpha \gg 1$, $L^t_{\alpha,\beta}(k,N)$ quickly goes to zero for $k \gg \lambda$ so that the dominant contribution comes from the small-k regime, where $k \ll \lambda$. Expanding

$$\left(1+\left(\frac{k}{\lambda}\right)^2\right)^{\alpha/2} \cos\left(\alpha \arctan \frac{|k|}{\lambda}\right) \simeq \left[1+\frac{\alpha}{2}\left(\frac{|k|}{\lambda}\right)^2\right]\left[1-\frac{\alpha^2}{2}\left(\frac{|k|}{\lambda}\right)^2\right]$$

$$\simeq 1 + \frac{1}{2}\frac{\alpha(1-\alpha)}{\lambda^2}k^2,$$

and

$$\tan\left(\alpha \arctan \frac{|k|}{\lambda}\right) \simeq \tan\left(\alpha \frac{|k|}{\lambda}\right) \simeq \alpha \frac{|k|}{\lambda},$$

we find that the asymmetry contribution to (4.98) may be neglected and $L^t_{\alpha,\beta}(k,N)$ is given by

$$\ln L^t_{\alpha,\beta}(k,N) = -\frac{1}{2}\sigma^2(N)k^2 \quad (N \text{ large}), \tag{4.102}$$

where

$$\sigma^2(N) = N \frac{2C\pi(1-\alpha)}{\Gamma(\alpha)\sin\pi\alpha}\lambda^{\alpha-2} > 0 \quad \text{for } 0 < \alpha < 2 (\alpha \neq 1) \tag{4.103}$$

with $C = c_+ = c_-$. Therefore, $L^t_{\alpha,\beta}(k,N)$ becomes Gaussian in the small-k limit.

On the other hand, if $N\lambda^\alpha \ll 1$, all k-values contribute. In addition to the Gaussian behavior for small k we can also consider $k \gg \lambda$. Formally, this corresponds to the limit $\lambda \to 0$. Since $\arctan(|k|/\lambda) \to \pi/2$ and $c_0 \to 0$ for $\lambda \to 0$, we obtain (4.11), i.e.,

$$\ln L^t_{\alpha,\beta}(k,N) \simeq -Nc|k|^\alpha\left[1+i\beta\frac{k}{|k|}\tan\left(\frac{\pi\alpha}{2}\right)\right]. \tag{4.104}$$

This shows that the pure Lévy character of (4.98) emerges if both λ and N are small.

The minimal N value required to enter the Gaussian regime can be estimated by the following argument. Let us take $\beta = 0$. Then, the dominant finite N correction to the Gaussian is determined by the kurtosis, $\hat{\kappa}_4$, of $p(\ell)$ (see (2.45)) and scales as $\hat{\kappa}_4/N$. Therefore, the correction is negligible if

$$N \gg N_\times := \hat{\kappa}_4 \quad (\hat{\kappa}_4 > 0), \tag{4.105}$$

where [18]

$$\hat{\kappa}_4 = \frac{(2-\alpha)(3-\alpha)\Gamma(\alpha)}{2C\pi}\frac{\sin\pi\alpha}{1-\alpha}\lambda^{-\alpha}. \tag{4.106}$$

Equation (4.106) shows that the kurtosis is the larger, the smaller λ is. A small λ value shifts the truncation into the far wings of the Lévy distribution. Then, $L^t_{\alpha,\beta}(x,N)$ has a pronounced Lévy character and the convergence to the Gaussian is slowed down.

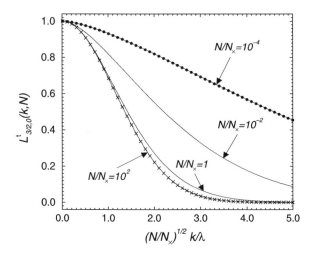

Fig. 4.3 Illustration of the slow crossover from Lévy to Gaussian behavior for the truncated Lévy flight (4.98). In this example, $\alpha = 3/2$ and $\beta = 0$ (*solid lines*), which implies that the α-dependent prefactors of (4.107) and (4.109) are 3/8 and -1, respectively. The crossover from the pure Lévy distribution (4.104) (*filled circles*) to the Gaussian (4.102) (*crosses*) is studied as a function of $N/N_\times$. If $N/N_\times = 10^{-4}$ or $N/N_\times = 10^2$, (4.98) is indistinguishable from (4.104) or (4.102), respectively

Illustration Using (4.105) and (4.106), we can rewrite the Gaussian limit (4.102) in the following way

$$\ln L^t_{\alpha,\beta}(k,N) = -\frac{1}{2}(2-\alpha)(3-\alpha)\frac{N}{N_\times}\left(\frac{k}{\lambda}\right)^2. \tag{4.107}$$

This suggests studying the crossover from Lévy to Gaussian behavior as a function of the scaled variable

$$\bar{k} = \sqrt{\frac{N}{N_\times}}\frac{k}{\lambda}. \tag{4.108}$$

In terms of this variable, (4.98) reads for $\beta = 0$

$$\ln L^t_{\alpha,\beta}(k,N) = \frac{(2-\alpha)(3-\alpha)}{\alpha(1-\alpha)}\frac{N}{N_\times}\left[1-\left(1+\frac{N_\times}{N}\bar{k}^2\right)^{\alpha/2}\right.$$
$$\left.\times \cos\left(\alpha \arctan\sqrt{\frac{N_\times}{N}}|\bar{k}|\right)\right]. \tag{4.109}$$

As an example, let us choose $\alpha = 3/2$. This is shown in Fig. 4.3. Consider now a fixed value of $\bar{k}$, say, $\bar{k} = 2$. If $N/N_\times$ is very small, k/λ must be large to realize this value, whereas it has to be small when $N/N_\times$ is large. According to our previous discussion, the first case gives rise to Lévy behavior and the second to Gaussian behavior. Therefore, we expect a crossover from a Lévy-dominated to a Gaussian regime if $N/N_\times$ increases. This trend is clearly visible in Fig. 4.3.

Addendum Since Lévy flights exhibit superdiffusive behavior, whereas a finite-jump-length, but Lévy-like waiting-time distribution can give rise to subdiffusion, it is plausible that the divergence of moments may also be avoided by considering non-separable, continuous-time random walks. An example is the so-called *Lévy walk* defined by [105, 187]

$$\Psi(\ell, \tau) = \psi(\tau|\ell)p(\ell) = \delta\left(\tau - \frac{|\ell|}{v(\ell)}\right)p(\ell), \qquad (4.110)$$

where the velocity $v(\ell)$ is associated with each step. Contrary to Lévy flights, a 'Lévy walker' does not simply jump an arbitrary length in the same time, but has to move with a given velocity from his starting to the end point. For the simplest case of constant velocity, it is immediately clear that large jumps require a longer time than shorter ones. The overall consequence of this spatio-temporal coupling is a finite mean-square displacement for all values of α [105]. The Lévy walk is one specific example of a non-separable, continuous-time random walk. Another interesting model is a generalization of the Weierstrass random walk with space-time coupling [111, 112]. Several further approaches are discussed in [145].

4.5 Summary

A characteristic signature of Brownian motion is that a typical displacement, X_N, scales with the number of steps N as

$$X_N^2 \sim N$$

if N is large. To derive this result, one has to assume that the step length and the time needed to perform a step are on average finite. With these assumptions the diffusive behavior, $X_N^2 \sim N$, is a direct consequence of the central limit theorem (CLT). The theorem can be expressed as follows: If there are

1. N random variables $\{\ell_n\}_{n=1,\ldots,N}$ which are independent and identically distributed, $p(\ell_1) = \cdots = p(\ell_n) = p(\ell)$, and have
2. finite first and second moments, $\langle \ell \rangle < \infty$, $\langle \ell^2 \rangle < \infty$,

then the sum variable

$$S_N = \sum_{n=1}^{N} \ell_n$$

has a Gaussian distribution in the limit $N \to \infty$ and a variance

$$X_N^2 = \langle S_N^2 \rangle \sim \langle \ell^2 \rangle N \quad (\text{if } \langle \ell \rangle = 0).$$

The remarkable feature of this theorem resides in the fact that the details of $p(\ell)$ are of only minor importance. Provided the first and second moments exist, the sum variable is always normally distributed, no matter what the actual shape of the individual distribution looks like. All details of $p(\ell)$ are lumped into two relevant quantities, $\langle \ell \rangle$ and $\langle \ell^2 \rangle$, which 'survive' in the limit $N \to \infty$.

These properties can change either if N is not truly infinite or if the premises of the CLT are violated. For finite N, there are corrections to the Gaussian behavior which depend on higher-order moments of $p(\ell)$ and are thus sensitive to the tails of the distribution. If the tails are broad, but decay still sufficiently fast to maintain a finite variance, the convergence to the normal distribution is strongly slowed down. We have encountered an example of this behavior in Sect. 4.4.

On the other hand, truly non-Gaussian behavior results as soon as premises 1 and 2 are violated:

- A violation of premise 2 can be brought about if $p(\ell)$ decays so slowly for $\ell \to \pm\infty$ that even the lowest-order moments diverge. Mathematically, this violation is well understood. It gives rise to stable distributions, of which the normal distribution is a special case.

  These distributions find important applications in the theory of continuous-time random walks (CTRW). A CTRW is characterized by the probability $\Psi(\ell, \tau)$ that a random walker performs a step of length $[\ell, \ell + d\ell]$ in the time interval $d\tau$ after having waited for a time τ. In this chapter we have treated only separable CTRW, i.e.,

$$\Psi(\ell, \tau) = p(\ell)\psi(\tau).$$

If we assume symmetric walks ($\langle \ell \rangle = 0$), we can distinguish the following cases:

– the first moment of $\psi(\tau)$ and the second moment of $p(\ell)$ are finite,

$$\langle \ell^2 \rangle = \int_{-\infty}^{+\infty} d\ell \, \ell^2 p(\ell) < \infty,$$

$$\langle \tau \rangle = \int_0^\infty d\tau \, \tau \psi(\tau) = \tau_0 < \infty.$$

This leads to ordinary diffusion, i.e., Brownian motion.
– $\tau_0 < \infty$ but $\langle \ell^2 \rangle = \infty$ gives rise to superdiffusion, i.e.,

$$X_N^2 \sim N^{2/\alpha} \quad (0 < \alpha < 2)$$

with logarithmic corrections when $\alpha = 1$. Superdiffusive behavior is characteristic of Lévy flights, i.e., random processes in continuous space with a Lévy-like jump-length distribution, but finite (average) jump time.
– $\tau_0 = \infty$ but $\langle \ell^2 \rangle < \infty$ yields subdiffusion, i.e.,

$$X^2(t) \sim t^\alpha$$

if $0 < \alpha < 1$. For $1 \leq \alpha < 2$, τ_0 is finite and one obtains ordinary diffusion, again with logarithmic corrections, when $\alpha = 1$.

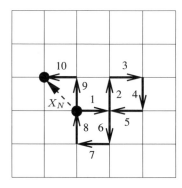 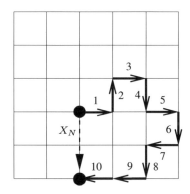

Fig. 4.4 Schematic picture of a random walk (*left panel*) and of a self-avoiding walk (*right panel*) on a square lattice. Both walks consist of $N = 10$ steps. Whereas a random walk can intersect its path arbitrarily often, such as where step 5 returns to the site already visited by the first step (see *left panel*), a self-avoiding walk may occupy each site only once. Sites already visited therefore act like a repulsive potential for the continuation of the walk. This leads to an increase in the distance between the start and the end, X_N (*dashed lines*), compared to the random walk

– A violation of premise 1 means that the random variables are correlated. These correlations also change the normal diffusive behavior. The previous chapter briefly mentioned one example, the self-avoiding walk (see Fig. 4.4). In this process the random walker has to keep track of the whole history of his path while he moves along, since he is not allowed to visit a site twice. Intuitively, this 'long-range repulsive interaction' along the path should make the overall displacement grow stronger with increasing N than in the case of the random walk. A considerable amount of difficult analysis is necessary to support this expectation, but the final result is (see (3.34))

$$X_N^2 \sim N^{2\nu} \quad (\nu = 0.589 \text{ in 3 dimensions}),$$

i.e., superdiffusive behavior. On the other hand, if the walker is allowed to revisit sites after having propagated self-avoidingly for a certain number of steps, the correlation is only short ranged. Such short-range correlations do not alter the normal diffusive behavior. They can be accounted for by defining a new (increased) elementary step length, the persistence length $\ell_p^2 > \langle \ell^2 \rangle$. All of these processes, random walks, walks with short-range interactions and self-avoiding walks, are important concepts in polymer physics [35, 179].

4.6 Further Reading

The discussion of this chapter has been compiled from many sources. We briefly list our main references and make suggestions for further reading:

Books Our presentation of the mathematics of stable distributions closely follows the discussion in the book of Gnedenko and Kolmogorov [62]. Furthermore, we have considerably profited from the books of Feller [52] and Breiman [20] (and from Appendix B of [17]). Besides this mathematical literature the book by Bardou et al. [7] is also very commendable. It provides a concise introduction to Lévy processes and applies them to obtain a better understanding of laser cooling. A very good overview of the theoretical description of anomalous diffusion can be found in the book by Coffey, Kalmykov and Waldron [28].

Review Articles There are many excellent review articles on the theoretical background and applications of Lévy processes. The most extensive ones which we have used are the articles of Bouchaud and Georges [17], Montroll and West [147], Montroll and Shlesinger [145], and West and Deering [213]. These articles are very pedagogically written throughout and comprehensively discuss many aspects which we have not considered at all. There are also some shorter reviews, for instance, [105, 187], which elaborate the main ideas very clearly. An approach from the perspective of fractional calculus may be found in the review articles by Metzler and Klafter [143] and by Hilfer [79]. Furthermore, a compilation of lecture notes on Lévy processes has been published as a conference proceedings [189]. The topics of the lectures range from mathematical properties to applications in biology. The editors emphasize that the proceedings is 'the first book for physical scientists devoted to Lévy processes' and hope that the 'essays will make Lévy's work known to a wider audience and inspire further developments.' In this chapter we have endeavored to do the same.

Chapter 5
Modeling the Financial Market

In 1900, a Ph.D. student of Henri Poincaré finished his dissertation. The name of the student was Louis Bachelier. In his report on the thesis, Poincaré praised the cunning way by which Bachelier had derived the Gaussian distribution. But he also pointed out that the topic had been very different from the typical ones that his other students presented [145].

This was certainly the case. Bachelier named his thesis 'Théorie de la spéculation' [3]—Theory of speculation—and suggested in it a probabilistic description for price fluctuations on the financial market. Essentially, he developed the mathematics of Brownian motion—five years before Einstein published his famous paper [43]—to model the time evolution of asset prices.

Perhaps it was because of the application to the financial market that Bachelier's work was not recognized by the scientific community at that time. It seems as if it fell into complete oblivion until the early 1940s. In 1944, Itô used it as a motivation to introduce his calculus and a variant of Brownian motion, geometric Brownian motion.

Geometric Brownian motion, in turn, became an important model for the financial market. Its economic significance was recognized in the work of Paul A. Samuelson (1915–2009) from 1965 onwards [182]. For his contributions Samuelson received the 1970 Nobel Prize in Economics [153]. In 1973, Fischer Black (1938–1995) and Myron Scholes [14] and, independently, Robert Merton [138] used the geometric Brownian motion to construct a theory for determining the price of stock options. This theory represents a milestone in the development of mathematical finance, and the resulting valuation formulas have become indispensable tools in today's daily capital market practice. These achievements were also honored by the Nobel Prize in Economics (Scholes and Merton 1997) [154]. Though widely applied, the valuation formulas are not perfect. An important input parameter, the volatility, has to be adjusted empirically to obtain usable predictions. This discrepancy between theory and application is well documented in the financial literature [82, 215].

Beginning in the 1990s, some physicists have tried to understand these deviations with very diverse approaches, ranging from statistical analyses of the time evolution of asset prices to microscopic trader models. The perception of the financial market as a complex many-body system offers an interesting challenge for testing well-established physical concepts and methods in a new field. If some of these tools can be successfully transferred, improved insight into the underlying mechanisms of the market should be gained. This new interdisciplinary field of physics is called *econophysics* [131].

The aim of the present chapter is to sketch some of the steps of this development. We start with an introduction to financial markets. This first section is meant to present basic concepts and to explain why the theory of stochastic processes finds important applications in this non-physical field. It sets the stage for the following two sections, in which we discuss a 'classical' example, the Black-Scholes theory of option pricing, and some of the 'non-classical' approaches suggested by the econophysics community.

5.1 Basic Notions Pertaining to Financial Markets

If trained in natural sciences, one is usually not well acquainted with the language of the financial market. Learning new vocabularies and relevant concepts is the first hurdle. Providing respective help is the main aim of this section.

Economists would define a *market* as a 'location', where buyers and sellers meet to exchange 'products'. At every time instant t, the products have a price, the *spot price* $S(t)$. The spot price is determined by the interplay of *supply and demand* in a free-market economy. If demand in a given product increases and supply correspondingly decreases, $S(t)$ increases, and vice versa.

A special kind of market is the *financial market*, where the product traded is, loosely speaking, money. On the financial market, large sums of money are lent, borrowed and invested in commodities, like metals or corn, and securities, such as bonds or stocks ('shares' in British English). In principle, there are two ways how an investment can take place: either on the *stock exchange* or by *over-the-counter* trading.

The stock exchange is the official trading place for securities, foreign currency exchange and (some) commodities. The task of the stock exchange is to comprehensively provide the traders with all pertinent information, to promote reliable transactions between them, and to determine the price of the *assets* traded on the basis of resulting supply and demand. To accomplish this aim many stringent rules are imposed. Only approved assets may be traded, the credit worthiness of the involved parties to meet their obligations is thoroughly tested ('credit rating'), and transactions are standardized. These means are meant to make trading more efficient and safer. But, they also limit the freedom.

5.1 Basic Notions Pertaining to Financial Markets

Market participants[1] who do not want to renounce that freedom can resort to over-the-counter (OTC) trading. This implies that a contract is established between two parties, the features of which are adapted to the individual needs of the partners. It may involve assets and rules which are not approved by the stock exchange. Such tailor-made deals can be very speculative and therefore much more risky than standardized contracts of a regulated exchange. But, traders are willing to accept (a certain amount of) risk. If this were not the case, they could deposit all their money on a bank account and pocket the interest rate granted by the bank.[2] This return is certain, but low compared to the possible outcome of a highly profitable transaction. This expected gain is the main driving force for the market participants to trade at all and to deliberately tolerate a certain exposure to risk.

However, everybody obviously wants to make money at the end of the day. Nobody wants to lose it. The aim of modern *risk management* consists in finding the sources of potential losses and in devising strategies to limit the embedded risk as much as possible [36]. Risk management starts with a classification of risk. Basically, there are three different kinds of risk: *credit risk*, *operational risk* and *market risk*.

Credit Risk Credit risk means that the counter party cannot or can only partially meet the obligations incurred when signing the contract. The possibility of such a failure is particularly present in OTC transactions, but largely reduced on an exchange. If a failure should still happen in spite of the strict trading conventions and the severe credit rating, the exchange guarantees clearance of the debts.

The preceding paragraph summarizes the market theory as we stated it in the first edition (1999) of this book. At the time of writing this second edition (2012), the financial markets around the world have tumbled into the most severe crisis since the world-wide depression in the 1930s. Sadly so, the regulations about the credit ratings imposed on transactions at stock exchanges were not worth the paper they were printed on. In 2008 one of the largest investment banks of the world, Lehman Brothers, went bankrupt, and it became clear that only a small fraction of the financial transactions such an institution is involved in, is actually backed by securities. Nobody believed that such a big bank could actually go bankrupt in adverse market conditions, and so everybody believed the credit risk, incurred in transactions with such a company, to be basically zero. However, the big leverage effect[3] that these

---

[1] Market participants is a general expression for individuals or groups who are active in the market, such as banks, investors, investment fonds, traders (for their own account). Often, we use the term 'trader' as a synonym for 'market participant'.

[2] Here and in the following the term 'bank' signifies the 'safest place' for an investment. In the reality of capital markets, it would correspond to the State so that the risk-free interest rate equals the yield on a 'Bundesanleihe' in Germany or on a 'US Government Treasury Bond' in the United States, for instance.

[3] In most speculative investments a bank or other financial institution embarks on, only a few percent of the invested money is owned by them, the rest is borrowed. This creates a leverage effect when a gain—or loss—is measured as return on invested own capital.

institutions employ to achieve large gains in good market conditions also seems to increase the amplitude of fluctuations in the market (see Sect. 5.3.4). When the size of the fluctuations is governed by the leveraged money involved, even the balance sheet of a big bank, like Lehman Brothers, is small, and it is easily brought down by the fluctuations which it helped to generate itself. Unfortunately, Lehman Brothers is only the most visible case. Basically all big banks around the world were at the edge of bankruptcy and needed to be rescued by state intervention. We will discuss market fluctuations and econophysics ideas on how to quantify and understand them later in this chapter. At his point we would only like to conclude that the development of appropriate methods to quantitatively assess credit risk is a very important issue in modern risk management [36].

Operational Risk Operational risk is a fairly new term [89, 90]. It has been given a definition in the process of finding international regulations for the banking sector within the so-called Basel II convention from which we quote:

> *Operational risk is defined as the risk of loss resulting from inadequate or failed internal processes, people and systems or from external events [8]. This definition includes legal risk, but excludes strategic and reputational risk.*

Possible sources of operational risk include: hidden errors in computer programs, which only pop up after a sequence of rare events, lack of controls to prevent inappropriate or unauthorized transactions, simple human errors, such as misspellings which lead to faulty accounting, etc. These few examples show that operational risk comprises factors which are very specific to the individual institutions and business transactions. This makes it difficult to quantify and control operational risk.

Market Risk Market risk means the hazard of losing money due to the unavoidable fluctuations of the asset prices. The fluctuations can be caused by the intrinsic market dynamics of supply and demand or by external influences, such as political decisions or changes of the interest rates.

Figure 5.1 presents a typical example of these fluctuations. It shows the time evolution of the *Standard & Poor's Composite Index* (S&P500) from 1950 to 1999. Quite generally, a *stock index* is a suitably normalized average of the spot prices of stocks traded on an exchange. The weight of a stock in the average is usually determined by its market capitalization (i.e., spot price of one stock × number of shares currently traded). The S&P500, for instance, comprises 500 different stocks: 400 industrials, 40 financial institutions, 40 utilities and 20 transportation companies. It represents about 80 % of the market capitalization of all stocks listed on the New York Stock Exchange. There are many other indices like the S&P500 traded on all exchanges of the world, the oldest and certainly the most well-known being the Dow Jones (Industrial Average) Index (first published 1887). Since they are representative averages over all kinds of stocks, indices serve as sensible indicators of the market situation.

The spot price $S(t)$ of the S&P500 in Fig. 5.1 has two characteristic features. First, its time evolution is not smooth, but rather irregular with random up and down shifts. Second, $S(t)$ gradually 'drifts' to larger values with increasing time. Based

5.1 Basic Notions Pertaining to Financial Markets

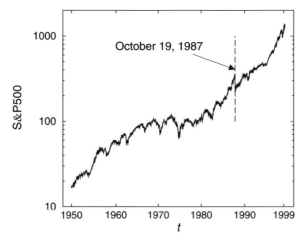

Fig. 5.1 Time evolution of the daily closing price of the Standard & Poor's Composite Index (S&P500) from 1950 until 1999. The price is given in points. The years on the abscissa correspond to the following days: $1950 = 3.1.1950$, $1960 = 4.1.1960$, $1970 = 2.1.1970$, $1980 = 2.1.1980$, $1990 = 2.1.1990$, $1999 = 27.4.1999$. The vertical dashed line indicates the crash of October 19, 1987. On that day, the S&P500 went down by about 20 % compared to the closing price of the previous day. The data are a time series obtained from http://quote.yahoo.com

on these observations, it is very tempting to model the time evolution of the index price as a stochastic process. But what kind of stochastic process? A fundamental assumption about the market, the *efficient market hypothesis*, suggests that it should be a Markov process.

Efficient Market Hypothesis   A market is called 'efficient' [46]

- if the participants quickly and comprehensively obtain all information relevant to trading.
- if it is *liquid*. This means that an investor can easily buy or sell a financial product at any time. The more liquid a market is, the more secure it is to invest. The investor knows that he can always cash-in his assets. This easy exchange between money and financial products raises the attractivity of the market. On a 'mature' liquid market, the myriad transactions efficiently balance the decision of a single investor (or of a small group of investors) so that individual purchases or sales are possible at any time without destabilizing the asset prices.
- if there is low *market friction*. Market friction is a collective expression for all kinds of trading costs. These include trader provisions, transaction costs, taxes, bid–ask spreads, i.e., differences in the prices that an investor obtains ('bid-price') when selling or has to pay ('ask-price') when buying, etc. The sum of these costs is negligible compared with the transaction volume if the market friction is low.

The efficient market hypothesis states that a market with these properties 'digests' the new information so efficiently that all the current information about the market development is at all times completely contained in the present prices. No advantage is gained by taking into account all or part of the previous price evolution. This amounts to a Markov assumption.

The rationale behind this hypothesis rests on the following argument: Imagine that a time series exhibits a structure from which the rise of an asset price could be predicted in the near future. Certainly, investors would buy the asset now and sell it later in order to pocket the difference. However, an efficient market immediately responds to the increased demand by increasing the price. The profitable opportunity vanishes due to competition between the many active traders. This argument limits any correlations to a very short time range and advocates the random nature of the time series.

In fact, this Markov assumption seems very plausible and has a long history. After a detailed analysis of the French market, Bachelier drew the same conclusion at the end of the 19th century [3, 145]. He suggested a model for the time series $S(t)$ which we would today call a Wiener process with a drift term to account for the gradual increase of $S(t)$. His model suffers from the unrealistic property that it allows negative asset prices. Prices, however, are by definition always positive. This fact led to a modified version of Brownian motion—the *geometric Brownian motion*—which underlies many modern approaches to pricing derivative securities, such as options.

Geometric Brownian Motion To get an idea of how to define a sensible model for the evolution of $S(t)$ let us first assume that at $t = 0$ we deposit a sum $S(0)$ in a bank. The bank grants a *risk-free* interest rate r for the deposit. If the interest is paid once at the end of time t, the initial sum has grown by [109]

$$S(t) = S(0) + rtS(0) = S(0)(1 + rt).$$

On the other hand, if it is paid twice, we have

$$S(t) = \left(S(0) + \frac{rt}{2}S(0)\right) + \left(S(0) + \frac{rt}{2}S(0)\right)\frac{rt}{2} = S(0)\left(1 + \frac{rt}{2}\right)^2.$$

Iterating this reasoning for n payments in time t, we finally obtain

$$S(t) = S(0)\left(1 + \frac{rt}{n}\right)^n, \qquad (5.1)$$

which yields in the limit of continuously compounded interest

$$S(t) = S(0)\left(1 + \frac{rt}{n}\right)^n \xrightarrow{n \to \infty} S(0)\exp(rt). \qquad (5.2)$$

Now, we imagine the asset price to be similar to a bank deposit, but perturbed by stochastic fluctuations. Therefore, the price change dS in the small time interval dt should consist of two contributions: a deterministic and a random one. In analogy to (5.2) a reasonable ansatz for the deterministic part is

$$dS = \mu S(t)dt, \qquad (5.3)$$

5.1 Basic Notions Pertaining to Financial Markets

where μ is called *drift* and measures the average growth rate of the asset price. One can expect μ to be larger than r. No sensible investor would assume the risk of losing money on the market if a bank account yielded the same or an even better profit.

The second contribution models the stochastic nature of the price evolution. It should be Markovian according to the efficient market hypothesis. A possible choice is an ansatz symmetric to (5.3)

$$dS = \sigma S(t) dW(t). \quad (5.4)$$

Equation (5.4) introduces a second phenomenological parameter, the *volatility* σ. The volatility measures the strength of the statistical price fluctuations.

Combining both contributions, we obtain the following result

$$dS = \mu S(t) dt + \sigma S(t) dW(t) \quad (S(0) > 0). \quad (5.5)$$

This stochastic differential equation defines a variant of Brownian motion, which is called *geometric Brownian motion*. It is a specific case of an Itô process

$$dS = a(S,t) dt + b(S,t) dW(t),$$

where the functions $a(S,t)$ and $b(S,t)$ are proportional to the random variable $S(t)$ (see (2.119)–(2.121)).

Equation (5.5) suggests that not the absolute change $dS = S(t + dt) - S(t)$, but the *return* $dS/S(t)$ in the time interval dt is the relevant variable. Financially, this is sensible. An absolute change of $dS = \$10$ in time dt is much more significant for a starting capital of $S(t) = \$100$ than for one of $S(t) = \$10000$. The return dS/S clearly expresses this difference [36, 82, 215].

The interpretation of dS/S as the relevant variable suggests that (5.5) should be rewritten in terms of $\ln S(t)$. Using Itô's formula (2.121)

$$df(S,t) = \left(\frac{\partial f}{\partial t} + a(S,t) \frac{\partial f}{\partial S} + \frac{1}{2} b(S,t)^2 \frac{\partial^2 f}{\partial S^2} \right) dt + b(S,t) \frac{\partial f}{\partial S} dW(t)$$

with $f(S,t) = \ln S$, $a(S,t) = \mu S$ and $b(S,t) = \sigma S$, we find

$$d \ln S = \left(\mu - \frac{1}{2} \sigma^2 \right) dt + \sigma dW(t). \quad (5.6)$$

Therefore, it is the logarithm of the asset price, and not the price itself, as Bachelier assumed, which performs a Wiener process with a (constant) drift. The discussion in Sect. 2.2.2 implies that $d \ln S$ is normally distributed with a mean and variance of (see also Appendix F)

$$E[d \ln S] = \left(\mu - \frac{1}{2} \sigma^2 \right) dt, \quad \text{Var}[d \ln S] = \sigma^2 dt, \quad (5.7)$$

and that the transition probability from (S, t) to (S', t') is given by

$$p(S', t'|S, t) = \frac{1}{\sqrt{2\pi(\sigma S')^2(t'-t)}} \exp\left(-\frac{[\ln(S'/S) - (\mu - \sigma^2/2)(t'-t)]^2}{2\sigma^2(t'-t)}\right). \quad (5.8)$$

The price $S(t)$ thus has a log–normal distribution. This distribution and the geometric Brownian motion from which it results underly the Black-Scholes theory of option pricing, which we will discuss in the next sections.

Geometric Brownian motion introduces two parameters, μ and σ. The first describes the expected gain, the second quantifies the fluctuations around the average behavior. If the fluctuations are large, the asset is very 'volatile', and an investment is considered risky. A prudent investor wants to avoid this risk and to protect himself against possible losses. How can he achieve that?

Certainly, it is advisable not to place all of one's money on a single asset, but instead to have a *portfolio* containing a large number of assets from different, uncorrelated sectors of the market. For instance, an investor could hold shares in the General Motors Corp. and in the Coca-Cola Co. Even if the customers do not buy cars, they might still drink Coca-Cola. Eventual losses in General Motors could thus be compensated by gains in Coca-Cola. If the portfolio comprises many mutually uncorrelated assets, the risk of the portfolio, measured, for instance, by the overall variance of its return, is largely reduced. This method of risk reduction is known as *diversification*.

But it is only possible to 'diversify away' the *specific* (or *non-systemic*) risk associated with a single asset (or market sector, such as cars), not the *non-specific* (or *systemic*) *market risk* mentioned above. The S&P500 can be thought of as a huge portfolio including stocks from all branches of the market. This diversification should have removed the specific risk. Thus, the remaining fluctuations, visible in Fig. 5.1, must reflect the systemic risk which affects all stocks, i.e., the market as a whole.

There is a way to cope with the market risk. The idea is to add strongly anti-correlated assets to the portfolio. If one asset decreases in value, the other increases, and vice versa. In ideal cases, this trick allows one to obtain a completely risk-free portfolio (for instance, in the Black-Scholes theory). The reduction of the risk by taking advantage of such anti-correlations is called *hedging* in financial jargon. There are special financial products, so-called *derivatives*, which can be used as versatile tools for hedging.

Derivatives: Options and Futures A *derivative* is a financial product whose value 'derives' from the price of an underlying asset. This basic asset is simply referred to as the *underlying* and the derivative is also called *contingent claim*. We can rephrase this definition in a more mathematical language by saying that if the underlying is characterized by its spot price $S(t)$ then a derivative is just a (mathematical) function of $S(t)$.

5.1 Basic Notions Pertaining to Financial Markets

Two examples for derivatives are *options* and *futures*. The basic idea of a future is identical to that of a so-called *forward* contract. A forward contract, or more briefly a forward, is a contract between two parties in which

- one partner agrees to buy an underlying asset ('long position')
- at a certain specified time T in the future, known as the *delivery date*,
- for a prescribed price K, known as the *delivery price*,
- and the other partner pledges to sell the underlying at time T for price K ('short position').

In financial jargon, the party who buys assumes a *long position*, whereas the other who sells assumes a *short position*. The main difference between a forward and a future is that the details of the future cannot be negotiated individually. Futures are standardized and traded on an exchange. The exchange requires from both parties a security deposit, the so-called *margin*. Margins serve to guarantee that equalizing payments accruing from the contract can actually be made. The level of the required margins is fixed by the exchange. If the value of the future contract rises during the trading day, the profit is credited to the margin account. Otherwise, it is debited. This daily clearing of profits and losses is called 'daily settlement'. If the balance of a margin account decreases below the level required, the party has to transfer money to its account or the account is closed. These stringent trading rules of an exchange are meant to eliminate the credit risk, in contrast to an OTC forward.

The basic motivation to sign a contract, such as a forward or a future, is to remove the market risk of the unpredictable price evolution. Both parties can adjust their further transactions and investments accordingly. As an example, imagine a farmer who grows corn. If he knows at the time of sowing that he can sell a certain quantity of his harvest at a prescribed price, he is able to plan his work and expenses much more efficiently. It is therefore not astonishing that forwards and futures have a long history. Futures on commodities have been traded on organized exchanges since the middle of the 19th century, especially in the United States. In 1848, the Chicago Board of Trade (CBOT) was founded, and other exchanges opened in subsequent years, for instance, in New York and London. On the other hand, futures on financial products are rather recent. In 1972, a division of the Chicago Mercantile Exchange (CME), the International Money Market (IMM), started trading financial futures. In England, the London International Financial Future Exchange (LIFFE) was established in 1982, and in Germany trading started as late as 1990 with the foundation of the Deutsche Terminbörse (DTB).

Futures are contracts that bind both parties. Contrary to that, *options* are contracts in which only one partner assumes the obligation, whereas the other obtains a right. There are many different kinds of options. The simplest are *European options*. A European option is a contract between two parties in which

- the seller of the option, known as the *writer*,
- grants the buyer of the option, known as the *holder*,
- the right to purchase (= *call option*) from the writer or to sell (= *put option*) to him an underlying with current spot price $S(t)$.

- for a prescribed price K, called the *exercise* or *strike price*
- at the *expiry date T* in the future.

The key property of an option is that only the writer has an obligation. He must sell or buy the underlying asset for the strike price at time T. On the other hand, the holder has the *possibility* to exercise his option or to not exercise it. He will only exploit his right if he gains a profit, i.e., if $S(T) > K$ for a call option. Otherwise, he can buy the underlying for a cheaper price, $S(T) < K$, on the market. Of course, the writer does not incur this liability without requiring compensation. The central question, therefore, is how much an option should cost. The Black-Scholes theory gives an answer to this question for European options.

Historically, the first major use of option-like contracts occurred in the Netherlands in the 17th century. At that time, there was an immense passion for tulips. The tulip growers wanted to protect themselves against market-price fluctuations. Therefore, they purchased contracts which entitled them to sell tulip bulbs at a minimum price if the market price fell under this threshold. On the other hand, the writers of the contracts expected the prices to increase so that the growers would refrain from their right, and then they could pocket the premium of the contract. However, the tulip market crashed in 1637. Since the writers could not meet their obligations to buy, the Netherlands experienced a serious economic crisis.

After that, such option-like contracts were defamed in Europe for a long time. Nevertheless, an organized trading of options gradually developed in London in the 18th century. Yet, large losses were not uncommon due to the lack of a sufficient legal framework. Strict laws for option trading were only created in the 1930s. But it was not before 1973 that options began to win the importance that they have today.

In 1973, the world's first option exchange, the Chicago Board Options Exchange (CBOE), opened in Chicago. Before that, only OTC trading was possible. Today, options are traded on all major world exchanges. The simplest (European) call and put options are now very common. Colloquially, they are called *plain vanilla* options to emphasize that they are ubiquitous and of simple structure. Many further types of options have been invented. Among them are *American* and also so-called *exotic* options. An American option deviates from the European style only in that the holder may exercise his right at *any* time prior to expiry. Exotic options, however, are very different. They can have complicated payoffs, depend on more than one underlying, or their value can be determined by the *whole* time evolution of the underlying and not only by its value at the end. Pricing these options correctly is a great challenge [215]. We will only be concerned with European options in the remainder of this chapter.

The market for options and futures has extended so much during the last twenty years because they offer unique possibilities to different types of traders. They can be very speculative, but also serve as a versatile tool for hedging the market risk.

Let us illustrate the latter possibility by the following example. Imagine that a US company has to pay 1000000 Euro to a German supplier in three months. The company faces a risk exposure due to the uncertain development of the foreign exchange rate. Suppose that the Euro/US-dollar exchange rate is €1 = \$1.05 today, but increases to €1 = \$1.1 during the next three months. If the company waits until the money is due, it has to pay $1100000 instead of $1050000 and so loses $50000. There are two ways to hedge this risk by using derivatives:

- The company could sign a forward contract with a bank which fixes the exchange rate to the present value of €1 = $1.05. Thus, the company and the bank agree to buy and to sell for that rate, respectively. The agreement is binding for both parties. If the exchange rate rises to €1 = $1.1 as assumed, the company ends up $50000 better by the forward. However, if the rate drops to €1 = $1, the company still loses $50000 because the contract obliges that the purchase be made for €1 = $1.05. This example shows that hedging through a forward does not necessarily improve the company's situation. It merely makes the future development predictable.
- If the company does not like the restrictive features of the forward contract, it might consider hedging its currency risk using a call option. To this end, it has to find a bank which grants the right to buy 1000000 Euro at €1 = $1.05 in three months. If the rate rises during this time, the company exercises its right. Otherwise, it lets the contract expire and exchanges the money for the current, cheaper rate. Therefore, such a call option seems to eliminate the risk for the company and to transfer it completely to the bank. But of course, the bank does not generously take over the risk without compensation. It charges a premium, the call price, for the right conferred. This premium partly transfers the risk back to the company. The call price has to be well chosen. The bank will not take the risk if the price is too low. On the other hand, a price that is too large might prevent the company from buying the option at all, so no transaction is made. The key question therefore is: What is a fair price for an option?

Finding the 'correct' price of options and other derivatives is a challenging and very important problem. Practitioners constantly seek and invent new complex financial products. Exotic options are an example. If priced incorrectly, large losses can result. It is therefore indispensable to understand the structure of these products and to assess the risk involved properly. The next section introduces how to solve this problem for the by now well-established case of European options.

5.2 Classical Option Pricing: The Black-Scholes Theory

A European option confers on its holder the right to buy from a writer (call option) or to sell to him (put option) an underlying asset for the strike price K at a future time T. This right has a value, the option price. In general, we denote the option price by $\mathcal{O}$, unless call or put options are considered explicitly. In such cases, we use the notation $\mathcal{C}$ and $\mathcal{P}$ for call and put prices, respectively.

The option price must be a function of the current value S of the underlying and of time t, i.e., $\mathcal{O} = \mathcal{O}(S, t)$. Furthermore, it depends on parameters, such as the strike price K and the expiry time T. At the expiry time, we can calculate $\mathcal{O}(S, T)$ by the following simple argument: Consider a call option. If $S(T) < K$, the holder will not exercise the option, because he can obtain a better price on the market. Thus, $\mathcal{C}(S, T) = 0$. In the opposite case, however, it would be profitable to exercise a call. By paying an amount K, the holder receives an asset worth $S(T)$, which provides him with a gain $S(T) - K$ if he sells it immediately. This gain is the maximum price that the holder would be willing to pay for the option. On the other hand, the writer requires at least $\mathcal{C}(S, T) = S(T) - K$, since he has to buy the asset for $S(T)$ on the market, and only receives an amount K. So they agree on a fair price of $\mathcal{C}(S, T) = S(T) - K$. Putting the cases $S(T) < K$ and $S(T) > K$ together, the call price at expiry, the so-called *payoff*, must be

$$\mathcal{C}(S, T) = \max\bigl(S(T) - K, 0\bigr). \tag{5.9}$$

A similar argument yields for the payoff of a put option:

$$\mathcal{P}(S, T) = \max\bigl(K - S(T), 0\bigr). \tag{5.10}$$

Given these limiting values, what are fair option prices when signing the contract, i.e., at times $t < T$? In 1973, Black and Scholes, and, independently, Merton, proposed an answer to this question. An entertaining account of the history of how this answer was found is given in [13]. The next sections summarize and discuss the results of the theory.

5.2.1 The Black-Scholes Equation: Assumptions and Derivation

The Black-Scholes analysis is based on the following assumptions:

1. There is no credit risk, only market risk.
   The price of an option does not depend on any assessments of the reliability of the counter party. It is only influenced by the market fluctuations of the underlying. This is (supposed to be) a good approximation for products traded on an exchange (but see our comments on the world economic crisis as of 2008 in the paragraph on credit risk).
2. The market is maximally efficient, i.e., it is infinitely liquid and does not exhibit any friction.
   This implies that all relevant information is instantaneously and comprehensively available and at all times fully reflected by the current prices, that purchases and sales can be readily performed at any time, and that there are no additional costs, such as trader provisions, transaction costs, bid–ask spreads, etc. This assumption may also be reasonably well realized by trading on a (very liquid) exchange.
3. Continuous trading is possible.
   Continuous trading means that the time interval Δt between successive quotations of the price of the underlying tends to zero. Certainly, this is an idealization. On the one hand, exchanges are closed on weekends and bank holidays, and on trading days successive quotations are separated by a finite time interval of the

order of seconds to minutes. On the other hand, continuous trading would entail infinitely high costs in practice. So, the previous assumption is a necessary prerequisite for this one.
4. The time evolution of the asset price is stochastic and exhibits geometric Brownian motion.
   Mathematically, this assumption is expressed by (5.5),
   $$dS = \mu S(t)dt + \sigma S(t)dW(t) \quad (S(0) > 0).$$
   Since geometric Brownian motion is a Markov process, this assumption can also be considered as a realization of assumption 2, i.e., of the efficient market hypothesis.
5. The risk-free interest rate r and the volatility σ are constant.
   This assumption can be relaxed if r and σ are known functions of time. The Black-Scholes valuation formulas for European call and put options ((5.48) and (5.49)) remain valid with r and σ replaced by [215]
   $$r \to \frac{1}{T-t}\int_t^T r(t')dt', \qquad \sigma^2 \to \frac{1}{T-t}\int_t^T \sigma^2(t')dt'. \qquad (5.11)$$
   But r and σ should not be stochastic. Such models, however, exist. They were proposed as extensions of the Black-Scholes analysis and are known as *stochastic interest rate* and *stochastic volatility models* [82, 215].
6. The underlying pays no *dividends*.
   A *dividend* is usually a part of a company's profit which is paid to the shareholders. This assumption can also be dropped if the dividend $\bar{D}$ is a known function of time, $\bar{D} = \bar{D}(t)$. Then, the stock price has to be replaced by [36, 215]
   $$S \to S\exp\left(-\int_t^T \bar{D}(t')dt'\right) \qquad (5.12)$$
   in the Black-Scholes formulas (5.48) and (5.49).
7. The underlying is arbitrarily divisible.
   This means that the amount of underlying in the portfolio need not be an integer. It can be any real number.
8. The market is *arbitrage-free*.
   The absence of *arbitrage* possibilities is a fundamental hypothesis with regard to the properties of the market. Let us illustrate this concept by the following thought experiment (see Fig. 5.2): Suppose that an investor borrows the amount B_0 from a bank, which he has to pay back, plus interest, after a time T. At that time, he owes the bank $B_0 e^{rT}$. Suppose, furthermore, that he knows an investment strategy on the market, which yields a guaranteed, risk-less gain whose growth parameter μ_0 is larger than the interest rate required by the bank, i.e., $\mu_0 > r$. Then, he could clear his debts at the bank and pocket the difference $B_0(e^{\mu_0 T} - e^{rT}) > 0$. Motivated by this success, he would certainly repeat this cycle over and over again and would finally become infinitely rich, even if he had no money to begin with. Such a risk-less profit without any costs by exploit-

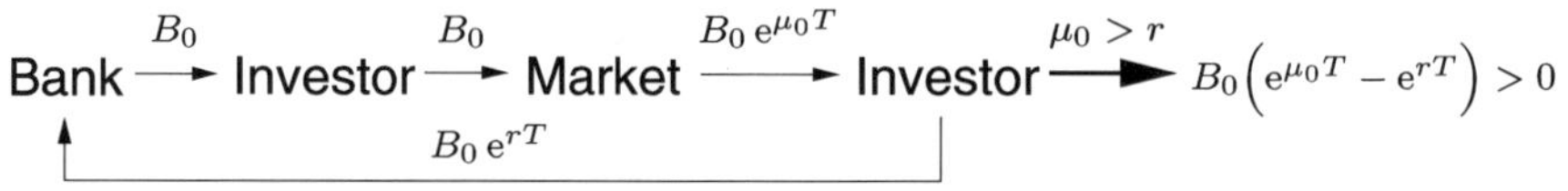

```
      B₀              B₀          B₀ eμ₀T           μ₀ > r
Bank ────► Investor ────► Market ────► Investor ────► B₀(eμ₀T − erT) > 0
  ▲                    B₀ erT
  └──────────────────────┘
```

Fig. 5.2 Thought experiment to motivate the concept of arbitrage. A bank lends an amount B_0 to an investor who exploits the market as a 'money-pump' to obtain a risk-free return, $B_0 e^{\mu_0 T}$, larger than the amount, $B_0 e^{rT}$, required by the bank after time T. The risk-free profit, $B_0(e^{\mu_0 T} - e^{rT}) > 0$, is called arbitrage

ing price disparities ($\mu_0 \neq r$) on different markets ('market' and 'bank') is called *arbitrage* or a *free lunch* in financial jargon.

Such arbitrage opportunities cannot persist for a long time. If this were possible, *arbitrageurs*, such as the smart investor in our thought experiment, could extract unlimited quantities of risk-less gains from the market. The market would gradually lose its liquidity and finally fall out of equilibrium. Usually, this does not happen—crashes being an exception—because the market is inhabited by arbitrageurs who constantly try to spy out and to take advantage of mispricings. It is the competition between the arbitrageur community which quickly leads to price adjustments. Therefore, a reasonable, though idealized, assumption is to claim that there is no risk-less growth rate other than the risk-free interest rate granted by the bank, i.e., $\mu_0 = r$, and that the market is arbitrage-free at any time. Referring to the above thought experiment this hypothesis can be expressed in the following way:

> There is no periodically working financial process which generates a risk-free profit from nothing.

In this formulation, the no-arbitrage concept appears to play the same role in finance as the law of energy conservation does in physics. In fact, it is very fundamental and powerful. A realistic, mathematically tractable model of the financial market can be built from it. Furthermore, arbitrage arguments suffice to deduce bounds on prices of options and other derivatives or to derive relations between them. This is called the *arbitrage-pricing technique*. An application of this technique (see Appendix E) leads to the following relation between call and put prices of (European) options,

$$\mathcal{P}(S,t) = \mathcal{C}(S,t) - S + Ke^{-r(T-t)}. \tag{5.13}$$

This relation is called *put–call parity*.

These assumptions can be used to derive the Black–Scholes pricing formulas. The first step is to realize that the option price $\mathcal{O}(S,t)$ is a function of the stochastic variable S which performs the Itô process (5.5). Its change in the time interval dt is therefore given by Itô's formula (2.121),

$$\begin{aligned}d\mathcal{O} &= \left(\frac{\partial \mathcal{O}}{\partial t} + \mu S(t)\frac{\partial \mathcal{O}}{\partial S} + \frac{1}{2}(\sigma S(t))^2 \frac{\partial^2 \mathcal{O}}{\partial S^2}\right)dt + \sigma S(t)\frac{\partial \mathcal{O}}{\partial S}dW(t) \\ &= \left(\frac{\partial \mathcal{O}}{\partial t} + \frac{1}{2}(\sigma S(t))^2 \frac{\partial^2 \mathcal{O}}{\partial S^2}\right)dt + \frac{\partial \mathcal{O}}{\partial S}dS, \end{aligned} \tag{5.14}$$

5.2 Classical Option Pricing: The Black-Scholes Theory

in which (5.5) was inserted to obtain (5.14). Equation (5.14) reflects the response of the option price caused by the stochastic time evolution of the underlying. All assumptions except the last two, 7 and 8, have been used to derive this equation.

In order to take the remaining assumptions into account, we adopt the role of the writer. When signing the contract, the writer faces the risk that the odds are against him and he has to sell the underlying at the expiry date below the market price to the holder. How can he hedge this risk?

Certainly, it is advisable to own a fraction $\Delta(t)$ ($0 \leq \Delta(t) \leq 1$) of the underlying at any time t. This fraction should be adjusted depending on the changes of the asset price S. Imagine that the writer has sold a call option. If S rises, the option becomes more likely to be exercised, so $\Delta(t)$ should be increased, and vice versa. For these adjustments he needs money. Thus, it is also advisable to have a cash amount $\Pi(t)$ which he could in turn invest to increase its value. Certainly, this investment should be risk-less to avoid introducing further uncertainties. Taking advantage of both forms of advice, the writer should possess the 'wealth' $\mathcal{W}(t)$ at time t:

$$\mathcal{W}(t) = \Delta(t)S(t) + \Pi(t). \tag{5.15}$$

Given the above, a reasonable hedging strategy is to make the holder provide $\mathcal{W}(t)$ through the option price $\mathcal{O}(t)$. The writer should therefore require

$$\mathcal{O}(t) \stackrel{!}{=} \mathcal{W}(t) = \Delta(t)S(t) + \Pi(t). \tag{5.16}$$

Equation (5.16) must hold at any time, and thus also for the variation in the small interval dt from the present time t to a time $t + dt$ in the near future. This means

$$d\mathcal{O} = \Delta(t)dS + d\Pi = \Delta(t)dS + r\Pi dt. \tag{5.17}$$

The first term of the right-hand side assumes that $\Delta(t)$ does not change in the time interval. A heuristic argument to justify this assertion is that a change of Δ should be interpreted as a reaction to a price fluctuation. Therefore, the amount of underlying can only be adjusted *after* a price variation has occurred. The second term expresses the growth of the cash amount $\Pi(t)$. It reflects the requirement that the market is arbitrage-free. In such a market, there is no risk-less investment strategy which yields a better return than that granted by the bank.

Equations (5.14) and (5.17) can only be consistent if the coefficients of dS and dt agree with each other. Therefore, we have

$$\Delta(t) = \frac{\partial \mathcal{O}}{\partial S} \quad \text{(delta-hedge)} \tag{5.18}$$

and

$$r\Pi = \frac{\partial \mathcal{O}}{\partial t} + \frac{1}{2}[\sigma S(t)]^2 \frac{\partial^2 \mathcal{O}}{\partial S^2}. \tag{5.19}$$

If the writer continuously adjusts the amount of underlying according to (5.18) during the lifetime of the option (called *delta-hedge*), he can eliminate the risk completely by calculating the option price from (5.19). This equation is deterministic.

The stochastic character, contained in both terms of the right-hand side individually, has to vanish upon addition, since the left-hand side is non-stochastic. Inserting (5.16) and (5.18) in (5.19), we obtain the *Black-Scholes equation* for European options

$$\frac{\partial \mathcal{O}}{\partial t} + \frac{1}{2}(\sigma S)^2 \frac{\partial^2 \mathcal{O}}{\partial S^2} + rS\frac{\partial \mathcal{O}}{\partial S} - r\mathcal{O} = 0. \tag{5.20}$$

Note that this equation is *independent* of the drift parameter μ and valid for any derivative which satisfies the assumptions compiled above, in particular for call and put options. The difference between a call and a put option resides in the boundary conditions. For a call we have

$$\begin{aligned} t = T: \quad & \mathcal{C}(S,T) = \max\bigl(S(T) - K, 0\bigr), \\ S = 0: \quad & \mathcal{C}(0,t) = 0, \\ S \to \infty: \quad & \mathcal{C}(S,t) \sim S, \end{aligned} \tag{5.21}$$

and for a put we have

$$\begin{aligned} t = T: \quad & \mathcal{P}(S,T) = \max\bigl(K - S(T), 0\bigr), \\ S = 0: \quad & \mathcal{P}(0,t) = K\exp\bigl[-r(T-t)\bigr], \\ S \to \infty: \quad & \mathcal{P}(S,t) \to 0. \end{aligned} \tag{5.22}$$

The boundary conditions at $t = T$ are just (5.9) and (5.10). The others can be explained as follows: If ever $S = 0$, the price of the underlying can never change due to (5.5) and remains zero at any later time. Thus, a call has no value, $\mathcal{C} = 0$, whereas a put is certain to be exercised. It must therefore be worth the strike price, discounted by the risk-less interest rate, in an arbitrage-free market. On the other hand, if S tends to infinity, we definitely have $S > K$. Then, a put is worthless, and the value of a call is determined by the price of the underlying.

This finishes the derivation of the Black-Scholes equation for European call and put options. In the subsequent section, we turn to its solution. However, before proceeding, we want to add a remark which allows a different, physically inspired interpretation of risk-less hedging.

Riskless Hedging and Legendre Transformation

If the delta-hedge (5.18) is inserted into (5.16), the left-hand side of (5.23) is obtained

$$\Pi(\Delta,t) = \mathcal{O} - \left(\frac{\partial \mathcal{O}}{\partial S}\right)S \quad \longleftrightarrow \quad f(T,v) = u - \left(\frac{\partial u}{\partial s}\right)s. \tag{5.23}$$

Portfolio		Option	Free energy		Internal energy
(Δ,t)	$\longleftrightarrow$	(S,t)	(T,v)	$\longleftrightarrow$	$s(s,v)$

5.2 Classical Option Pricing: The Black-Scholes Theory

The cash amount Π can be interpreted as the value of a portfolio of a trader who bought one option and sold an amount $\partial \mathcal{O}/\partial S$ of an asset with price S. According to our discussion above, this portfolio is risk-less.

Mathematically, it has the structure of a Legendre transform. Legendre transforms often occur in physics. In particular, they are very common in thermodynamics. The right-hand side of (5.23) provides an example. Here, $f(T, v)$ is the (intensive) free energy which depends on two natural variables: temperature T and (specific) volume v. It is the thermodynamical potential of the canonical ensemble. The free energy is related to the thermodynamical potential of the microcanonical ensemble, the (intensive) internal energy u, which has as natural variables (intensive) entropy s and volume v, by a Legendre transform. The Legendre transform is used to substitute the variables entropy and temperature *without any loss of information*. All of the information contained in u with regard to the thermodynamic system is transferred to the free energy by this mathematical operation.

In a similar way, we can interpret the risk-less delta-hedging in the Black-Scholes framework as a complete transfer of 'information', which is contained in the dependence of the option on the dynamics of the underlying, to the portfolio, whose independent variables are time and the fraction of underlying owned. Riskless hedging therefore corresponds to an information-conserving variable transformation from an 'option ensemble' with natural variables (S, t) to a 'portfolio ensemble' with natural variables (Δ, t).

5.2.2 The Black-Scholes Equation: Solution and Interpretation

This section deals with the solution and the interpretation of the Black-Scholes equation for European call and put options. Our presentation closely follows the excellent discussion in Sect. 5.4 of [215].

Consider a call option. Its price $\mathcal{C}(S, t)$ is the solution of the Black-Scholes equation

$$\frac{\partial \mathcal{C}}{\partial t} + \frac{1}{2}(\sigma S)^2 \frac{\partial^2 \mathcal{C}}{\partial S^2} + rS \frac{\partial \mathcal{C}}{\partial S} - r\mathcal{C} = 0, \qquad (5.24)$$

subject to the following boundary conditions (see discussion of (5.22)):

$$\begin{aligned} t = T: \quad & \mathcal{C}(S, T) = \max\bigl(S(T) - K, 0\bigr), \\ S = 0: \quad & \mathcal{C}(0, t) = 0, \\ S \to \infty: \quad & \mathcal{C}(S, t) \sim S. \end{aligned} \qquad (5.25)$$

Solution

When trying to solve an equation like (5.24), the first steps should be: Find suitable scales to render the equation dimensionless, and simplify it as much as possible by a clever change of variables.

The obvious scale for the price is K. However, for the time we could either use r, σ^2 or T. We choose the volatility, but also eliminate T by letting time evolve in reverse direction, starting at the expiry date. This transforms the boundary condition at $t = T$ into an initial condition at the time origin. Furthermore, the structure of the Black-Scholes equation suggests that $\ln S$ is a better variable than S because it removes the dependence of the coefficients on the asset price. Putting these ideas together we make the ansatz

$$C = K f(x, \tau), \qquad S = K e^x, \qquad t = T - \frac{\tau}{(\sigma^2/2)}, \tag{5.26}$$

which allows us to rewrite (5.24) and (5.25) as

$$\frac{\partial f}{\partial \tau} = \frac{\partial^2 f}{\partial x^2} + (\kappa - 1) \frac{\partial f}{\partial x} - \kappa f \quad (\kappa = 2r/\sigma^2), \tag{5.27}$$

$$\begin{aligned} \tau = 0: & \quad f(x, 0) = \max(e^x - 1, 0), \\ x \to -\infty: & \quad f(x, \tau) \to 0, \\ x \to +\infty: & \quad f(x, \tau) \sim e^x. \end{aligned} \tag{5.28}$$

This shows that the Black-Scholes theory only depends on one single parameter, κ. All other, initially present parameters have been adsorbed in the transformation of variables.

Equation (5.27) resembles a diffusion equation. It would be identical to the diffusion equation if the two last terms of the right-hand side were absent. This can be achieved by another clever change of variables. Inserting the ansatz

$$f(x, \tau) = e^{ax + b\tau} g(x, \tau) \quad (a \text{ and } b \text{ are real and arbitrary}) \tag{5.29}$$

into (5.27) yields

$$\frac{\partial g}{\partial \tau} = \frac{\partial^2 g}{\partial x^2} + [2a + (\kappa - 1)] \frac{\partial g}{\partial x} + [a^2 + (\kappa - 1)a - \kappa - b] g. \tag{5.30}$$

By choosing the undetermined constants a and b as

$$a := -\frac{1}{2}(\kappa - 1), \qquad b := a^2 + (\kappa - 1)a - \kappa = -\frac{1}{4}(\kappa + 1)^2, \tag{5.31}$$

the coefficients of $\partial g/\partial x$ and g vanish, and we obtain the diffusion equation

$$\frac{\partial g}{\partial \tau} = \frac{\partial^2 g}{\partial x^2} \tag{5.32}$$

5.2 Classical Option Pricing: The Black-Scholes Theory

with the following boundary conditions

$$\tau = 0: \quad g(x,0) = \max\left(e^{(\kappa+1)x/2} - e^{(\kappa-1)x/2}, 0\right)$$

$$\Rightarrow g(x,0)e^{-\alpha x^2} \xrightarrow{|x|\to\infty} 0 \quad (\alpha > 0),$$

$$\tau > 0: \quad g(x,\tau) \xrightarrow{|x|\to\infty} e^{(\kappa+1)x/2 - (\kappa+1)^2\tau/4} \quad (5.33)$$

$$\Rightarrow g(x,\tau)e^{-\alpha x^2} \xrightarrow{|x|\to\infty} 0 \quad (\alpha > 0),$$

where α is an arbitrary, real, positive constant. The second line in the cases $\tau = 0$ and $\tau > 0$ illustrates that g is 'well-behaved' for $|x| \to \infty$, and so a unique solution exists.

To find this solution, we use the Green's function method. This involves the following steps: The function $g(x,\tau)$ has a meaning only for $\tau > 0$. However, if we introduce

$$\bar{g}(x,\tau) := \Theta(\tau)g(x,\tau), \quad (5.34)$$

where $\Theta(\tau)$ is the Heavyside function

$$\Theta(\tau) = \begin{cases} 0 & \text{for } \tau < 0, \\ 1 & \text{for } \tau \geq 0, \end{cases} \quad (5.35)$$

the time variable can be extended to $\tau < 0$. The definition (5.34) turns (5.32) into an inhomogeneous differential equation,

$$\left(\frac{\partial}{\partial \tau} - \frac{\partial^2}{\partial x^2}\right)\bar{g}(x,\tau) = g(x,\tau)\delta(\tau) = \bar{g}(x,0)\delta(\tau), \quad (5.36)$$

which is solved by

$$\bar{g}(x,\tau) = \int_{-\infty}^{+\infty} dy\,\bar{g}(y,0)p(x,\tau|y,0) \quad (5.37)$$

if the integral kernel (Green's function) satisfies

$$\left(\frac{\partial}{\partial \tau} - \frac{\partial^2}{\partial x^2}\right)p(x,\tau|y,0) = \delta(x-y)\delta(\tau). \quad (5.38)$$

This is just the partial different equation for the diffusion (Gaussian) propagator

$$p(x,\tau|y,0) = \frac{1}{\sqrt{4\pi\tau}}\exp\left(-\frac{(x-y)^2}{4\tau}\right). \quad (5.39)$$

Combining these results, we see that the solution of (5.32) can be written as

$$g(x,\tau) = \int_{-\infty}^{+\infty} dy\,g(y,0)p(x,\tau|y,0) \quad (5.40)$$

$$= \frac{1}{\sqrt{4\pi\tau}} \int_{-\infty}^{+\infty} dy\, g(y,0) \exp\left(-\frac{(x-y)^2}{4\tau}\right). \quad (5.41)$$

Equation (5.40) will become important again in the next section, which sketches a different approach in mathematical finance, the so-called 'risk-neutral valuation'.

When inserting the boundary condition for $g(x,0)$ into (5.41), the maximum function limits the integration interval to positive y because $(e^{(\kappa+1)y/2} - e^{(\kappa-1)y/2}) = e^{(\kappa+1)y/2}(1 - e^{-y}) < 0$ for $y < 0$. Thus, we find

$$g(x,\tau) = \frac{1}{\sqrt{4\pi\tau}} \int_0^\infty dy \left(e^{(\kappa+1)y/2} - e^{(\kappa-1)y/2}\right) \exp\left(-\frac{(x-y)^2}{4\tau}\right) \quad (5.42)$$

and by introducing a new integration variable $z = (y-x)/(2\tau)^{1/2}$, this becomes

$$g(x,\tau) = \int_{-x/\sqrt{2\tau}}^\infty \frac{dz}{\sqrt{2\pi}} \left(e^{(\kappa+1)(\sqrt{2\tau}z+x)/2} - e^{(\kappa-1)(\sqrt{2\tau}z+x)/2}\right) e^{-z^2/2}$$

$$= e^{(\kappa+1)x/2 + (\kappa+1)^2\tau/4} \int_{-x/\sqrt{2\tau}}^\infty \frac{dz}{\sqrt{2\pi}} e^{-\frac{1}{2}(z-\sqrt{2\tau}(\kappa+1)/2)^2}$$

$$- e^{(\kappa-1)x/2 + (\kappa-1)^2\tau/4} \int_{-x/\sqrt{2\tau}}^\infty \frac{dz}{\sqrt{2\pi}} e^{-\frac{1}{2}(z-\sqrt{2\tau}(\kappa-1)/2)^2}$$

$$= e^{(\kappa+1)x/2 + (\kappa+1)^2\tau/4} N(d_1) - e^{(\kappa-1)x/2 + (\kappa-1)^2\tau/4} N(d_2), \quad (5.43)$$

where

$$d_1 = \frac{x}{\sqrt{2\tau}} + \frac{1}{2}(\kappa+1)\sqrt{2\tau} = \frac{\ln(S/K) + (r+\sigma^2/2)(T-t)}{\sigma\sqrt{T-t}}, \quad (5.44)$$

$$d_2 = \frac{\ln(S/K) + (r-\sigma^2/2)(T-t)}{\sigma\sqrt{T-t}} = d_1 - \sigma\sqrt{T-t}, \quad (5.45)$$

and

$$N(x) = \frac{1}{\sqrt{2\pi}} \int_{-\infty}^x dz\, \exp\left(-\frac{z^2}{2}\right). \quad (5.46)$$

The function $N(x)$ is the probability that the normally distributed variable z adopts a value smaller than x. It is the cumulative probability distribution for the Gaussian distribution.

The solution of the Black-Scholes equation is almost finished. All that is left to do is to express the call price, using (5.26), (5.29) and (5.31), as

$$C = Ke^{-\frac{1}{2}(\kappa-1)x - \frac{1}{4}(\kappa+1)^2\tau} g(x,\tau) \quad (5.47)$$

and to restore the original variables using (5.26). This gives

$$C(S,t) = SN(d_1) - Ke^{-r(T-t)} N(d_2), \quad (5.48)$$

5.2 Classical Option Pricing: The Black-Scholes Theory

and for the put option, using the put–call parity (5.13)

$$\mathcal{P}(S,t) = -S[1 - N(d_1)] + Ke^{-r(T-t)}[1 - N(d_2)]. \tag{5.49}$$

Interpretation

The Black-Scholes equations (5.48) and (5.49) tell the writer which price he should charge for an option at time t, at which the contract is signed. This price depends on the parameters K and T of the contract and on the market characteristics r and σ.

In addition to that, the equations also provide the necessary information on how to eliminate the risk. The writer's portfolio only stays risk-less if he continuously adjusts the amount of underlying, $\Delta(t)$, according to (5.18). For a call option, this means

$$\begin{aligned}\Delta(t) &= \frac{\partial \mathcal{C}}{\partial S} \\ &= N(d_1) + \left[S \frac{\partial N(d_1)}{\partial d_1} - Ke^{-r(T-t)} \frac{\partial N(d_2)}{\partial d_2} \right] \frac{\partial d_1}{\partial S} \\ &= N(d_1) \quad (0 \leq N(d_1) \leq 1), \end{aligned} \tag{5.50}$$

since the expression in square brackets [·] vanishes if (5.44) and (5.45) are inserted. Similarly, one obtains for a put option

$$\Delta(t) = \frac{\partial \mathcal{P}}{\partial S} = -[1 - N(d_1)] \quad (-1 \leq N(d_1) - 1 \leq 0). \tag{5.51}$$

Therefore, the first terms in (5.48) and (5.49) can be interpreted as the fraction of underlying which the writer should buy (call option) or sell (put option) to maintain a risk-less position.

Using these results and (5.16), the second terms of (5.48) and (5.49) may be identified with the cash amount $\Pi(t)$. For instance, for a call we have

$$\begin{aligned}\mathcal{C}(S,t) &= SN(d_1) - Ke^{-r(T-t)}N(d_2) \\ &= S\Delta(t) + \Pi(t)\end{aligned}$$

so that

$$\Pi(t) = -Ke^{-r(T-t)}N(d_2), \tag{5.52}$$

and similarly for a put

$$\Pi(t) = Ke^{-r(T-t)}[1 - N(d_2)]. \tag{5.53}$$

The factor $Ke^{-r(T-t)}$ is the strike price discounted to the present time, the so-called *present value*. For a call option, $\Pi(t)$ is given by the present value of the payment, due when the call is exercised, multiplied by the probability $N(d_2)$ that the call will be exercised. The minus sign indicates that this amount of money must be borrowed.

Fig. 5.3 Solution of the Black-Scholes equation for a call option. The call price C/K (*solid lines*) and the delta-hedge Δ (*dashed lines*) are shown as a function of S/K for three different time intervals to expiry: $T - t = 0, 1$ year, 2years. *All curves are calculated for the risk-free interest rate $r = 10\,\%$/year and the volatility $\sigma = 20\,\%/\sqrt{\text{year}}$. These are realistic values in practice*

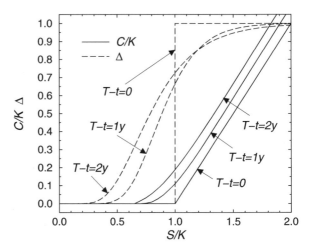

On the other hand, (5.53) specifies the cash amount which should be invested to replicate a put option. It is given by the present value of the payment, received when the put is exercised, multiplied by the probability, $1 - N(d_2)$, that the put will be exercised.

The solution of the Black-Scholes equation for a call option is illustrated in Fig. 5.3. The figure shows the call price in units of the strike price, C/K (5.48), and the delta-hedge (5.50) versus the *moneyness* S/K for three different times: the expiry date T, a year before expiry and two years before. One or two years are the longest times to maturity of an option contract. Standard contracts are usually shorter.

At expiry, the call only has a value if $S > K$. The option is then said to be 'in-the-money', whereas it is 'out-of-the-moncy', if $S < K$. In this case, the call price is zero, and there is no need to stock the underlying in the portfolio. However, as soon as the option is 'at-the-money' (i.e., $S = K$) or in-the-money, the holder is likely to exercise his right, and the writer must be prepared to deliver the underlying.

At times before expiry, the call price is more expensive and non-zero, even if the option is still out-of-the-money. Furthermore, hedging has to set in much earlier to maintain a risk-less portfolio. On the other hand, it is not necessary to keep exactly one underlying as soon as S becomes larger than K. These properties are caused by the interplay of the volatility and risk-less compounding, both of which are absent at $t = T$ (see (5.44) and (5.45)).

5.2.3 Risk-Neutral Valuation

Although the Black-Scholes theory is based on geometric Brownian motion of the underlying, an important parameter of the underlying's dynamics does not appear in the final result, the drift μ. The option prices are identical for all assets evolving with the same volatility, no matter what μ is.

5.2 Classical Option Pricing: The Black-Scholes Theory

This is a surprising property. Intuitively, one would have expected the opposite. If μ is large, $S(t)$ rises steeply and probably exceeds K in the future. A call option is thus likely to be in-the-money and a put option out-of-the-money. So, the option prices should be adjusted to the value of μ. However, this is not true in the Black-Scholes model. Our intuition is misleading because it does not take into account the possibility of constructing a portfolio which eliminates the risk completely.

The fact that the option price is independent of μ allows an interesting interpretation. In the financial literature, it is sometimes said that the drift parameter μ is a hallmark of the 'risk preferences' of the investors. Investors require a larger return if they believe a stock to be very risky. The extent of their risk aversion is reflected by the value of μ. The Black-Scholes analysis formally replaces μ by the risk-free interest rate r. This can be interpreted to mean that traders would not demand an extra premium for investing in a risky asset. They would seem to be *risk neutral*.

This interpretation may appear a little far-fetched because it confers a 'subjective' character on the drift. The drift appears to be qualitatively different from the other 'objective' parameters, like the volatility. This is certainly not true. However, the interpretation points to an important development in mathematical finance, the *risk-neutral valuation* [11, 109]. The basis of this is that the price of any derivative is given by the *discounted expectation value under an equivalent martingale measure*. The remainder of this section tries to explain what is meant by this sentence. We use the Black-Scholes theory as a vehicle to expose the main idea.

Example: Black-Scholes Price of a Call Option

Consider again (5.47) and replace $g(x, \tau)$ by (5.42). The call price is then given by

$$C = K e^{-(\kappa-1)x/2 - (\kappa+1)^2 \tau/4}$$
$$\times \frac{1}{\sqrt{4\pi\tau}} \int_{-\infty}^{+\infty} dy \max\left(e^{(\kappa+1)y/2} - e^{(\kappa-1)y/2}, 0\right) e^{-(x-y)^2/4\tau}. \quad (5.54)$$

If we write

$$\max\left(e^{(\kappa+1)y/2} - e^{(\kappa-1)y/2}, 0\right) = e^{(\kappa-1)y/2} \max\left(e^y - 1, 0\right),$$

$$\frac{1}{4}(\kappa+1)^2 \tau = \frac{1}{4}(\kappa-1)^2 \tau + \kappa\tau,$$

and substitute x, τ and κ by (5.26) and (5.27), we obtain

$$C = e^{-r(T-t)} \int_0^\infty dS_T \max(S_T - K, 0)$$
$$\times \left[\frac{1}{\sqrt{2\pi(\sigma S_T)^2(T-t)}} \exp\left(-\frac{(\ln(S_T/S) - (r - \sigma^2/2)(T-t))^2}{2\sigma^2(T-t)}\right)\right]$$

$$=: e^{-r(T-t)} \int_0^\infty dS_T \max(S_T - K, 0) p^*(S_T, T|S, t) \tag{5.55}$$

$$\equiv e^{-r(T-t)} E^*\big[\max(S_T - K, 0)|(S_{t'})_{t' \in [t,T]}\big]. \tag{5.56}$$

Equation (5.56) shows that the call price can be expressed as the conditional expectation value of its pay-off, $\max(S_T - K, 0)$, at expiry, discounted by the risk-free interest rate.

Intuitively, this result seems reasonable. If we deposited an amount $B(t) = \mathcal{C}(t)$ at time t in a bank, it would grow by the continuously compounded interest to

$$B(T) = B(t) \exp\big[r(T-t)\big]$$

in the remaining time before expiry. This is an investment alternative to the option contract. In an arbitrage-free market, no investment strategy can yield a better certain return than a bank account. The return from the option contract is its pay-off which is, however, not certain due to the stochastic evolution of the underlying. To eliminate this uncertainty, it seems sensible to identify $B(T)$ with the expectation value of the call's pay-off and to postulate

$$\mathcal{C}(S, t) = e^{-r(T-t)} E\big[\mathcal{C}(S_T, T)\big].$$

Equation (5.56) differs from this educated guess in two respects:

- Instead of a simple average, the call price is the conditional expectation value of its pay-off, which depends on the stochastic process $(S_{t'})_{t' \in [t,T]}$ of the underlying in the time interval $[t, T]$.
- The conditional expectation is *not* calculated by the 'real historic' probability (5.8), but by a 'fictitious' distribution

$$p^*(S_T, T|S, t)$$
$$= \frac{1}{\sqrt{2\pi (\sigma S_T)^2 (T-t)}} \exp\left(-\frac{[\ln(S_T/S) - (r - \sigma^2/2)(T-t)]^2}{2\sigma^2 (T-t)}\right), \tag{5.57}$$

where the drift μ is replaced by the risk-less interest rate r. This new 'risk-neutral' distribution is generated in the Black-Scholes theory by setting up a risk-free portfolio which replicates the option price completely. In mathematical language, the distribution (5.57) represents the unique *equivalent martingale probability* for the geometric Brownian motion of the stock-price process.

Geometric Brownian Motion and Martingales

In Sect. 2.2.1 we defined a martingale as a specific stochastic process, in which the best estimate for the future value, based on all information provided by the preceding

5.2 Classical Option Pricing: The Black-Scholes Theory

process, is the current value. The Wiener process is an example of a martingale, since

$$\begin{aligned}
\mathrm{E}\big[W(t+\mathrm{d}t)|(W_{t'})_{t'\le t}\big] &= \mathrm{E}\big[W(t+\mathrm{d}t) - W(t) + W(t)|(W_{t'})_{t'\le t}\big] \\
&= \mathrm{E}\big[\mathrm{d}W(t)|(W_{t'})_{t'\le t}\big] + W(t) \\
&= \mathrm{E}\big[\mathrm{d}W(t)\big] + W(t) = W(t),
\end{aligned} \quad (5.58)$$

where the last line holds because the increments $\mathrm{d}W(t) = W(t+\mathrm{d}t) - W(t)$ are symmetrically distributed and statistically independent of the past process $(W_{t'})_{t'\le t}$.

A comparison of (5.56) with (5.58) shows that the structure of both equations becomes identical when discounting the call price by e^{rt}, i.e.,

$$\frac{C(t)}{e^{rt}} = \mathrm{E}^*\left[\frac{C(T)}{e^{rT}} \,\bigg|\, (S_{t'})_{t'\in[t,T]}\right]. \quad (5.59)$$

The only essential difference is that not the real, but a different distribution has to be used for the calculation because geometric Brownian motion is, due to the finite drift ($\mu \ne 0$), not a martingale, unlike the Wiener process. Since the distribution used, p^*, is equivalent to the historic one (i.e., they have the same set of impossible events, see Sect. 2.1.3) and makes the discounted price process a martingale, it is called the 'equivalent martingale probability' [11, 109].

In the mathematical analysis of the Black-Scholes model, the equivalent martingale probability of the geometric Brownian motion can be found without setting up a risk-less portfolio and solving a partial differential equation. The main steps of the reasoning are as follows [11]: Consider two price processes—a *numéraire*, i.e., an (almost surely) strictly positive price process, for which we take a bank account $B(t)$, and geometric Brownian motion

$$\mathrm{d}B = rB(t)\mathrm{d}t, \quad B(0) = 1,$$
$$\mathrm{d}S = S(t)\big[\mu\mathrm{d}t + \sigma\mathrm{d}W(t)\big], \quad S(0) > 0.$$

Define the discounted asset-price process $\tilde{S}(t)$ as

$$\tilde{S}(t) := \frac{S(t)}{B(t)} = S(t)e^{-rt} \quad (5.60)$$

and use Itô's formula to derive

$$\mathrm{d}\tilde{S} = \tilde{S}(t)\big[(\mu - r)\mathrm{d}t + \sigma\mathrm{d}W(t)\big]. \quad (5.61)$$

The discounted price process exhibits a geometric Brownian motion with a reduced drift. However, it is not a martingale because $(\mu - r)$ is finite. To transform (5.61) into a martingale, it is necessary to change the probability from the original p to an equivalent p^* such that the drift vanishes. This can be achieved by an application of the following theorem [11, 109]:

Theorem 5.1 (Girsanov) *Let* $(\gamma(t) : 0 \leq t \leq T)$ *be a measurable process which satisfies Novikov's condition, i.e.,*

$$E\left[\exp\left(\frac{1}{2}\int_0^T \gamma^2(t)dt\right)\right] < \infty.$$

Furthermore, define two processes $(L(t) : 0 \leq t \leq T)$ *and* $(W^\mu(t) : 0 \leq t \leq T)$ *by*

$$L(t) = \exp\left(-\int_0^t \gamma(t')dW(t') - \frac{1}{2}\int_0^t \gamma(t')^2 dt'\right), \tag{5.62}$$

$$W^\mu(t) = W(t) + \int_0^t \gamma(t')dt', \tag{5.63}$$

where $W(t)$ *is a Wiener process with respect to the measure* ν. *Then,* $L(t)$ *is a martingale and* $W^\mu(t)$ *is also a Wiener process under the equivalent probability measure* μ *with Radon-Nikodým derivative*

$$\frac{d\mu}{d\nu} = L(T). \tag{5.64}$$

If $\gamma(t)$ were independent of t, the transformation of the measure by the Radon-Nikodým derivative would simply shift the drift while keeping the Wienerian character of the stochastic fluctuations. This is exactly what we need. By Girsanov's theorem, we have

$$dW(t) = dW^\mu(t) - \gamma(t)dt. \tag{5.65}$$

Inserting this equation into (5.61), we find

$$d\tilde{S} = \tilde{S}(t)\{[\mu - r - \sigma\gamma(t)]dt + \sigma dW^\mu(t)\}, \tag{5.66}$$

which becomes a martingale if we choose

$$\mu - r - \sigma\gamma(t) \stackrel{!}{=} 0 \Rightarrow \gamma(t) = \gamma = \frac{\mu - r}{\sigma} \tag{5.67}$$

so that

$$d\tilde{S} = \tilde{S}(t)\sigma dW^*(t), \tag{5.68}$$

where $W^*(t)$ is the Wiener process under the equivalent measure μ^* for which $\gamma(t)$ was chosen according to (5.67). This choice leads to a unique martingale measure. Using Itô's formula again, we can transform back to the non-discounted price dynamics

$$dS = S(t)[rdt + \sigma dW^*(t)], \tag{5.69}$$

for which the conditional probability density is given by (5.57).

5.2 Classical Option Pricing: The Black-Scholes Theory

This excursion to mathematical finance was meant to show that risk-neutral valuation does *not* simply correspond to a mere replacement of μ by r. In an arbitrage-free market, there are two distributions to characterize an asset: the historic probability p, which describes the observable stochastic fluctuations of the asset, and the martingale probability p^*, which is equivalent to p and is used for derivative pricing. A priori, both probabilities have a different functional form. They happen to coincide for geometric Brownian motion, for which the martingale probability can be constructed from the historic one by exchanging μ and r. In general, one can only state the conditions under which such an equivalent martingale measure exists and prove the following important theorem [11, 109]:

Theorem 5.2 (Risk-Neutral Valuation Formula) *If the market is arbitrage-free, the (arbitrage) price process of any attainable derivative $\mathcal{D}(t)$, i.e., of a derivative, which can be replicated by a portfolio, is given by the risk-neutral valuation formula*

$$\mathcal{D}(S,t) = B(t) \mathrm{E}^* \left[\frac{\mathcal{D}(S_T, T)}{B(T)} \,\Big|\, (S_{t'})_{t' \in [t,T]} \right], \tag{5.70}$$

where $B(t)$ is a numéraire.

The Black-Scholes pricing formula (5.59) is an example of a special application of this theorem.

5.2.4 Deviations from Black-Scholes: Implied Volatility

The Black-Scholes prices for put and call options depend on four parameters: the risk-free interest rate r, the volatility σ, the expiry time T and the strike price K. Of these parameters, T and K are part of the option contract and r is either quoted in or can be estimated from (the yields and bond prices published in) financial newspapers. These parameters are therefore readily accessible.

On the other hand, the volatility is more difficult to determine. Being directly related to the dynamics of the underlying, a natural way to estimate σ is by using a time series analysis. Suppose that we have $(N+1)$ observations of the stock price $\{S_k\}_{k=0,\dots,N}$, which are separated by a constant time interval Δt. Let

$$\bar{m} = \frac{1}{N \Delta t} \sum_{k=0}^{N-1} \ln(S_{k+1}/S_k) \tag{5.71}$$

be the average return in the studied time interval $N \Delta t$. Then a good estimate for the volatility should be provided by the variance of the time series

$$\sigma_{\mathrm{his}}^2 = \frac{1}{(N-1)\Delta t} \sum_{k=0}^{N-1} \big[\ln(S_{k+1}/S_k) - \bar{m}\big]^2. \tag{5.72}$$

This quantity is often called *historic volatility*. However, the historic volatility is usually not used as an estimate for σ in practice because it changes with time. These fluctuations can be very strong and persist over long periods, so σ_{his} depends on the portion of the time series which is analyzed. A pertinent value of the volatility for option pricing is therefore hard to determine in this way.

On the other hand, many options are regularly traded in the market and represent liquid assets. The liquidity reduces the bid–ask spreads sufficiently to allow an accurate price to be defined by the interplay of supply and demand. This suggests a clever alternative by which to obtain σ. Even if the volatility defies all attempts to be extracted by time-series analysis, the market seems to 'know' it. We can extract the information contained in the market prices of the options by inverting either of the Black-Scholes formulas (5.48) and (5.49). For instance,

$$\mathcal{C}(S, t; K, T, \sigma_{imp}) \stackrel{!}{=} \text{current market price of the call.} \tag{5.73}$$

The inversion cannot be done analytically, but numerically. The resulting value of σ is called the *implied volatility*.

Since σ_{imp} is derived from the price for an option, it need not coincide with the historic volatility. However, if the Black-Scholes theory were rigorously correct, we should find $\sigma_{imp} = \sigma_{his}$. Therefore, the implied volatility can be used to quantify deviations between the market and Black-Scholes prices. Several analyses of this kind were done in the late 1970s. A very detailed empirical test for call options was published by MacBeth and Merville in 1979 [120]. They found the following results:

- The implied volatility depends on both the strike price and the time to expiry, i.e., $\sigma_{imp} = \sigma_{imp}(K, T - t)$.
- The implied volatility is fairly constant for call options which have at least 90 days left until expiry and are at-the-money (i.e., current asset price $S(t) =$ strike price K). MacBeth and Merville therefore assumed that the Black-Scholes model correctly prices these options and that the corresponding σ_{imp} is the 'true' volatility which should be used to calculate option prices.
- Using this assumption, they predicted Black-Scholes (BS) prices for in-the-money and out-of-the-money options. These prices deviate from the observed market prices in the following manner:

$$\text{BS price} \begin{cases} < \text{market price}, & S(t) > K \text{ (in-the-money)}, \\ > \text{market price}, & S(t) < K \text{ (out-of-the-money)}. \end{cases} \tag{5.74}$$

- The deviations between the Black-Scholes price and the market price are the more pronounced,
  - the further the option is in-the-money or out-of-the-money,
  - the smaller the remaining time to expiry is.

This early empirical study reveals that even the implied volatility is not constant across different strike prices and maturities—in contrast to the assumption of the

5.3 Models Beyond Geometric Brownian Motion

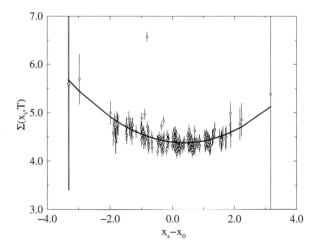

Fig. 5.4 Implied volatility $\sigma_{\text{imp}}(K, T)$ (denoted by Σ) as a function of $K - S(0)$ (denoted by $x_s - x_0$ with $x_0 = S(0)$ and $x_s = K$). Data points (*circles*) represent the prices quoted on April 26, 1995, of all options with a maturity of $T = 1$ month. A fit to the average curvature by (5.120) (*solid line*) yields an 'implied kurtosis' for this T. Reproduced with permission from [18]

Black-Scholes theory. In particular, σ_{imp} depends on the difference between the current and the strike price, $S(t) - K$. Since the volatility monotonously increases with the Black-Scholes price for the call, (5.74) suggests that $\sigma_{\text{imp}}(S(t) - K)$ should decrease as a function of K, if K increases towards $S(t)$, then level off for $K \approx S(t)$, and eventually decrease again, as $K > S(t)$.

This expectation is essentially confirmed by more recent studies. An example is shown in Fig. 5.4 [18]. Note how the implied volatility of options deeply in-the-money is significantly larger than for those at-the-money, but levels off or even increases for options out-of-the-money. Since this dependence on K resembles a 'smile', this curve is commonly called the *volatility smile*, although different shapes, like a 'skew' [21] or also a 'frown', may be observed depending on the market conditions.

The implied volatility is an efficient means by which to reconcile the imperfections of the Black-Scholes model with the reality of the market. When recorded as a function of the strike price and the remaining time to expiry for options on the same underlying, a so-called *volatility surface* is deduced. Volatility surfaces obtained from liquid options can then be used to price illiquid or OTC options on the same underlying. Therefore, the Black-Scholes model serves as an important guide in practice. However, it is obviously not perfect. Possible corrections are discussed in the financial literature [82]. More recently, physicists have also tried to find the flaws in the model and to suggest improvements. The following sections discuss some of these attempts.

5.3 Models Beyond Geometric Brownian Motion

An important factor for the pricing of derivatives is an accurate description of the time evolution of the underlying assets. The Black-Scholes theory assumes Brownian motion for the returns of the underlying. This hypothesis seems very plausible on a liquid and efficient market. However, it remains a postulate as long as it is not verified by empirical tests.

A significant test requires high-frequency data over a long period of time so that also the wings of the price distributions can be clearly resolved. Such data have become available only as of the 1990s and can be analyzed due to the rapid development of computing facilities. These statistical analyses challenge the validity of geometric Brownian motion for the movement of asset prices. The next section constitutes a selection of these studies.

The purpose of the statistical approach is to reveal empirical laws that characterize the financial market. Ultimately, it is hoped that the observations can be combined to explain the properties of the market by a model for the 'microscopic interactions' between the market constituents. Several models of this type have been proposed as speculations in recent years. Some borrow well-established concepts from statistical physics and reinterpret them in financial terms. The last sections will describe two of these physically inspired models.

5.3.1 Statistical Analysis of Stock Prices

The modeling of the dynamical evolution of asset prices as geometric Brownian motion rests on several assumptions:

- Trading is continuous, i.e., the time interval between successive quotations, Δt, tends to zero.
- The time evolution of the prices is a stochastic process whose fundamental random variables are the infinitesimal variations of the return

$$\Delta \ln S(t) = \ln S(t + \Delta t) - \ln S(t) \xrightarrow{\Delta t \to 0} d \ln S(t). \tag{5.75}$$

These variables are independent and identically distributed. They have a finite mean and variance.

Since Δt tends to zero, the number of quotations, N, in a finite time interval T diverges ($N = T/\Delta t \to \infty$), so that the difference, $\ln S(T) - \ln S(0)$, is normally distributed due to the central limit theorem.

The statistical analysis of financial time series aims to test these assumptions and eventually to improve them. Of course, it is immediately evident that continuous trading is an idealization. Even for the most active markets, the interval Δt between successive quotations lasts at least several seconds and is thus finite. This implies that the time evolution of the assets should rather be modeled as a stochastic process in discrete time, $(S_n)_{n=0,...,N}$, with $S_n = S(t_n)$ being the spot price at time $t_n = n \Delta t$.

The statistical properties of this process have also been studied extensively in the physics community. Important issues of these investigations included the search for correlations in the time series and for an adequate description of the distribution of price increments.

5.3 Models Beyond Geometric Brownian Motion

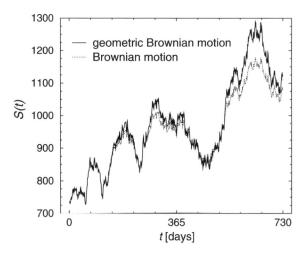

Fig. 5.5 Comparison of geometric Brownian motion (*thick solid line*) with Brownian motion (*thin dotted line*) for a daily updated spot price, $S(t)$ (measured in points). Time runs from 1 to 730 days (= two calendar years ≈ three trading years). *Both curves* are results from a Monte Carlo simulation with: $\mu = 10\,\%/\text{year}$, $\sigma = 20\,\%/\sqrt{\text{year}}$, $S(0) = 737.01$ points and the same set of Gaussian random numbers. The initial spot price was taken as the closing value of the S&P500 on January 2, 1997. Furthermore, the 'seed' (= 47911) of the Gaussian random-number generator (Box-Muller method [174]) was chosen so that the simulation approximately mimics the real evolution of the S&P500 as of 1997

Prices Versus Returns

Contrary to usual practice in finance, several of the physical studies did not choose (5.75) as the random variable, but rather the price increment

$$\Delta S_n \equiv \Delta S(t_n) = S(t_n + \Delta t) - S(t_n) \quad (t_n = n\Delta t), \tag{5.76}$$

where $n = 0, \ldots, N - 1$. Since the logarithm is bounded by (see [1], Eq. (4.1.33), p. 68)

$$\frac{\Delta S_n/S_n}{1 + \Delta S_n/S_n} < \ln\frac{S(t_n + \Delta t)}{S(t_n)} < \frac{\Delta S_n}{S_n}, \tag{5.77}$$

this choice is justified if $\Delta S_n \ll S_n$ and S_n varies only slowly with time, so that

$$\ln\frac{S(t_n + \Delta t)}{S(t_n)} \approx \frac{\Delta S_n}{S_n} \approx \frac{\Delta S_n}{S_0}. \tag{5.78}$$

These conditions should hold for short times and 'normal' periods during which prices do not change vehemently (no crashes).

Figure 5.5 supports this expectation. It compares the results of a Monte Carlo simulation for geometric Brownian motion

$$S(t_n + \Delta t) = S(t_n) + S(t_n)(\mu\Delta t + \sigma X_n\sqrt{\Delta t}), \tag{5.79}$$

with those of Brownian motion

$$S(t_n + \Delta t) = S(t_n) + S(0)(\mu \Delta t + \sigma X_n \sqrt{\Delta t}), \tag{5.80}$$

using the same drift (10 % per year) and volatility (20 % per year) and an identical set of normally distributed random numbers X_n. Over the period of two (calendar) years, the curves almost agree with one another, especially at the beginning of the time series. Therefore, it is often reasonable to use the 'additive' model

$$S(T) = S(0) + \sum_{n=0}^{N-1} \Delta S_n \quad (T = N \Delta t), \tag{5.81}$$

where the increments of the prices add up, instead of the 'multiplicative' model

$$\ln S(T) - \ln S(0) = \sum_{n=0}^{N-1} \Delta \ln S(t_n) \quad (T = N \Delta t), \tag{5.82}$$

where the ratios of the successive prices, S_{n+1}/S_n, multiply.

Distribution of Asset Prices

In 1963, Benoit Mandelbrot pointed out that the real distribution of asset prices differs from the Gaussian model in a very characteristic way [123, 125]: Large price changes occur much more frequently than predicted by the Gaussian law. The real distribution has 'fat tails'. To model this leptokurtic[4] character, he suggested a stable distribution with a characteristic exponent, $\alpha < 2$.

Mandelbrot's suggestion seems to suffer from a serious drawback. The variance of Lévy distributions with $\alpha < 2$ diverges, whereas financial time series have a well-defined variance. Thus, a Lévy flight cannot be a viable description for the *whole* distribution of price changes. However, a truncated Lévy flight could be. As discussed in Sect. 4.4, the distribution of this process preserves the typical self-similar Lévy scaling over an extended range before the truncation becomes effective and imposes a finite variance. Since Mandelbrot only had approximately 2000 data points at his disposal, it is possible that the wings of the distribution could not be sampled sufficiently to resolve the onset of the truncation.

This point of view was advocated in a seminal paper by Mantegna and Stanley [129]. Based on all records of the S&P500 between 1984 and 1989, they determined the distribution $p_{\Delta t}(\ell)$ of the indexes variations, $\ell = \Delta S = S(t + \Delta t) - S(t)$, for $\Delta t = 1, \ldots, 10^3$ min. These times are so short that one can expect the distribution to be symmetric. This expectation is, in fact, borne out. Figure 5.6 reproduces the

---

[4]Possible etymological origin: *leptos (Greek)* = slender, slight; *kyrtos (Greek)* = curved, bulging. Quite generally, a distribution is called 'leptokurtic' if its tails are broader than those of a Gaussian.

5.3 Models Beyond Geometric Brownian Motion

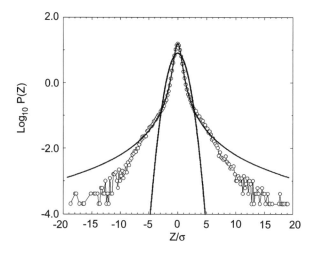

Fig. 5.6 Probability distribution $p_{\Delta t}(\ell)$ of the price variations $\ell\ [=S(t+\Delta t)-S(t)]$ for the S&P500, determined from all records between 1984 and 1989 (about 1.5 million records). $p_{\Delta t}(\ell)$ (denoted by $P(Z)$) is plotted versus ℓ/σ for $\Delta t = 1$ min, where $\sigma\ (= 0.0508)$ is the standard deviation calculated from the data points. The Gaussian distribution (*thick solid line*) corresponding to this value of σ is compared to the best fit with a Lévy distribution (*thin solid line*). The Lévy distribution (with $\alpha = 1.4$ and $c = 3.75 \times 10^{-3}$, see (5.83)) gives a much better representation of the data for $\ell/\sigma \leq 6$. If $\ell/\sigma \geq 6$, the distribution of the index decays approximately exponentially. Reproduced with permission from [129]

result for $\Delta t = 1$ min from [129]. It shows $p_{\Delta t}(\ell)$ as a function of ℓ/σ, where σ is the standard deviation of the S&P500. The central part of $p_{\Delta t}(\ell)$ is well represented by a symmetric Lévy distribution,

$$p_{\Delta t}(\ell) = L_\alpha(\ell, \Delta t) = \frac{1}{2\pi} \int_{-\infty}^{+\infty} dk \exp\left(-ik\ell - c\Delta t |k|^\alpha\right), \qquad (5.83)$$

with $\alpha \simeq 1.4$. It extends over about 3 orders of magnitude in probability and up to 6σ in ℓ. For larger variations, the distribution decays almost exponentially. The exponential truncation is visible for $\Delta t < 10$ min, whereas the data set for larger time increments is too limited to exhibit pronounced deviations from the Lévy behavior [130]. These results support Mandelbrot's original suggestion.

A further important property of the price increments is their correlation in time. Geometric Brownian motion assumes ΔS to be uncorrelated, even if Δt tends to zero. A critical test of this hypothesis consists in monitoring the price-price autocorrelation function

$$\Phi_{\Delta S}(t) = \frac{\langle \Delta S(t_n + t) \Delta S(t_n)\rangle - \langle \Delta S(t_n + t)\rangle \langle \Delta S(t_n)\rangle}{\langle \Delta S(t_n)^2\rangle - \langle \Delta S(t_n)\rangle^2}, \qquad (5.84)$$

which decays from 1 (for $t = 0$) to 0 (for $t \to \infty$). This function should be zero for all $t > 0$ if the hypothesis were true. In practice, however, one finds substantial correlations at short times. The form of $\Phi_{\Delta S}(t)$ is often well approximated by an exponential, $\Phi_{\Delta S}(t) \approx \exp(-t/\tau_{\Delta S})$, with a typical relaxation time $\tau_{\Delta S}$ of the order of a few minutes [65]. This implies that a complete decorrelation of successive price variations, i.e., $\Phi_{\Delta S}(t) \approx 0$, is only realized at about $t \geq 15$ min. Thus, in order to work with independent price increments, one has to choose $\Delta t \geq \Delta t^* = 15$ min [15, 65].

Prompted by these results, it is tempting to suggest the following model [18, 30]: As soon as the time increment exceeds Δt^*, the price variations ΔS can be considered as independent random variables which are identically distributed according to an exponentially truncated, symmetric Lévy flight (note that Δt^* here corresponds to the choice $N = 1$ in (4.98)),

$$p_{\Delta t^*}(\ell) = L_\alpha^t(\ell, \Delta t^*)$$
$$= \int_{-\infty}^{+\infty} \frac{dk}{2\pi} e^{-ik\ell} \exp\left\{\frac{c}{\cos \pi \alpha/2}\left[\lambda^\alpha - (k^2 + \lambda^2)^{\alpha/2} \cos\left(\alpha \arctan \frac{|k|}{\lambda}\right)\right]\right\}, \tag{5.85}$$

where λ is the exponential cutoff parameter and c is given by (see (4.17))

$$c = \frac{2\pi C \cos(\pi\alpha/2)}{\alpha \Gamma(\alpha) \sin(\pi\alpha)} \quad (C = \text{const.}). \tag{5.86}$$

A pivotal test of this model was performed by Bouchaud and Potters [18]. If $\Delta S_n^* = S(t_n + \Delta t^*) - S(t_n)$ were independent and identically distributed random variables, the probability distribution of the sum

$$x = S(T) - S(0) = \sum_{n=0}^{N^*-1} \Delta S_n^* \quad (T - N^*\Delta t^*)$$

would necessarily be given by the convolution of $p_{\Delta t^*}(\ell)$

$$p_T(x) = p_{\Delta t^*}(\ell) * \cdots * p_{\Delta t^*}(\ell) \quad (N^* \text{ factors}). \tag{5.87}$$

Figure 5.7 shows a comparison of the cumulative probability distribution corresponding to (5.87), i.e., (see Sect. 2.1.4)

$$\text{Prob}[x \geq \Delta S] = \int_{\Delta S}^{\infty} dx\, p_T(x), \tag{5.88}$$

with the empirically determined distribution of the S&P500 for $T = 1$ h, ..., 5 days. Several points should be noted:

- The elementary distribution $p_{\Delta t^*}(\ell)$ is well described by a truncated, symmetric Lévy flight with $\alpha = 3/2$. In the actual fit, the characteristic exponent was kept fixed at this value and only the remaining parameters, c and λ, were optimized. The choice $\alpha = 3/2$ is compatible with the results of [65, 129].

5.3 Models Beyond Geometric Brownian Motion

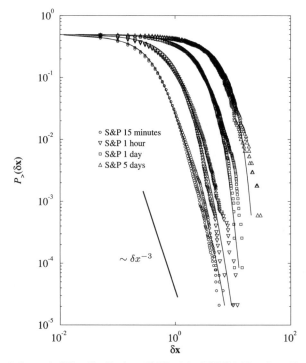

Fig. 5.7 Cumulative probability distributions (5.88) of the S&P500 (data from November 1991 to February 1995) versus the price variations ΔS (denoted by δx in the figure) for $\Delta t^* = 15$ min (○) and for $T = 1$ h (▽), 1 day (□) and 5 days (△). Two data points are shown for each time; they correspond to $\text{Prob}[x \geq \Delta S]$ for $\Delta S > 0$ and to $\text{Prob}[x \leq -\Delta S] = F(-\Delta S)$ for $\Delta S < 0$ (see (2.30)). The empirical results are compared to the hypothesis of a convolution (5.87) of a truncated Lévy flight (*solid lines*). The parameters of the truncated Lévy flight (5.85) are optimized for $\Delta t^* = 15$ min after fixing $\alpha = 3/2$: $\lambda^{-1} = 2.21$ and $c = 0.17$. The splaying-out of the probabilities for positive and negative ΔS could partly result from statistical inaccuracies of rare events, but could also indicate an asymmetry of the distribution perhaps due to drift in the index. The power law $\text{Prob}[x \geq \Delta S] \sim (\Delta S)^{-3}$ (see (5.99)) is indicated (*thick solid line*). This power law is not contained in the original figure which we reproduce with permission from [18]

- The convolution (5.87) is a viable approximation for the probability distributions at larger times $T \gg \Delta t^*$. As T increases, the shape of $p_T(x)$ progressively deforms towards a Gaussian. Using (4.105), the cross-over time to the Gaussian behavior can be estimated: $T_\times = \hat{\kappa}_4 \Delta t^* \approx 195$ min (≈ 0.4 (trading) day),[5] where $\hat{\kappa}_4 \approx 13$ is the kurtosis of $p_{\Delta t^*}(\ell)$.

However, the hypothesis of a simple convolution is not perfect. On the one hand, there are systematic deviations between (5.87) and the empirical distribution. As T increases, the financial data tend to cross the theoretical curve, lying slightly below it at the beginning of the curvature and above it farther down in the tails. On the

---

[5] A typical trading day lasts between 6 and 8 hours and a month has a about 21 trading days.

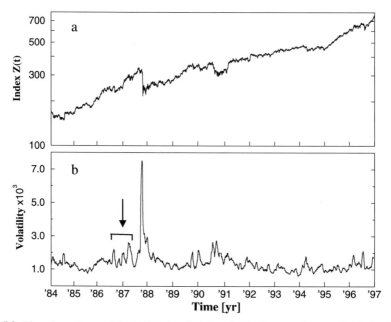

Fig. 5.8 Time dependence of the S&P500 and of its volatility between January 3, 1984 and December 31, 1996. (**a**) Semi-log plot of the S&P500 (denoted by $Z(t)$) versus time. Successive records are separated by an interval, $\Delta t = 1$ min. The overall increase can be fitted by a straight line, yielding a drift of $\mu \approx 15$ %/year [119]. Note the large drop (= crash) on October 19, 1987. (**b**) Volatility of the S&P500 versus time. The volatility, $v_T(t)$, is defined by (5.89) with a time window, $T \simeq 1$ month (8190 min = 21 days × 390 min), a sampling interval, $\Delta t = 30$ min, and a total number of trading days, $N_{\text{td}} = 3309$ in the 13-year period studied. The sampling interval is so large that $\Phi_{\Delta S}(\Delta t) \simeq 0$, i.e., all correlations between price increments have completely vanished. Note the large burst around October 19, 1987, which decays only gradually over a period of several months and which is preceded by precursors, indicated by an arrow. Reproduced with permission from [27]

other hand, the time of the convergence to Gaussian behavior is strongly retarded compared to the estimate of about half a day. It is rather of the order of several days to weeks, or even longer, depending on the market considered (see [65] for a detailed discussion of this point).

Stochastic Behavior of the Volatility

Contrary to our assumption, the volatility is not constant in reality. It fluctuates with time. This is nicely demonstrated in Fig. 5.8, which contrasts the time evolution of the S&P500 to that of its volatility [27]. In this study, the 'volatility', $v_T(t_v)$, at a given time t_v is defined as the average of the absolute value of the normalized returns over some time window of length T ($= N \Delta t$), i.e.,

$$v_T(t_v) := \frac{1}{N} \sum_{n=v}^{N-1+v} |g(t_n)| \quad (t_v = v\Delta t, t_n = n\Delta t), \tag{5.89}$$

5.3 Models Beyond Geometric Brownian Motion

with

$$g(t_n) = \frac{\ln[S(t_n + \Delta t)/S(t_n)]}{\frac{1}{N_{td}} \sum_{i=1}^{N_{td}} |\ln[S_i(t_n + \Delta t)/S_i(t_n)]|}, \tag{5.90}$$

where N_{td} denotes the total number of trading days in the entire time series and $S_i(t_n)$ is the spot price of the index at time t_n on day i.

The normalization by the denominator in (5.90) is supposed to remove intra-day fluctuations of the volatility. Over the day, the market activity is large at the beginning and at the end, presumably due to the presence of many 'information traders' during the opening hours and of many 'liquidity traders' near the closing hours, but exhibits a broad minimum around noon [119]. Since the volatility measures the magnitude of the market activity, this intra-day pattern should be eliminated to obtain an unbiased average over all t_n in time T.

Figure 5.8 clearly shows that the volatility depends on time and that periods of high volatility tend to be correlated. This is particularly pronounced for the stock market crash of October 19, 1987. The sharp drop of the S&P500 leads to a peerlessly high value of the volatility, which decreases to the normal level only after several months. This correlation is called *volatility clustering*: large changes of the asset price tend to follow one another, but they do not necessarily occur in the same direction.

Therefore, time-displaced correlation functions, which are sensitive to the magnitude of price changes, such as

$$\Phi_{|g|}(t) = \frac{\langle |g(t_n + t)||g(t_n)|\rangle - \langle |g(t_n + t)|\rangle\langle |g(t_n)|\rangle}{\langle |g(t_n)|^2\rangle - \langle |g(t_n)|\rangle^2}, \tag{5.91}$$

should decay much more slowly than the price-price autocorrelation function (5.84), which depends on both sign and magnitude. This expectation is borne out. Figure 5.9 shows that the disparity in the relaxation rate between $\Phi_{|g|}(t)$ and $\Phi_g(t)$ (defined analogously to (5.91) for g) is several orders of magnitude. $\Phi_{|g|}(t)$ is represented well by a slowly decaying power law, i.e.,

$$\Phi_{|g|}(t) \sim \frac{1}{t^\gamma} \quad (\gamma \approx 0.3, t \gg 1 \text{ min}), \tag{5.92}$$

contrary to the fast exponential decrease of the price-price correlation function.

A Simple Model

The previous discussion reveals a clear separation of the decorrelation times between the direction and the amplitude of the price variations. This difference suggests the following simple model [15, 18, 19]: A price variation can always be split into two factors—magnitude and sign. They measure the amplitude and the direction of the fluctuations, respectively. Therefore, they are random variables which evolve on well-separated time scales, the sign being the fast degree of freedom and the magnitude, i.e., the 'volatility', the slow degree of freedom.

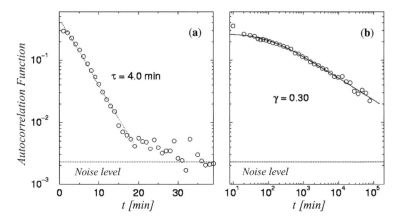

Fig. 5.9 Time dependence of the correlation functions (**a**) for $g(t)$ and (**b**) for $|g(t)|$. The correlation function of $g(t)$ decays almost exponentially, $\Phi_g(t) \approx \exp(-t/\tau_g)$, with a small time constant $\tau_g \approx 4$ min (*dashed line* in (**a**)). However, the correlation function of $|g|$ exhibits a very slow power-law relaxation, i.e., $\Phi_{|g|}(t) \approx (1+t^\gamma)^{-1}$, with $\gamma = 0.30 \pm 0.08$ (*solid line* in (**b**)), which extends over several days to months. A trading day corresponds to 390 min and there are about 21 trading days per month. All data refer to the S&P500 recorded between January 3, 1984, and December 31, 1996. Reproduced with permission from [119]

Let us choose Δt^* as the basic time lag between successive quotations, i.e., the sign is uncorrelated, and consider an intermediate time interval, Δt, which is much larger than Δt^*, but also much smaller than the time scale on which the magnitude varies appreciably. This implies

$$\Delta t = n^* \Delta t^* \ (n^* \gg 1) \quad \text{and} \quad v_\nu = \frac{1}{n^*} \sum_{n=\nu}^{n^*-1+\nu} |\Delta S_n^*| \simeq |\Delta S_n^*|. \quad (5.93)$$

The price variation between time t_ν and $t_\nu + \Delta t$ can then be written as

$$S(t_\nu + \Delta t) - S(t_\nu) = \sum_{n=\nu}^{n^*-1+\nu} \Delta S_n^* \quad (t_\nu = \nu \Delta t^*)$$

$$= \sum_{n=\nu}^{n^*-1+\nu} |\Delta S_n^*| \left(\frac{\Delta S_n^*}{|\Delta S_n^*|} \right)$$

$$\simeq v_\nu \sum_{n=\nu}^{n^*-1+\nu} \left(\frac{\Delta S_n^*}{|\Delta S_n^*|} \right) =: v_\nu \varepsilon_\nu. \quad (5.94)$$

When calculating the distribution of $S(t_\nu + \Delta t) - S(t_\nu)$ from a time series, t_ν is increased in steps of Δt^* and the difference is determined assuming that the series is stationary. However, this assumption neglects the variation of v_ν in the series.

5.3 Models Beyond Geometric Brownian Motion

Therefore, the distribution of $S(t_\nu + \Delta t) - S(t_\nu)$ does not depend on Δt only, but also on t_ν.

A simple way to account for this dependence could be as follows: As long as the drift is negligible, the sign ε_ν is a sum of the independent random variables $\Delta S_n^* / |\Delta S_n^*|$, which can be either $+1$ or -1 with equal probability. Thus, ε_ν can be interpreted as the distance between the start and the end of a one-dimensional, symmetric random walk. According to the discussion in Sect. 1.2, the corresponding distribution is Gaussian in the limit $n^* \gg 1$. Now, we make the following assumption: Since the sign is the fast variable, shifting t_ν throughout the time series can sample the full distribution of ε_ν before a significant change of v_ν occurs. This suggests that the functional form of the sign distribution should remain stationary along the series, whereas its parameter v_ν evolves with time, i.e.,

$$\text{Prob}[\varepsilon \leq \varepsilon_\nu \leq \varepsilon + d\varepsilon] = p_\nu(\varepsilon) d\varepsilon \stackrel{!}{=} \frac{1}{v_\nu} p_\pm\left(\frac{\ell}{v_\nu}\right) d\ell$$

$$= \frac{1}{\sqrt{2\pi n^* v_\nu^2}} \exp\left(-\frac{\ell^2}{2n^* v_\nu^2}\right) d\ell. \quad (5.95)$$

Since the volatility is also a random variable characterized by

$$\text{Prob}[v \leq v_\nu \leq v + dv] = p_{|\Delta S|}(v) dv, \quad (5.96)$$

the probability distribution of the price increments ℓ can be written as

$$p_{\Delta t}(\ell) = \int_0^\infty dv\, p_{|\Delta S|}(v) \frac{1}{v} p_\pm\left(\frac{\ell}{v}\right). \quad (5.97)$$

In order to calculate $p_{\Delta t}(\ell)$, we have to know $p_{|\Delta S|}(v)$. In [119], it was found that the central part of $p_{|\Delta S|}(v)$ can be fitted by a log–normal distribution, whereas the tail for v larger than the maximum position, $v_{\max}$, of $p_{|\Delta S|}(v)$ exhibits a power-law decay,

$$p_{|\Delta S|}(v) \sim \frac{1}{v^4} \quad \text{for } v > v_{\max}. \quad (5.98)$$

Inserting (5.95) and (5.98) into (5.97), we obtain

$$p_{\Delta t}(\ell) \sim \int_{v_{\max}}^\infty dv\, \frac{1}{v^5} \exp\left(-\frac{\ell^2}{2n^* v^2}\right)$$

$$= \sqrt{\frac{2(n^*)^3}{\pi}} \frac{1}{\ell^4} \left[1 - \left(1 + \frac{\ell^2}{2n^* v_{\max}^2}\right) \exp\left(-\frac{\ell^2}{2n^* v_{\max}^2}\right)\right]$$

$$\sim \frac{1}{\ell^4}, \quad (5.99)$$

if $\ell^2/2n^* v_{\max}^2 > 1$. This is compatible with the empirical results of the S&P500 for large price variations (see Fig. 5.6 and [65]).

Note that (5.99) is not necessarily in contradiction to the presented analysis in terms of truncated Lévy flights. Equation (5.99) is applicable in an interval of large price variations which belongs to the crossover regime from the Lévy-dominated to the truncation-dominated part of the price distribution. Therefore, the power law interpolates between a slow Lévy-like decay and a fast exponential decay. It is then not unreasonable to find an effective exponent much larger than the theoretically allowed values of Lévy distributions, since

$$p_{\Delta t}(\ell) \sim \exp(-\lambda \ell) = \lim_{\alpha \to \infty} \left(1 + \frac{\lambda}{\alpha}\ell\right)^{-\alpha}. \quad (5.100)$$

This might also explain why characteristic exponents larger than 3/2, i.e., $1.6 \leq \alpha \leq 1.8$, are often found when fitting price distributions by a pure Lévy law without truncation only [18, 65, 123].

In the last years the exact form of the tails in the price distribution has been much argued about. Especially the group of H.E. Stanley performed several studies on different financial markets around the world utilizing large databases [65, 169, 199]. Different from Fig. 5.6 these studies looked at the distribution of returns, $\ln[S(t_n + \Delta t)/S(t_n)]$ ($\approx \Delta S_n/S_n$, see "Prices Versus Returns" on p. 193), and came to the conclusion that the tail of the cumulative distribution of the return is best described by a power-law decay with exponent ≈ 3 (seemingly universal for mature markets) outside the Lévy-stable regime. On time scales where the variation of the spot price is negligible ($S(t_n) \approx S(0)$), this predicts $p_{\Delta t}(\ell) \sim \ell^{-4}$, in agreement with (5.99). From that perspective, the truncated Lévy distribution just yields some approximation to this power law generally found for mature markets.

The simple model proposed above also allows us to rationalize why the tails of the empirical price distributions $p_T(x)$ are broader than the result of the convoluted truncated Lévy flights, although Δt^* was chosen as the basic time lag (see Fig. 5.7) [18]. Due to the stochastic character of the volatility, the kurtosis $\hat{\kappa}_{4,T}$ of $p_T(x)$ decreases more slowly with increasing T than expected for independent variables. The kurtosis is defined as (remember $T = N^* \Delta t^*$)

$$\hat{\kappa}_{4,N^*} = \frac{\overline{\sum_{i,j,k,l=1}^{N^*} \langle \Delta S_i^* \Delta S_j^* \Delta S_k^* \Delta S_l^* \rangle}}{\overline{\langle (\sum_{i=1}^{N^*} \Delta S_i^*)^2 \rangle}^2} - 3, \quad (5.101)$$

where $\langle \cdot \rangle$ and $\overline{}$ represent the average over the sign and the volatility, respectively. Using the fact that the signs are uncorrelated on the time scale Δt^*, i.e.,

$$\langle (\Delta S_i^*)^n \Delta S_j^* \rangle = \langle (\Delta S_i^*)^n \rangle \langle \Delta S_j^* \rangle = 0,$$

5.3 Models Beyond Geometric Brownian Motion

we can write

$$\overline{\sum_{ijkl} \langle \Delta S_i^* \Delta S_j^* \Delta S_k^* \Delta S_l^* \rangle}$$

$$= \sum_{i=1}^{N^*} \overline{\langle (\Delta S_i^*)^4 \rangle} + 3 \sum_{i \neq j} \overline{\langle (\Delta S_i^*)^2 \rangle \langle (\Delta S_j^*)^2 \rangle}$$

$$= (3 + \hat{\kappa}_4) \sum_{i=1}^{N^*} \overline{\langle (\Delta S_i^*)^2 \rangle}^2 + 3 \sum_{i \neq j} \overline{\langle (\Delta S_i^*)^2 \rangle \langle (\Delta S_j^*)^2 \rangle},$$

where $\hat{\kappa}_4$ is the kurtosis of the elementary variable ΔS_i^* and

$$\overline{\left(\left(\sum_i \Delta S_i^* \right)^2 \right)^2} = \left(N^* \overline{\langle (\Delta S^*)^2 \rangle} \right)^2 = \sum_{i,j=1}^{N^*} \overline{\langle (\Delta S_i^*)^2 \rangle \langle (\Delta S_j^*)^2 \rangle},$$

so that (5.101) becomes

$$\hat{\kappa}_{4,N^*} = \frac{(3+\hat{\kappa}_4) \sum_{i=1}^{N^*} \overline{\langle (\Delta S_i^*)^2 \rangle}^2 + 3 \sum_{i \neq j} \overline{\langle (\Delta S_i^*)^2 \rangle \langle (\Delta S_j^*)^2 \rangle}}{\sum_{i,j=1}^{N^*} \overline{\langle (\Delta S_i^*)^2 \rangle \langle (\Delta S_j^*)^2 \rangle}} - 3$$

$$= \frac{1}{N^*} \left[\hat{\kappa}_4 + \frac{3}{N^*} \left(\frac{\overline{\langle (\Delta S_i^*)^2 \rangle^2}}{\overline{\langle (\Delta S_i^*)^2 \rangle}^2} - 1 \right) \sum_{i \neq j} \Phi_{(\Delta S^*)^2}(|i-j|) \right], \quad (5.102)$$

with

$$\Phi_{(\Delta S^*)^2}(|i-j|) = \frac{\overline{\langle (\Delta S_i^*)^2 \rangle \langle (\Delta S_j^*)^2 \rangle} - \overline{\langle (\Delta S_i^*)^2 \rangle}^2}{\overline{\langle (\Delta S_i^*)^2 \rangle^2} - \overline{\langle (\Delta S_i^*)^2 \rangle}^2}. \quad (5.103)$$

To obtain this result, we have used the stationarity of the process so that the averages in (5.103) do not explicitly depend on i, but only on the time difference $|i-j|$. Since

$$\sum_{i \neq j}^{N} f(|i-j|) = 2 \sum_{i=1}^{N-1} \sum_{j=i+1}^{N} f(|i-j|)$$

$$= 2 \sum_{i=1}^{N-1} \sum_{k=1}^{N-i} f(k)$$

$$= 2 \sum_{k=1}^{N-1} (N-k) f(k),$$

and

$$\overline{\langle(\Delta S)^2\rangle^2} = \overline{\langle(v\varepsilon)^2\rangle^2} = \overline{v^4} = \overline{\langle(\Delta S)^4\rangle},$$

we, finally, find

$$\hat{\kappa}_{4,N^*} = \frac{1}{N^*}\left[\hat{\kappa}_4 + 6(\hat{\kappa}_4 + 2)\sum_{k=1}^{N^*-1}\left(1 - \frac{k}{N^*}\right)\Phi_{(\Delta S^*)^2}(k)\right]. \quad (5.104)$$

This equation shows that the typical result for independent variables, $\hat{\kappa}_{4,N^*} = \hat{\kappa}_4/N^*$, is only obtained if $\Phi_{(\Delta S^*)^2} = 0$. However, if $\Phi_{(\Delta S^*)^2}$ decreases slowly according to a power law (see (5.92)),

$$\Phi_{(\Delta S^*)^2}(k) \sim \frac{1}{k^\gamma},$$

the kurtosis is considerably enhanced, leading to a broadening of the tails of the probability distribution. Inserting the last equation into (5.104) and using

$$\sum_{k=1}^{n} k^a = \frac{n^{1+a}}{1+a} + O(n^a),$$

the kurtosis depends on N^* as follows:

$$\hat{\kappa}_{4,N^*} \stackrel{N^*\gg 1}{=} \frac{\hat{\kappa}_4}{N^*} + \frac{6(\hat{\kappa}_4 + 2)}{(1-\gamma)(2-\gamma)}\frac{1}{(N^*)^\gamma} \sim \frac{1}{(N^*)^\gamma}. \quad (5.105)$$

Empirically, it is possible to fit $\hat{\kappa}_{4,N^*}$ by a power law with an exponent $0.3 \le \gamma \le 0.6$, depending on the market considered (see Fig. 5.12 as an example). This range of γ-values is compatible with the behavior of $\Phi_{(\Delta S^*)^2}$ [15, 18, 30].

Addendum

Many of the results described in this section were not discovered by physicists for the first time. The presence of broad tails in the price distribution, the time dependence of the volatility and its long-lived correlations, etc., are well documented in the financial literature (see [159], for instance). In order to account for the fluctuations of the volatility, several (Gaussian) models have been invented both in continuous ('stochastic volatility models') and in discrete time ('GARCH' = generalized autoregressive conditional heteroskedasticity)[6][82]. Some of the models in discrete time have been compared with the analyses presented above [65, 129].

---

[6]In financial jargon, the term *heteroskedasticity* is sometimes used as a synonym for the time dependence of volatility. Possible etymological origin: *hetero (Greek)* = different; *skédasis (Greek)* = dispersion.

5.3.2 The Volatility Smile: Precursor to Gaussian Behavior?

The discussion in the preceding section showed that the real distribution of price variations is, in general, non-Gaussian. To a first approximation, it is well modeled by the convolution of truncated Lévy distributions for times larger than Δt^*. The distinguishing feature of this underlying process is the presence of broad tails, which strongly delays the crossover to Gaussian behavior. Even for very liquid markets, Gaussian behavior begins to emerge only on the scale of several days, and a full convergence presumably takes many weeks to months.

On the other hand, the typical duration of a short option contract also lasts a few months. Therefore, the question arises of whether the deviations from the Black-Scholes theory, which one tries to compensate by introducing the implied volatility, could not be explained by finite-time corrections to the central limit theorem. This idea was proposed by Potters, Cont and Bouchaud [171], and we pursue their discussion in this section.

Drift and Volatility for Short Maturities

The shortest option contracts expire after about 100 days. This time to maturity is so short that the drift can usually be neglected, except for periods of vehement market activities, like crashes. If we assume typical values for the drift and volatility, such as $\mu = 10\,\%/\text{year}$ and $\sigma = 20\,\%/\sqrt{\text{year}}$, geometric Brownian motion yields a 'signal-to-noise' ratio (*Sharpe ratio*) of

$$\frac{\mu S T}{\sigma S \sqrt{T}} = \frac{\mu \sqrt{T}}{\sigma} \approx 0.3 \quad (T = 100 \text{ days}). \tag{5.106}$$

For short times, the price evolution is dominated by stochastic fluctuations. The contribution of the drift can be neglected. This is also shown by Fig. 5.10, which compares the results of a Monte Carlo simulation for geometric Brownian motion with and without drift. As long as the time is smaller than about 100 days, the difference between the two curves is exceedingly small. Therefore, it is reasonable to omit the drift term and the risk-less interest rate when calculating option prices for short maturities.

Black-Scholes Versus Bachelier

If the time lapse between signing and expiry of the option is short, the Black-Scholes pricing formulas can be simplified. Consider a call option, for instance. The call price is given by (see (5.55) and (5.57))

$$\mathcal{C}_{\text{BS}} = e^{-rT} \int_0^\infty dS_T \max(S_T - K, 0)$$
$$\times \left[\frac{1}{\sqrt{2\pi (\sigma S_T)^2 T}} \exp\left(-\frac{[\ln(S_T/S_0) - (r - \sigma^2/2)T]^2}{2\sigma^2 T} \right) \right]. \tag{5.107}$$

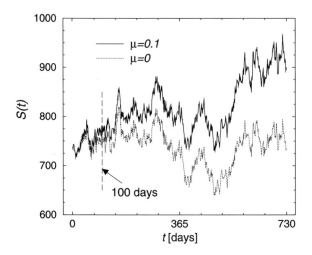

Fig. 5.10 Comparison of geometric Brownian motion with drift (*thick solid line*) and without drift (*thin dotted line*). The drift parameter is $\mu = 10$ %/year in the first and $\mu = 0$ in the second case. *Both curves* are results from a Monte Carlo simulation of (5.79), with $\sigma = 20\,\%/\sqrt{\text{year}}$, $S(0) = 737.01$ points and the same set of Gaussian random numbers. Time runs from 1 to 730 days (= two (calendar) years). Note that the curves almost coincide if time is smaller than about 100 (calendar) days (*dashed vertical line*)

For small T, we have $(r - \sigma^2/2)T \ll 1$ and $S_T \simeq S_0$, so that

$$\frac{[\ln(S_T/S_0) - (r - \sigma^2/2)T]^2}{2\sigma^2 T} \simeq \frac{(S_T - S_0)^2}{2(\sigma S_0)^2 T}, \tag{5.108}$$

which implies

$$\mathcal{C}_{\text{BS}} \simeq \mathcal{C}_{\text{B}} = \int_0^\infty dS_T \frac{\max(S_T - K, 0)}{\sqrt{2\pi (\sigma S_0)^2 T}} \exp\left(-\frac{(S_T - S_0)^2}{2(\sigma S_0)^2 T}\right) \tag{5.109}$$

$$= \sqrt{\frac{(\sigma S_0)^2 T}{2\pi}} \exp\left(-\frac{u_K^2}{2}\right) - \frac{1}{2}(K - S_0)\text{erfc}\left(\frac{u_K}{\sqrt{2}}\right), \tag{5.110}$$

in which

$$u_K = \frac{K - S_0}{\sqrt{(\sigma S_0)^2 T}}. \tag{5.111}$$

Equation (5.109) shows that the call price can be approximated as an average of the option's pay-off with respect to a Gaussian distribution. This is exactly the suggestion made by Bachelier in 1900 (therefore the index 'B'). Figure 5.11 shows a comparison of Bachelier's approximation with the Black-Scholes theory for short maturities. The difference is indeed fairly small, so we choose to work with (5.109) in the following.

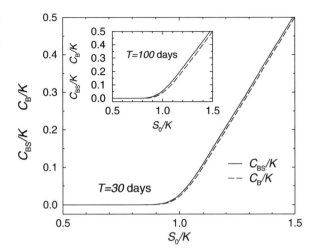

Fig. 5.11 Comparison of the call prices from the Black-Scholes theory (5.107) (*solid lines*) and from Bachelier's approximation (5.110) (*dashed lines*). In both cases the volatility is $\sigma = 20\,\%/\sqrt{\text{year}}$. Furthermore, $r = 10\,\%/\text{year}$ for the Black-Scholes formula. Two different expiration times are shown: $T = 30$ days (*main figure*) and $T = 100$ days (*inset*)

The Volatility Smile

In the limit $N \to \infty$, the central limit theorem holds for the probability distribution of the sum of independent and identically distributed random variables if the variance exists. For finite N, there are corrections to Gaussian behavior which depend on the tails of the distribution of the random variables (see Sect. 2.1.6). Since the price process is modeled well by a truncated Lévy flight, the pre-asymptotic effects are very pronounced. Can these finite-time effects give rise to the 'volatility smile'?

If we choose $\Delta t = \Delta t^*$ and set $\mu = 0$, the price increments are (to first order) independent of one another and symmetrically distributed. Then, the price process is a martingale, and the historic distribution coincides with the equivalent martingale distribution. So, we can write

$$
\begin{aligned}
C &\simeq \int_K^\infty dS_T (S_T - K) p(S_T, T|S_0, 0) \\
&= (S_T - K) \left[\int_\infty^{S_T} dS p(S, T|S_0, 0) \right]\bigg|_K^\infty - \int_K^\infty dS_T \left[\int_\infty^{S_T} dS p(S, T|S_0, 0) \right] \\
&= \int_K^\infty dS_T \left[\int_{S_T}^\infty dS p(S, T|S_0, 0) \right],
\end{aligned}
\qquad (5.112)
$$

where we used an integration by parts in the second line and chose ∞ as the low bound of the integral to obtain a finite result. Introducing the new variable

$$
u = \frac{S - S_0}{\sqrt{(\sigma S_0)^2 T}}, \qquad (5.113)
$$

and remembering (2.45), i.e.,

$$\int_{u_T}^{\infty} du\, p(u, T) = \int_{u_T}^{\infty} \frac{du}{\sqrt{2\pi}} e^{-u^2/2}$$
$$+ \frac{1}{\sqrt{2\pi}} e^{-u_T^2/2} \left(\frac{Q_1(u_T)}{N^{1/2}} + \frac{Q_2(u_T)}{N} + \cdots \right) \quad (5.114)$$

with

$$Q_1(u_T) e^{-u_T^2/2} = \frac{\hat{\kappa}_3}{6} \frac{d^2}{du_T^2} e^{-u_T^2/2}, \quad (5.115)$$

$$Q_2(u_T) e^{-u_T^2/2} = -\frac{\hat{\kappa}_4}{24} \frac{d^3}{du_T^3} e^{-u_T^2/2} - \frac{\hat{\kappa}_3^2}{72} \frac{d^5}{du_T^5} e^{-u_T^2/2}, \quad (5.116)$$

Eq. (5.112) becomes

$$\mathcal{C} \simeq \int_K^{\infty} dS_T \int_{S_T}^{\infty} \frac{dS}{\sqrt{2\pi(\sigma S_0)^2 T}} \exp\left(-\frac{(S-S_0)^2}{2(\sigma S_0)^2 T}\right)$$
$$+ \sigma S_0 \sqrt{T} \left[\int_{u_K}^{\infty} \frac{du_T}{\sqrt{2\pi}} e^{-u_T^2/2} \left(\frac{Q_1(u_T)}{N^{1/2}} + \frac{Q_2(u_T)}{N} + \cdots \right) \right]$$
$$\simeq \mathcal{C}_B + \frac{\hat{\kappa}_4 \sigma S_0 \sqrt{T}}{24 N \sqrt{2\pi}} \exp\left(-\frac{(K-S_0)^2}{2(\sigma S_0)^2 T}\right) \left[\frac{(K-S_0)^2}{(\sigma S_0)^2 T} - 1 \right], \quad (5.117)$$

since $\hat{\kappa}_3 = 0$ for a symmetric process. If this formula is to give rise to the implied volatility, it must be possible to write the second term as a correction of the variance σ used in $\mathcal{C}_B$. To test this idea, let us require

$$\mathcal{C} - \mathcal{C}_B \stackrel{!}{=} \frac{d\mathcal{C}_B}{d\sigma} d\sigma. \quad (5.118)$$

Using (5.110), we find

$$\frac{d\mathcal{C}_B}{d\sigma} = S_0 \sqrt{\frac{T}{2\pi}} \exp\left(-\frac{(K-S_0)^2}{2(\sigma S_0)^2 T}\right). \quad (5.119)$$

Equation (5.119) must be inserted into (5.118) and the result compared to (5.117). The comparison shows that the leading-order finite-time correction to $\mathcal{C}_B$ can indeed be reproduced by Bachelier's pricing formula after replacing σ via the following K- and T-dependent implied volatility

$$\sigma_{\text{imp}}(K, T) := \sigma + d\sigma = \sigma \left[1 + \frac{\hat{\kappa}_{4,N}}{24} \left(\frac{(K-S_0)^2}{(\sigma S_0)^2 T} - 1 \right) \right], \quad (5.120)$$

where $\hat{\kappa}_{4,N} = \hat{\kappa}_4 / N$.

This is the famous 'volatility smile': The volatility of the underlying, extracted from the option market, depends on the difference between the strike and the current price in a parabolic fashion, and this dependence becomes weaker with increasing time. Both predictions are possible scenarios in practice [82].

Equation (5.120) suggests that the 'smile effect' is caused by the broad tails of the underlying's price distribution. However, we have based our analysis on the simplifying assumption of time-independent volatility. In reality, this is not true. The discussion of the previous section pointed out that the volatility gradually evolves with time, which leads to a slow power-law decay of the kurtosis $\hat{\kappa}_{4,N}$ instead of the $1/N$-behavior used in (5.120). A simple way to remove this drawback of our analysis could be to replace $\hat{\kappa}_{4,N}$ by the kurtosis calculated from the historic price series.

A test of this idea was performed in [171]. Instead of inserting the historic kurtosis into (5.120) and comparing the result with experimental data for the implied volatility the analysis proceeded vice versa. The implied volatility was fitted by (5.120) for different expiry dates, with the kurtosis as an adjustable parameter (see Fig. 5.4). The resulting 'implied kurtosis' was then compared to the kurtosis obtained from historic time series. Figure 5.12 shows the outcome of this analysis. Both kurtosis are indeed compatible with one another, which is further evidence in favor of the approach presented here.

5.3.3 Are Financial Time Series Stationary?

When attempting to extract statistical properties of a stochastic time series, like the records of the closing values of a financial index, the time series of earthquakes in a seismically active region, or climate data, we are faced with the problem that these series are always one of a kind. There is no possibility to rerun the experiment yielding the time series again. To get the desired information one generally assumes the series to be stationary. Here stationarity is interpreted in the sense that the series has no objective time origin. Then, it is justified to perform a gliding average over starting times in order to extract statistical properties with sufficient accuracy from the time series. This approach has been taken in Sect. 5.3.1, for instance to obtain the distributions of the price variations in Figs. 5.6 and 5.7 or the correlation functions of Fig. 5.9. However, it is doubtful whether stationarity is a safe assumption. For financial markets this problem has been recognized early in the physics literature [18, 131] and recently the assumption of stationarity has again been strongly challenged by McCauley and collaborators [132, 134, 135]. Some of this criticism results from an analysis of the variance of the price fluctuations in terms of commonly employed theoretical concepts—geometric Brownian motion and martingales. In the following we describe and discuss these arguments.

In Sect. 5.1 we argued that the efficient market hypothesis suggests the price process to be Markovian and we introduced geometric Brownian motion as a viable

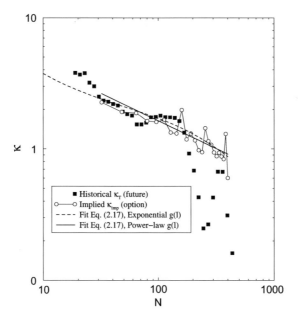

Fig. 5.12 Log-log plot of the kurtosis $\hat{\kappa}_{4,N}$ versus N, where $N = T/\Delta t$ with $\Delta t = 30$ min. All data refer to options on the German Bund future traded on LIFFE from 1993 to 1995. This is a very liquid market, so corrections to Gaussian behavior should be small. The figure compares the historic kurtosis, κ_T (■), calculated from the time series of the Bund contract, with the implied kurtosis, κ_{imp} (○), determined by fitting the volatility smile with (5.120) for different expiry times T (see Fig. 5.4 as an example). Furthermore, the results of (5.104) are included when assuming a power-law dependence for $\Phi_{(\Delta S)^2}(t)$ (denoted by $g(l)$), i.e., $\Phi_{(\Delta S)^2}(t) \sim t^{-\gamma}$ with $\gamma = 0.6$ (*solid line*) or an exponential decay for $\Phi_{(\Delta S)^2}(t)$ (*dashed line*). Both fits are acceptable. Reproduced with permission from [171]

mathematical model for this process. For the return $x(t) = \ln S(t)/S(0)$ this leads to (see (5.6))

$$dx = \left(\mu - \frac{1}{2}\sigma^2\right)dt + \sigma dW(t). \qquad (5.121)$$

This is Brownian motion with constant drift $(\mu - \sigma^2/2)$ and constant diffusion coefficient $(\sigma^2/2)$. The corresponding stochastic process is Gaussian with time-homogeneous increments (see Sect. 3.2.2), i.e.,

$$x(t + \Delta t) - x(t) = \text{function of } \Delta t \text{ only.} \qquad (5.122)$$

Due to this property the increments are often said to be stationary.[7] We also use this denomination in the following. As a consequence, the variance of the process only

---

[7]Note that the use of the term 'stationary' here is different from the definition given in Chap. 2. There, stationarity was defined as the time-translational invariance of the n-point probability density (cf. (2.66)) which implies (2.78) for Markov processes. Applying this criterion to Brow-

5.3 Models Beyond Geometric Brownian Motion

depends on the time difference and we know from (5.7) that it grows linearly with time:

$$\langle [x(t) - \langle x(t) \rangle]^2 \rangle \sim t^{2H} \quad \text{with } H = \frac{1}{2}. \tag{5.123}$$

Here we introduced what is often simply called the *Hurst exponent* H, although the original definition [126] referred to a related but different property of a stochastic time series. In fact, the exponent defined by (5.123) is one of the so-called *generalized Hurst exponents* [48], H_q, defined by

$$\langle [x(t) - \langle x(t) \rangle]^q \rangle \sim t^{q H_q}. \tag{5.124}$$

So for $q = 2$ we obtain the definition in (5.123). Furthermore, from (5.8) we see that the Gaussian distribution is centered at $\langle x(t) \rangle = (\mu - \sigma^2/2)t$. Thus the globally detrended variable

$$\tilde{x} = x - \left(\mu - \frac{1}{2}\sigma^2\right)t \tag{5.125}$$

is Gaussian distributed with zero average and a variance scaling as

$$\langle \tilde{x}^2(t) \rangle = \langle \tilde{x}^2(1) \rangle t^{2H}. \tag{5.126}$$

Every Gaussian distribution is completely determined by its first two moments. So all moments of $\tilde{x}$ must behave as ($n = 0, 1, \ldots$)

$$\langle \tilde{x}^{2n}(t) \rangle = t^{2nH} \langle \tilde{x}^{2n}(1) \rangle$$
$$\langle \tilde{x}^{2n+1}(t) \rangle = 0.$$

This defines a self-similar process for which one can also write

$$\tilde{x}(t) = t^H \tilde{x}(1), \tag{5.127}$$

meaning this equation to hold in distribution.

Quite generally, a self-similar (or scaling) process is characterized by the property that its moments obey the scaling relation

$$\langle x^n(t) \rangle = t^{nH} \langle x^n(1) \rangle. \tag{5.128}$$

Combining this with

$$\langle x^n(t) \rangle = \int_{-\infty}^{+\infty} dx \, x^n p(x, t),$$

---

nian motion reveals that the (position) process is non-stationary on the interval $(-\infty, \infty)$ (cf. Sect. 3.2.2) because the 1-point probability density vanishes in the limit $t \to \infty$.

we can conclude that the 1-point probability density depends on its two arguments in a scaling form

$$p(x,t) = t^{-H} \tilde{p}\left(\frac{x}{t^H}\right). \quad (5.129)$$

This scaling behavior is a property that all Lévy distributions share, as we have noted in (4.95). For a stochastic process with Lévy distributed (stationary) increments we thus expect a Hurst exponent of

$$H = \frac{1}{\alpha} \geq \frac{1}{2}, \quad (5.130)$$

with the equality holding for Gaussian distributed increments, i.e., for the Brownian motion discussed above. However, for $\alpha < 2$ stationarity of the increments can only hold for moments $n < \alpha$ (see (4.15)).

Assuming that the price fluctuations can be described by a detrendable process with stationary increments and that the detrended process is a self-similar Markov process, we have derived that the (1-point) probability density of the increments obeys a scaling property.

This result guides the analysis of financial time series in practice. As mentioned at the beginning of the section, one usually assumes the time series to be stationary and performs a gliding average over starting times to obtain the scaling behavior of the increment distribution or of its second moment, and from the latter the Hurst exponent. Typically, such an analysis gives a Hurst exponent which depends on time. The Gaussian behavior ($H = 1/2$) is only reached for large times. Before that, two time regimes exist: the *anti-persistent regime* with $H < 1/2$ at very short times and the *persistent regime* with $H > 1/2$ at longer times. This time dependence of H is part of the so-called *stylized facts* of financial markets [31] which also involve the broad-tailed fluctuations of Fig. 5.6, the volatility clustering of Fig. 5.8, and others.

From the discussion of Sect. 5.3.1 the presence of the anti-persistent regime is unexpected. A Hurst exponent of $H < 1/2$ lies outside the range of values compatible with stable distributions (cf. (5.130)). A possible interpretation of this (short-time) behavior employs the 'microscopic' dynamics of the market—the 'mechanics of order books' by which one denotes the details of the deterministic trading and price building process which is implemented at a stock exchange. We will explicate this point in more detail in Sect. 5.3.4. The persistent regime, on the other hand, is expected from the discussion of Sect. 5.3.1. It may be interpreted as a precursor to Gaussian behavior, arising from a truncated Lévy distribution of stationary price increments. However, as also pointed out in Sect. 5.3.1, this interpretation is not fully convincing. There are noticeable deviations between the empirical and predicted price distributions, possibly related to the time dependence of the volatility (cf. Fig. 5.8). This time dependence is not compatible with geometric Brownian motion, challenging one (or all) of the assumptions of the model, for instance the stationarity of the price increments.

That non-stationarity could be the reason why $H > 1/2$ in the persistent regime is supported by the work of McCauley [132]. Following [132] we make two assumptions about the price process: The process is detrendable and the detrended process

5.3 Models Beyond Geometric Brownian Motion

is a martingale. The martingale assumption is a weaker constraint on the process describing a financial time series than the Markov one. It only makes an assumption about the conditional expectations but not about the full joint distribution of the process. As discussed in Chap. 2, the martingale condition implies that the best estimate for the future return is the current value—this appears to be well justified by the efficient market hypothesis. Here we write this condition as (see (2.71))

$$E[x(t_0+t)|x(t_0)] = x(t_0) \quad \text{for } t > 0.$$

For the autocorrelation of the process $x(t)$ we thus obtain

$$\begin{aligned} E[x(t_0+t)x(t_0)] &= \int_{-\infty}^{+\infty} dx_1 \left[\int_{-\infty}^{+\infty} dx_2 x_2 p(x_2, t_0+t | x_1, t_0) \right] x_1 p(x_1, t_0) \\ &\stackrel{(2.40)}{=} E[E[x(t_0+t)|x(t_0)]x(t_0)] \\ &= E[x^2(t_0)]. \end{aligned} \quad (5.131)$$

It follows that

$$E[(x(t_0+t) - x(t_0))x(t_0)] = E[x(t_0+t)x(t_0)] - E[x^2(t_0)] = 0,$$

and with this we have

$$\begin{aligned} E[(x(t_0&+t) - x(t_0))(x(t_0) - x(t_0-t))] \\ &= E[(x(t_0+t) - x(t_0))x(t_0)] - E[(x(t_0+t) - x(t_0))x(t_0-t)] \\ &= E[x(t_0)x(t_0-t)] - E[x(t_0+t)x(t_0-t)] \\ &= E[x^2(t_0-t)] - E[x^2(t_0-t)] = 0. \end{aligned} \quad (5.132)$$

A martingale therefore has uncorrelated (not necessarily statistically independent) increments in non-overlapping time windows. For the variance of these increments we obtain

$$\begin{aligned} E[(x(t_0+t) - x(t_0))^2] &= E[x^2(t_0+t)] + E[x^2(t_0)] - 2E[x(t_0+t)x(t_0)] \\ &= E[x^2(t_0+t)] - E[x^2(t_0)]. \end{aligned} \quad (5.133)$$

This result depends on t_0; in general, it is thus non-stationary. However, if the variances in the right-hand side show scaling behavior with Hurst exponent H, (5.133) can be written as

$$E[(x(t_0+t) - x(t_0))^2] = \langle x^2(1) \rangle [(t_0+t)^{2H} - t_0^{2H}]. \quad (5.134)$$

A martingale therefore only has a stationary second moment of the increments if $H = 1/2$ because then the t_0-dependence drops out from (5.134). An analysis of the second moment of the increment distribution assuming martingale behavior and stationarity of the underlying process which yields $H \neq 1/2$ is therefore inconsistent. So one (or both) of the assumptions is (are) not fulfilled. For efficient markets it appears more likely that stationarity is violated.

Of course, this conclusion about the non-stationarity based on the second moment is not stringent. A stringent test on the stationarity of a stochastic process would require one to determine the n-point probability distribution (see Chap. 2). However, in practice the available time series data are barely sufficient for a reliable determination of the increment distribution. Additionally, one would need some repetitiveness of the time series, allowing for an assignment of blocks over which statistics can be generated [9, 27, 132, 133]. Also this is hard to realize in practice. Instead of trying to prove non-stationarity by such an approach, one can follow a different route. One can assume that the apparent non-stationary behavior is related to the simultaneous presence of more than one intrinsic time scale governing the stochastic process. In the simplest case this would just be one additional time scale and one could try to come to a stationary modeling by including an additional stochastic variable into the description. This approach is, e.g., followed in stochastic volatility models [82, 215] or in the modeling presented in Sect. 5.3.1 where we assumed different time scales for the change of the sign and the absolute values of the increments.

Finally, also the assumption that the process is detrendable is not as harmless as it may sound. There are approaches trying to perform an estimate of the local trend of a time series from the data themselves and to analyze locally detrended fluctuations [166]. However, this only removes the non-stationarity of the first moment of the stochastic variable under investigation and also seems susceptible to creating artifacts [22]. A way beyond all frequency based determinations of price increment probabilities may lie in the application of the extended logic approach to probabilities (cf. Sect. 2.1.2), but this path has not been followed yet.

5.3.4 Agent Based Modeling of Financial Markets

We saw in the last section that non-stationarity could be an important ingredient in the modeling of the time series of price fluctuations and that currently no method has been established how to derive such a non-stationary model from the existing time series. Under these circumstances, we can proceed in a way that is often followed in statistical physics and build on an analysis of microscopic models of the phenomenon, either by analytical or by numerical means. When we think of such a statistical-mechanical modeling of financial markets, agents or traders are the microscopic degrees of freedom of the market. Their decisions to buy or sell create the macroscopic variables supply and demand which in turn determine the changes in price as we will discuss below. A trader bases the decision to buy or sell or do nothing on different influences:

- His rational assessment of the current market situation, which is based on experience and economic knowledge. This is an individual ('idiosyncratic' [94]) criterion.

- The influence exerted by the environment. This influence is two-fold. A trader is in contact with a certain number of (selected) colleagues with whom he exchanges 'news and views'. Furthermore, he learns about the opinion of other traders by observing the price evolutions on the stock market. The first influence is direct, the second indirect. However, both have the effect that the trader tends to 'imitate' the decisions of the others.

A trader who always bases his decision on the first point and never follows the trend is called a *fundamentalist*, whereas one who only mimics the others is called a *trend chaser* or *noise trader*. In practice, no trader can afford to be a pure fundamentalist because he might ruin his fortune by continuously disregarding the signs of the market. To some degree, he has to follow the trend. Imitation is therefore an important factor in the traders' behavior. It also produces a strong feedback mechanism: When the traders follow the trend, their decisions influence the market prices, which in turn changes the trend and requires a renewed adjustment, and so on. Therefore, the stock market can be considered as a complex self-organizing system.

This description of the markets microstructure appears to be fairly vague compared to that usually developed in a physical approach. However, physics has the advantage that the interactions of the microscopic degrees of freedom and the equations of motion are precisely known. This is (hitherto) lacking in finance. But even if they were known, there is no guarantee that the theory would be mathematically tractable, due to the complexity of the system. Being faced with this situation, a legitimate alternative is to postulate simple models which incorporate the ingredients discussed above and which can be tackled either analytically or by computer simulations. Many such models have been proposed, see e.g. [5, 29, 193, 200] (and also Chap. 4 of [148] and [201]). In the following we will discuss two agent-based market models: a market model taking into consideration the working of an order book and a relatively abstract model of a trading process, involving only key generic features of this process.

An Order Book Based Market Model

Price formation at a stock exchange is realized through the interaction of buy and sell orders of traders mediated by the rules of the *order book*. The order book stores offers and demands of the various traders and, as an idealization (but see Sect. 5.3.1), enables a continuous trading which is called 'continuous double auction', i.e., an auction where both buyers and sellers quote their price ideas simultaneously. Prices are given as multiples of a minimum price change unit, the tick size. The highest offer price offered for buying is called the best bid-price, and the lowest price demanded for selling the best ask-price. The non-zero gap between these is the bid–ask spread. The market participants can enter two types of orders, limit orders and market orders. In limit sell orders for example, the market participant requests that a stock is sold for at least the price given as the limit price in the order. The same holds mutatis mutandis for a limit buy order. In contrast, market orders are given without a price limit, i.e., the market participant is willing to accept the best bid (if he submits a sell order) or best ask (if he submits a buy order) price currently listed on the order book. Typically, the price ideas and the requested trading volumes of buyers and sellers are not balanced. It is the task of the stock exchange and designated market makers (investment brokers charged with managing the trades of specific securities) to ensure that a continuous trading is upheld and that the volumes of buy and sell orders are matched. They perform this task by holding an inventory of money and securities and submitting buy and sell orders concurrently.

Based on this structure, which is basically present at every stock exchange, one can set up order book based market models, such as the one presented in [32, 172, 173, 190]. In this model, one imagines a total of N_A agents submitting only limit orders with a probability density decaying exponentially away from the midpoint between the best bid and the best ask price, both on the ask and the bid side. This exponential distribution for the order placement leads to a stationary distribution of orders in the book which is approximately log–normal to both sides of the midpoint between best bid and best ask price, which is a reasonable representation of real order books [172, 173]. The width of this stationary order distribution is called the order book depth. The decay length, λ_0, of the exponential order placement distribution characterizes the entry depth of the orders. The limit traders submit their orders with a rate α and with equal probability on the ask as well as the bid sides. Thus, these traders provide an influx per unit time of $N_A \alpha/2$ limit orders on the bid and ask sides. An equal number of N_A agents submits market orders with a rate μ, which are again buy or sell orders with equal probability. These market orders lead to an outflux of orders from the order book per unit time of $N_A \mu/2$ on the bid and ask sides. This model assumes a homogeneous trader population, i.e., all orders are submitted with a volume of one. The probability for in flowing limit orders has to be larger than the one for market orders so that there is always some liquidity present in the order book, i.e., incoming market orders find some limit orders they can be matched against. Market makers thus are not explicitly present in the model, but are represented by a constraint on the order flow rates (see below). On real markets limit orders are furthermore submitted with a finite life time and can also be canceled. This is modeled by deleting them with a rate δ from the order book. The number of limit orders in the order book then follows a simple rate equation

$$N(t+1) = \big(N(t) + \alpha N_A\big) - \big(N(t) + \alpha N_A\big)\delta - \mu N_A$$

with a stationary state

$$N_{eq} = \frac{N_A}{\delta}\big[\alpha(1-\delta) - \mu\big].$$

Stability of the model then requires $\alpha(1-\delta) > \mu$ as constraint on the order flux rates. As a matching algorithm for the orders, price-time-priority [172, 173] is implemented, as is usually found for most assets in real markets. This means that if, e.g., a market buy order could be matched with several limit orders at the best ask price, the limit order which has been in the order book for the longest time is selected.

5.3 Models Beyond Geometric Brownian Motion

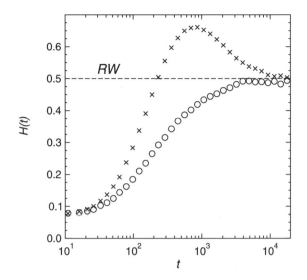

Fig. 5.13 Hurst exponent $H(t)$ as a function of time lag t for a multi-agent order book model of a financial market. The regime with $H < 1/2$ (anti-persistent regime) is created by the rules of an order book. For a stationary market (*circles*) this regime directly crosses over to the random walk regime (*dashed line* with label 'RW'). When one includes non-stationary order placement (*crosses*) a regime with $H > 1/2$ (persistent regime) is generated between the anti-persistent and random walk regimes

In such an order book model the spot price of the asset is always equal to the last trading price and the stochastic influx of orders generates its stochastic dynamics. When monitoring these price fluctuations one observes a variance increasing in time according to (5.123) with a time-dependent Hurst exponent, an example of which is shown in Fig. 5.13 by circles for a choice of parameters $N_A = 250$, $\alpha = 0.15$, $\mu = 0.025$, $\delta = 0.025$ and $\lambda_0 = 100$ (safely within the regime of stability of the model). Clearly, the homogeneous and stationary order book model we discussed so far reproduces the anti-persistent behavior at short times and the random walk regime at long times, but fails to generate an intermediate persistent regime. From our discussion in the previous section we expect that this could for instance be related to the missing non-stationarity of this model market on intermediate times.

Such non-stationarity can be motivated by the ubiquitous mood swings of market participants alternating between 'bullish' (the general sentiment is that the market will go up) and 'bearish' (the general sentiment is that the market will drop) phases. In such a phase, the market orders will be placed preferably on either the buy (bullish) or sell (bearish) sides. These bullish and bearish phases thus change the probability of market orders to be placed, e.g., on the buy side, which we denote by q in the following, away from its stationary value of $q = 1/2$. Mood swings exhibit an unpredictable time dependence, but the further they have driven the behavior of the market participants away from their well-balanced stationary behavior, the more risky the market becomes, and the stronger the tendency to return to the stationary state. One can model these mood swings by a mean-reverting random walk for $q(t)$ on the interval $[0, 1]$, where the probability to jump towards the average value of $\langle q(t) \rangle = 1/2$ is given by $1/2 + |q(t) - 1/2|$. At the same time, the traders providing liquidity to the market by their limit orders have to adjust the entry depth of their orders to prevent one side of the order book from running empty [173]. This can be modeled by a time-dependent order depth of the form

$$\lambda(t) = \lambda_0 \left[1 + C_\lambda \frac{|q(t) - 1/2|}{\sqrt{\langle (q(t) - 1/2)^2 \rangle}} \right].$$

The standard deviation of the mean-reverting random walk, $q(t)$, occurring in this equation as the natural scale for the deviations of $q(t)$ from its stationary value, can be determined in advance. When one introduces this non-stationary element into the simulations of the order book market model and chooses $C_\lambda = 10$ to affect a sufficient variation of $\lambda(t)$ keeping all other parameters constant, a time-dependent Hurst exponent as shown by the crosses in Fig. 5.13 results, which now reproduces the complete time dependence of the Hurst exponent found for real markets. Furthermore, this non-stationary order book model also generates broad-tailed price fluctuations in the time regime of the persistent behavior of the Hurst exponent [172, 173]. Such model simulations therefore allow the step-wise inclusion of agent behavior found in real markets and the identification of the consequences of such behavior. They give an explanation for the anti-persistent behavior of the Hurst exponent on short times and suggest a possible explanation for the occurrence of the persistent regime.

A Physicist's Market Model

The previous section gave an example for market models trying to incorporate detailed information about the mechanisms of price formation and the psychology of the market participants. Such 'realistic' models are nowadays very successful in explaining many of the stylized facts of financial markets. However, one can also go the opposite direction in model complexity by abstracting as much as possible from reality and retaining only a few features which are (assumed to be) essential for the market. This kind of 'generic' modeling has a long tradition in physics. We present such a physicist's market model in the following.

In Sect. 5.1 we introduced the idea that supply, s, and demand, d, determine the (spot) price, S. This idea could be called 'the zeroth law of economics'. In the economic literature it is usually assumed that there exists a monotonously decreasing function, $S(s)$, and a monotonously increasing function, $S(d)$, and that their uniquely determined intersection is the equilibrium price established on the market (this could be a financial market or any other market). One of the first physicists to become influential in his thinking about financial markets, M.F.M. Osborne [158], challenged this basic concept. He criticized that there is no experimental evidence in favor of this concept. In contrast, he argued that for each individual buyer and producer of a good the price determines the demand, $d(S)$, and supply, $s(S)$, respectively. Either demand drops to zero because the price becomes too high, or supply drops to zero because the price becomes too low. This is a valid characterization, e.g., of the action of an individual agent who provides limit orders in financial markets and is willing to buy at a certain price (or lower but not higher) or sell at a

5.3 Models Beyond Geometric Brownian Motion

certain price (or higher but not lower). But one has to take into account that markets involve a collection of participants leading to a distribution of individual price limits. Furthermore, there are also participants willing to accept the current best offers by submitting market orders, as discussed above. While it is no longer clear that, given this broad spectrum of ideas about the correct price of an asset, something like an equilibrium price exists [132], it seems reasonable—and is indeed a correct description of the price formation process on real markets—to state that *imbalances between supply and demand determine the direction of price changes*. One further fact about financial markets (and life in general) is that not everybody can participate with the same capacity. For a market participant this simply means that his current wealth determines how much he can invest.

These two basic facts—imbalances between supply and demand determine price changes and everybody invests proportional to his current wealth—are the only ingredients of the following minimal model [42] of the trading of two goods, called stock and money, respectively.

In each time step (Δt), each of N agents decides with equal probability to sell or to buy. Agent i invests a random fraction $x_i(t_n) \leq c$ (discrete time $t_n = n\Delta t$) of his wealth in stocks for selling or of his wealth in money for buying. As in (5.15), the current total wealth of the agent is given by $\mathcal{W}_i(t_n) = \Delta_i(t_n) S(t_n) + \Pi_i(t_n)$, where $\Delta_i(t_n)$ is the number of stocks owned by agent i, $\Pi_i(t_n)$ is his wealth in money and $S(t_n)$ is the spot price of the stock. By this double auction process a certain demand in stocks, $d(t_n) = \sum_i [x_i(t_n) \Pi(t_n)/S(t_n)]$, and supply of stocks, $s(t_n) = \sum_i [x_i(t_n) \Delta_i(t_n)]$, is generated (in both cases $[\cdot]$ denotes the largest integer smaller than the argument). The buy and sell orders at each individual time step t_n, which are given for the fixed spot price $S(t_n)$, are matched against each other as far as possible, while excess orders are discarded. This implements the second ingredient listed above.

The first ingredient is implemented in the following way: if there is an imbalance between supply, $s(t_n)$, and demand, $d(t_n)$, at time t_n, its impact on the price $S(t_n)$ of the stock is given by the following formula

$$S(t_n + \Delta t) = S(t_n) \left[1 + \frac{d(t_n) - s(t_n)}{d(t_n) + 2s(t_n)} \right]. \tag{5.135}$$

This price update rule is a possible realization of the concave price-impact functions of supply-demand imbalance found in real markets [47, 219]. The exact choice does not affect the qualitative behavior discussed in the following, nor does the exact choice of the fraction c as long as it is not too close to one.

Interestingly, when one starts such a market in a state which looks like a perfect equilibrium situation, i.e., all agents own the same amount of money and the same amount of stock, and the price is chosen such that wealth in money equals wealth in stock ($\Delta_i(0) S(0) = \Pi_i(0) \ \forall i$), one finds that this state is not stable. The self-organized stationary state of this simple model economy is characterized by the following properties [42] of the probability densities shown in Fig. 5.14:

- The wealth distribution is of Pareto type, i.e., a power law, as also found in the real world [160, 204] and, more specifically, for traders at a financial market [58]. The

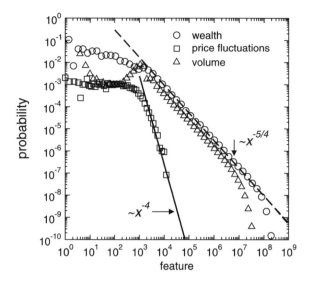

Fig. 5.14 *The open circles* depict the stationary Pareto-type wealth distribution of the minimal market model. This Pareto-type wealth distribution leads to a power law behavior of the volume traded at the market, shown here by *the triangles*. When each agent invests a maximum of half of his wealth in each step ($c = 0.5$, see text) a distribution of price fluctuations with a power law tail with exponent -4 results (*squares*), which was observed for real markets. For large values, finite size effects lead to an exponential cutoff of the power laws

exponent observed for the model differs from the one found for the real world, but the essential qualitative behavior is captured by the model. Furthermore, the trading process described by this model is not specific to financial transactions (stock and money), it could be fish sold for sea shells as well. This model therefore suggests that economic inequality among people developed as soon as the first equivalent of money was introduced into trading and the earlier barter economy abandoned.

- The distribution of trading volumes, $V(t_n) = S(t_n) \min[d(t_n), s(t_n)]$, is of power-law type, as found for real markets [170]. In [170] it was also argued that the power-law behavior in the distribution of trading volumes is a consequence of the one in wealth of the traders. Our simple model generates both of them concurrently from the basic trading process.
- The distribution of returns in the price time series develops broad power-law tails, which for the choice $c = 0.5$ agree well with the ones found for real markets [65] and calculated from the truncated Lévy flight model in (5.99).
- Although the available money per stock stays constant, the average price of stock increases compared to the starting value. This risk premium (see our discussion on the Black Scholes equation) often discussed for real stock prices therefore may be only a result of the price formation mechanism without any need for rationalization in any conscious decision of market participants.

In contrast to the behavior of the order book model, broad-tailed price fluctuations occur here also in the stationary state. But here they are generated by the inhomogeneity among traders, i.e., the power law distribution of their wealth and of the orders they submit, which were constrained to be of equal size in the order book model.

Before reaching its stationary state, the model exhibits a time window where the wealth distributions (money, value of stocks owned and total wealth) of the traders as well as the price fluctuations are Gaussian. However, this Gaussian state, which is often assumed to be the equilibrium state of real financial markets, is unstable. The instability is generated by the presence of a boundary at zero wealth. All traders accumulating there are temporarily taken out of the trading process as long as the current spot price is larger than their remaining money. They can no longer participate in this model economy. At the same time extremely wealthy traders appear and thus the market behavior is no longer determined by many equally important random decisions (Gaussian limit) but dominated by a few big orders or traders (Lévy or power-law limit). However, the model is still symmetric with respect to different traders. All traders move up and down the wealth scale in time—different from real life.

To come back to our critical discussion of the current (2012) crisis of financial markets, we can conjecture its origin to be the practice of leveraged speculations, which allowed the banking sector such huge profits in the years between 1990 and 2008. Leverage virtually shifts the upper limit of the wealth distribution in Fig. 5.14—in practice by one order of magnitude—and this in turn increases the fluctuations of the price. Only, these increased fluctuations are real and not virtual, so they can easily exceed the real capital of a market participant which in turn may lead to bankruptcy. The phenomenology of such huge market fluctuations, i.e., market crashes, and a way to analyze and maybe even to predict them will be discussed in the next section.

5.4 Towards a Model of Financial Crashes

The fear of every investor is a sudden and steep drop of asset prices: the occurrence of a *stock market crash*. Crashes are rare events that can happen even on mature markets. Prominent examples are the crashes on NYSE in October 1929 and October 1987, when prices fell by about 20 % to 30 % within a few days. Although being rare, crashes are far more probable than what could be expected for Gaussian price fluctuations, and even truncated Lévy flights cannot account for their frequency, as we will show in the next section.

Based on these findings it has been suggested that the market behavior can be characterized by qualitatively different phases: a normal trading phase with weak (or no) time correlations of price fluctuations—the one we have discussed up to now—and a run-up phase to a crash, where strong time correlations develop. This interpretation allows to draw an analogy between the market and a thermodynamic system undergoing a phase transition [49, 50, 94, 96, 97, 195, 196, 205, 206]. Normal market behavior is then associated with fluctuations in the one-phase region, while during a crash the market is close to a critical point.

The following discussion dwells upon this idea. We begin by presenting the phenomenology of crashes. Then, we discuss the analogy to critical behavior and give an introduction to the description of crashes proposed by Sornette, Johansen and coworkers [94, 96, 97, 195–197].

5.4.1 Some Empirical Properties

From the previous sections we know that the empirical distribution of asset price fluctuations has broad tails. The tails allow large price fluctuations to occur frequently. Therefore, a reasonable conjecture is to identify a crash with an extreme fluctuation, sampling the far edges of the tails.

Let us explore this hypothesis by modeling the price process as a truncated Lévy flight with finite variance. As shown in Fig. 5.6 the work by Mantegna and Stanley [127–129, 131] suggests that the central part of the distribution $p_{\Delta t}(\ell)$ of price fluctuations ℓ is well described by a Lévy distribution with $\alpha = 1.4$ up to $\ell/\sigma \approx 6$ (with σ being the standard deviation of the distribution). Let us therefore simulate the price process $S(t)$ as a truncated, symmetric (drift-free) Lévy flight. Its discretized equation of motion reads

$$S(t_n + \Delta t) = S(t_n) + \Delta L_{\alpha,6\sigma}(t_n) \quad (t_n = n\Delta t, \Delta t = 1), \tag{5.136}$$

where $\Delta L_{\alpha,6\sigma}$ denotes an increment drawn from a symmetric Lévy distribution with exponent α and cutoff parameter $\Delta L_{\text{cut}} - 6\sigma$ (scc (4.96)). For $\alpha = 1.4$ and the choice $c = 1$ for the scale factor (cf. (5.83)) one can determine σ numerically. This value is then taken to cut the distribution at $\Delta L_{\text{cut}} = 6\sigma = 13$. Lévy distributed increments can be generated following the algorithm of Chambers et al. [25]. For a symmetric distribution with $\alpha \neq 1$ this reads

$$\Delta L_{\alpha,6\sigma} = \frac{\sin \alpha u}{(\cos u)^{1/\alpha}} \left(\frac{\cos([1-\alpha]u)}{v} \right)^{(1-\alpha)/\alpha} \quad \text{for } |\Delta L_{\alpha,6\sigma}| \leq 6\sigma, \tag{5.137}$$

where u is a uniformly distributed random number in $[-\pi/2, \pi/2]$ and v is an exponentially distributed random number with mean value 1. Due to the cutoff in the increments, all moments of $p_{\Delta t}(\ell)$ exist and $S(t)$ performs Brownian motion for $t \to \infty$ (cf. Sect. 4.4).

The behavior of extreme excursions for this time series can be discussed in two ways. The first one is linked to our discussion of extreme value distributions and of the extreme excursion behavior of Brownian motion in Sect. 3.4. As the process $S(t)$ asymptotically behaves like Brownian motion we know that for long times its extreme excursion distribution is given by the distribution of its end points. For the chosen values of α and ΔL_{cut} this behavior is already found after about 100 steps

5.4 Towards a Model of Financial Crashes

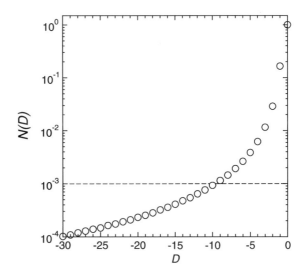

Fig. 5.15 Linear-log plot of the relative number $N(D)$ of drawdowns D for the truncated Lévy flight. The drawdowns are determined as described in the text. Percentage drawdown values are rounded to the next integer and the distribution is normalized to a maximum value of one

of the Lévy flight. So the specific non-Gaussian character of the price increments is rather quickly lost in this measure of extreme events [184].

A different measure for extreme events, which has been employed in the analysis of real markets (see below), is the distribution of drawdowns. A drawdown is defined as the percentage decrease from a relative maximum to the next relative minimum in the time series $S(t)$. This definition has the property that in contrast to the distribution of extreme excursions in a fixed time horizon, it averages over different time delays which can occur between these two extrema. To determine this distribution for our model Lévy flight, we have to fix a starting value $S(0)$ to set a scale for the relative drawdowns. This scale can be inferred from the standard deviation of the employed truncated Lévy distribution, which is given on an absolute scale and which is supposed to mimic daily price fluctuations on a 1 % level (on mature markets). This give an idea about the absolute scale for the price itself and we choose $S(0) = 200$. With this setting the distribution of drawdowns shown in Fig. 5.15 was determined.

Let us contrast the results from the truncated Lévy flight with the behavior found for the percentage decreases on the Dow Jones Index in the period from 1900 to 1994 shown in Fig. 5.16 [95]. For the Dow Jones Index, the authors identified two regimes in this distribution. In the regime where the drawdown is smaller than about 15 %, $N(D)$ was fitted by an exponential

$$N(D) = N_0 \exp\left(-\frac{D}{D_c}\right). \tag{5.138}$$

Using this exponential fit to predict the frequency of crashes on the 30 % level, one concludes that these occurred about 1000 times too often in the real data.

The simulation data for the truncated Lévy flight in Fig. 5.15 exhibit similar behavior in the small-D regime, the number of drawdowns at the 10 % level being

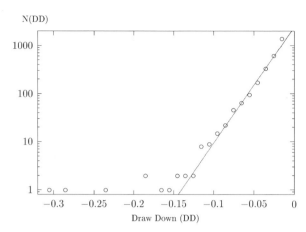

Fig. 5.16 Log-linear plot of the number of decreases $N(D)$ for the Dow Jones Industrial Average (commonly referred to as Dow Jones Index) versus the magnitude of a decrease D (referred to as 'draw down (DD)'). A decrease is defined as the percentage of the cumulative loss from the last local maximum to the next local minimum, using bins of size 1 %. If $D \lesssim 15$ %, $N(D)$ is fitted reasonably well by an exponential $N(D) = N_0 \exp(-D/D_c)$, with $N_0 \approx 2360$ and $D_c \approx 1.8$ %. The three largest drops are: NYSE, October 1987 (≈ 30.6 %), World War I (≈ 28.8 %) and NYSE, October 1929 (≈ 23.6 %). Each of these decreases lasted about three days. The analysis is based on the daily record of the Dow Jones Index from 1900–1994. Reproduced with permission from [95]

about three orders of magnitude smaller than on the 1 % level. However, they do not suggest an exponential dependence in this regime. Also contrary to Fig. 5.16, the simulation data continue to decrease smoothly for larger D without change of regime, as expected for the stationary increments employed. For drawdowns at the 30 % level, $N(D)$ reaches about 10^{-4} of the value at the percent level. Simulation of a Lévy flight based on the analysis of price fluctuations in (4.96) therefore predicts a much higher probability for the extreme drawdowns than the exponential fit would suggest. Nevertheless, a comparison of Figs. 5.15 and 5.16 shows that the frequency of drawdowns on the 30 % level is by a factor of about 10 larger for the Dow Jones Index than for the simulation data. This difference still corroborates the central conclusion from [95] that large drawdowns, i.e., crashes, are the result of different dynamic behavior. Originally, these rare extreme events were called 'outliers' in reference to their deviation from the 'normal' statistics. However, in recent years it has been recognized that the switching between a 'normal' dynamic regime and a regime characterized by huge fluctuations seems to occur in many different time series [198]. In reference to their potentially dramatic consequences, it has been suggested to call these rare events 'dragon-kings'.

If the dynamic behavior in the run-up phase is really different, it should be possible to identify specific and systematic features signaling the approach towards a crash. There is indeed evidence for this conjecture. Figure 5.17 presents an example. It shows the time series of the Dow Jones Index from 1927 to 1929 and from 1985 to 1987, respectively. In October of 1929 and October of 1987 severe stock mar-

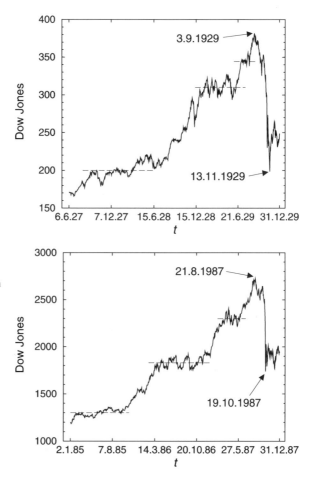

Fig. 5.17 Time evolution of the daily closing quotations of the Dow Jones Industrial Average close to the crashes of (**a**) October 1929 and (**b**) October 1987. The index is measured in points and the time in days. In both cases, time runs over 755 trading days ($\approx$ two to three years) before December 31, 1929, and December 31, 1987. The dates on the time axis are separated by 151 trading days. Both time series exhibit a step-like increase before the crash. The steps (*dashed lines*) shorten upon approaching the crash. The dates of the absolute minima and maxima close to the crash are specified in the figure. Data obtained from http://quote.yahoo.com

ket crashes occurred. Within a few days the index declined dramatically: by about 23 % in 1929 and by about 30 % in 1987. Both crashes were preceded by a gradual staircase-like increase which led to an unprecedented peak about two months before the market collapse. During this increase, the steps of the stairs and the time interval between successive steps became progressively shorter and eventually vanished close to the peak. Such an oscillatory structure with a shrinking period seems to be a characteristic precursor of crashes [94, 97].

5.4.2 A Market Model: From Self-organization to Criticality

The previous discussion suggests two characteristic features of a crash. First, it is preceded by oscillations whose period vanishes close to the actual crash time. Second, the magnitude of the price drop is so large that it cannot result from rare statistical fluctuations of the 'normal' dynamics.

One can argue that these 'normal' statistical fluctuations govern the behavior of the market when supply and demand are well balanced. However, the situation becomes different close to a crash. Then, an immense number of traders spontaneously decide to sell. This unusual 'cooperativity' shakes the liquidity of the market. There is no sufficient demand to defy the exploding supply. The prices drop, and the market falls out of equilibrium. This interpretation suggests that crashes are triggered by a spontaneous development of correlations among the traders like those occurring at a critical point in a thermodynamic system. In the following, we want to briefly sketch one model making this link to critical phenomena suggested by Cont and Bouchaud [18, 29].

The Market in Normal Periods: The Cont-Bouchaud Model

Let there be N_t traders in the market, all of which are assumed to have the same 'size', i.e., the same (average) trading volume. At every time instant t_n, a trader i can adopt three possible 'states' of action: He can buy ($\phi_i = 1$), sell ($\phi_i = -1$) or wait ($\phi_i = 0$). Thus, the difference between supply and demand is just the sum of ϕ_i over all traders, which in turn governs the price change

$$\Delta S(t_n) \propto \sum_{i=1}^{N_t} \phi_i(t_n), \qquad (5.139)$$

where it is assumed that price change is simply proportional to supply and demand imbalance (this approximation is valid only for small $\Delta S(t_n)$, see our discussion in Sect. 5.3.4). The proportionality constant is a measure for the *market depth*, i.e., how susceptible the market prices are to variations in the disparity of supply and demand. Consider now a market in which all traders 'communicate' with one another. Communication is established either by direct contact or indirectly by following the evolution of the stock prices. Eventually, this communication makes two traders adopt a common market position: They decide to buy, to sell or to be inactive. If this happens, a 'bond' is created between the traders. The creation of a bond should occur with a small probability

$$p_b = \frac{b}{N_t}, \qquad (5.140)$$

since any trader can bind to any other and the average number of bonds per trader, $(N_t - 1)p_b$, should be finite for a large market ($N_t \gg 1$). Thus, p_b is determined by a single parameter b, which characterizes the willingness of a trader to comply with his colleagues.

Since detailed information about the mechanism of bond formation is lacking, a 'maximally unbiased' assumption is to connect two randomly chosen traders with probability p_b. This divides the N_t traders into groups or 'clusters' of different sizes whose members are linked either directly or indirectly via a chain of intermediate traders. These groups are coalitions of market participants who share the same opinion about their activity. The decision of the group should be independent of its size and of the decision taken by any other cluster (see Fig. 5.18).

5.4 Towards a Model of Financial Crashes

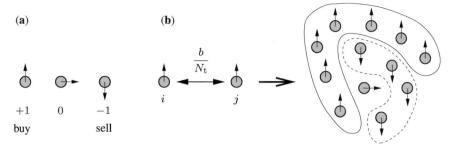

Fig. 5.18 Illustration of the Cont-Bouchaud model. The market consists of N_t traders. They are the 'microscopic degrees of freedom' of the market. Each trader can be in one of three states (see (**a**)): He can buy (+1; symbol: ↑), wait (0; symbol: →) or sell (−1; symbol: ↓). The traders are not independent of one another. Any two traders i and j can agree on a common market position with probability $p_b = b/N_t$ (see (5.140)). This 'interaction' divides the total number of traders into groups ('clusters') whose members adopt the same trading strategy: They sell (group encircled by a *dashed line* in (**b**)), buy (group encircled by a *solid line* in (**b**)) or are inactive (isolated trader in (**b**)). The resulting groups differ in size and adopt their market positions independently of the other groups. These assumptions lead to (5.141)

Let the state of a cluster be denoted by $\phi_c(t_n)$, with $\phi_c = \pm 1$ specifying an active and $\phi_c = 0$ an inactive position. If there are N_c clusters in total and the cth clusters has size s_c, then (5.139) can be rewritten as

$$\Delta S(t_n) \propto \sum_{c=1}^{N_c} s_c \phi_c(t_n). \tag{5.141}$$

So, the distribution of ΔS_n is determined by that of the sizes of the groups. However, the latter is well known in percolation theory [24]. It is the cluster-size distribution of an infinite-range bond percolation model, which is given by

$$p_c(s) \sim \frac{1}{s^{1+3/2}} \exp[-(1-b)^2 s] \quad \text{(for } 1 \ll s \ll N_t\text{)}, \tag{5.142}$$

if $b \lesssim 1$. This identifies $b = 1$ as the percolation threshold. As long as $b < 1$, all clusters are finite. This is the regime of 'normal' market fluctuations. However, for $b = 1$, one spanning cluster emerges which 'percolates' through the whole system. This is a critical phenomenon with, among others, strong temporal correlations in the dynamics, giving rise to a qualitatively different dynamic regime. Percolation is realized when every trader is on average connected to another. If $b > 1$, more and more traders join the spanning cluster, and the cluster begins to dominate the overall behavior of the system. This gives rise to a crash (if the members of the cluster decide to sell).[8] To avoid such a situation, b has to be smaller than one. On

---

[8]Otherwise, there would be a 'market boom' or a 'trading stop', if the cluster buys or becomes inactive. Both possibilities appear a bit strange and should be considered rather as an artifact than a realistic prediction.

the other hand, it must stay close to one to preserve the truncated Lévy behavior of (5.142). This behavior survives upon convolution. Thus, the simple model is capable of rationalizing the general form and the specific value $\alpha \simeq 3/2$ of the characteristic exponent which is often found in the analysis of empirical price distributions.

However, an important question is still open: Why should b remain near one? Cont and Bouchaud suggest that the intrinsic dynamics of the market is responsible for that. The nature of the market is supposed to be such that the system is driven towards $b \approx 1$ and kept in the 'critical' region $b \lesssim 1$ during normal periods. This argument is also at the heart of the model of Bak and coworkers [5].

Such a behavior is not unfamiliar in physics. Certain models gradually evolve towards a dynamical attractor without any fine-tuning of external control parameters (such as temperature). Close to the attractor, they behave 'critically', i.e., they exhibit power laws and broad-tailed distributions. Since it is the inherent dynamics and not an external field which pushes the system towards the critical point, this phenomenon is called *self-organized criticality* [4]. Therefore, the Cont-Bouchaud model considers the market as a self-organizing critical system.

Critical Crashes: The Sornette-Johansen Model

Self-organization is also a central concept in a model that Sornette, Johansen and coworkers have proposed to describe crashes (see [94, 97, 197] for reviews).

The basic motivation for this model stems from the following observations: The stock market crash of October 1987 was not limited to the United States. It shook all major markets around the world, although these markets are very diverse and only weakly correlated during normal times. Furthermore, crashes exhibit striking similarities. Certainly, the economy has changed a lot between 1929 and 1987. New financial products have been established (options, futures), trading techniques and exchange of information have altered (advent of computers), etc. But still, the crashes of 1929 and 1987 manifest via a sudden steep descent which is preceded by a gradual step-like increase. These features seem to be characteristic of crashes in general. When combining these observations, one could interpret a crash as a highly cooperative phenomenon, in which an otherwise non-existing strong correlation among the market constituents causes universal behavior.

Let us visualize this analogy between crashes and the physics of critical phenomena [12] in another way. Imagine a ferromagnetic material (see Fig. 5.19). At high temperatures, the thermal energy is much larger than the interaction between the elementary magnets. The magnets can easily flip and point on average in any direction so that the overall magnetization is zero (disordered paramagnetic phase). At low temperatures, however, the interaction dominates over the thermal energy. This makes the magnets align in the same direction, and the average magnetization becomes finite (ordered ferromagnetic phase). The transition from the disordered high-temperature phase to the ordered low-temperature phase occurs at a critical temperature, T_c, and is accompanied by the development of strong correlations among the magnetic degrees of freedom. Close to T_c, these correlated regions are

5.4 Towards a Model of Financial Crashes

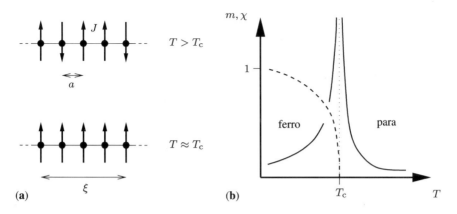

Fig. 5.19 Illustration of a ferromagnetic phase transition. Magnetic moments are symbolized by up- (↑) and down-arrows (↓). These 'spins' interact with their nearest-neighbors by a coupling constant J. A nearest-neighbor pair gains an energy of order J (> 0), if the spins point in the same direction, and loses an energy of order $-J$, if they do not (ferromagnetic interaction) (**a**). Above the critical temperature T_c, the thermal energy $k_B T$ is much greater than the ferromagnetic interaction. Therefore, the orientations of the spins are randomized, and the average magnetization m per spin is zero (**b**). The material is in the high-temperature paramagnetic phase. Below T_c, $k_B T \ll J$, and there is long-range ferromagnetic order, i.e., most spins point in the same direction (low-temperature ferromagnetic phase). The magnetization is then finite ($m > 0$) and tends to 1, as $T \to 0$ (*dashed line* in (**b**)). At T_c, the transition from the disordered paramagnetic to the ordered ferromagnetic phase occurs. The phase transition is manifest by the development of large clusters of size $\xi \gg a$ (a = lattice constant = typical range of the interaction J), whose spins point in the same direction. Close to T_c, the system is very 'susceptible' to external perturbations. Therefore, the magnetic susceptibility χ, i.e., the response of m to a change of the magnetic field, diverges, as T_c is approached from above and below (*solid lines* in (**b**)). See also [88] for illustrative pictures from Monte Carlo simulations

much larger than the typical microscopic distances and begin to dominate the physical behavior. At the critical temperature, the size of the regions diverges. This gives rise to universal and scale-invariant properties close to T_c, which is reflected by the emergence of power laws for many characteristic quantities.

Even if the ferromagnetic phase transition is only used as an analogy, these phase transitions and crashes are qualitatively different at least in one respect. There is no obvious external control parameter, such as temperature for the ferromagnet, which drives the market towards the crash. Furthermore, one can argue that such a parameter should not exist. An efficient market would have spotted it and informed its participants to avoid this risk. Effectively, this amounts to an elimination of the dangerous parameter. So, the stimulus of a crash must come from inside the market.

Sornette and coworkers suggest imitation as the prime source. When the prices increase, the attention of the traders is attracted. Eventually, some decide to buy, hoping that the rising trend persists. This increases the prices further, which, in turn, motivates more traders to speculate. Thus, a 'speculation bubble' is created. It is not unreasonable to assume that the bubble evolves in a breathing-like fashion. It is inflated by speculation, but the growth can also stagnate because some traders fear the trend will soon reverse and decide to sell. If they are the minority, the bubble keeps increasing and eventually reaches a level where many traders assume the prices will fall in the near future and place sell orders. This can trigger a crash, but does not inevitably cause it. There is always a finite chance for the bubble to deflate smoothly, since the market is not a deterministic system.

To model these ideas mathematically, Sornette et al. introduce a *hazard rate*, $h(t)$, which measures the probability per unit time that a crash will occur in the next time interval $[t, t + dt]$, provided it has not happened before. When the bubble inflates, the risk of a crash becomes more likely, and $h(t)$ increases. Close to a crash, speculation has driven the prices to such an exceedingly high level that many traders anticipate a prospective decline. They wait for the right moment to sell, while others still believe the rising trend will continue. Then, the market is 'nervous' and very susceptible to small perturbations. If some begin to sell, imitation quickly induces many others to follow. This creates an imbalance between supply and demand which cannot be instantaneously compensated by market makers and propagates through the hierarchy of traders of all sizes, ranging from the individual to large professional investors.

This leads to the following model of a crash: The market is a hierarchical system with traders of many different sizes. A crash corresponds to a critical point where the various levels of the system are highly correlated and the system becomes scale invariant. Then, a decision to sell which has been taken locally by some part within the hierarchy can easily propagate through all levels and create a macroscopically ordered state. While supply and demand are well balanced in normal times and 'disorder' dominates, 'order' prevails in times of crashes. Since the hierarchy consists of a various discrete levels, the analogy to the theory of critical phenomena suggests the following ansatz for the hazard rate:

$$h(t) = \frac{1}{(t_c - t)^\gamma}\left[B_0 + B_1 \cos\left(\omega \ln(t_c - t) + \psi\right)\right] \quad (t \lesssim t_c). \qquad (5.143)$$

Here, γ is a critical exponent which should be universal, i.e., the same for all crashes, B_0 and B_1 are constants, and the phase ψ can be related by $\psi = \omega \ln(1/t_0)$ to a time scale t_0. This time scale serves to match (5.143), which should be valid only asymptotically close to the crash, to the data of the price evolution.

Equation (5.143) exhibits two characteristic features. First, the dominant term has a power-law behavior which is a hallmark of scale invariance. Second, there are log–periodic corrections, the leading order of which is represented by the cosine term. We have already encountered such a combination of power-law and log–periodic behaviors for the Weierstrass random walk (see (4.32) and (4.34)). Quite generally, this is typical of all systems exhibiting a self-similar, but discrete hierarchical structure, i.e., the so-called *discrete scale invariance* [194].[9]

---

[9]Log–periodic corrections also occur in the formal solution of the renormalization group equation for the free energy of a critical system exhibiting continuous scale invariance, such as a ferromagnet [152]. In this case, however, the log–periodic oscillations depend on the coarse-graining length,

5.4 Towards a Model of Financial Crashes

The last missing link in modeling crashes is to establish a connection between the hazard rate and the actual price evolution. To this end, Sornette et al. construct a price dynamics, from which they derive a relationship between $h(t)$ and the drift $\mu(t)$ by invoking a martingale hypothesis. This approach has been criticized in [84] and defended again in [97]. Instead of discussing this debate, we want to present a simple—though perhaps more ad hoc—argument which yields the same result.

The examples of Fig. 5.14 suggest the impression that the evolution of the Dow Jones Index can be decomposed into a systematic ascent and superimposed stochastic fluctuations. If we neglect the fluctuations and only attempt to model the systematic increase by geometric Brownian motion, it suffices to concentrate on the deterministic contribution

$$dS_{det} = \mu(t) S_{det}(t) dt.$$

This approximation amounts to assuming that the long-term dynamics before the crash is dominated by the drift term with time-dependent μ. Economically interpreted, the drift reflects the reward demanded by the traders for investing in risky assets. The riskier an investment becomes, the larger μ must be. Since the hazard rate can be thought of as a measure of the risk evolution when a crash is approached (from below, i.e., for $t \to t_c^-$), it is reasonable to assume

$$\mu(t) = \kappa h(t) \quad (\kappa > 0), \tag{5.144}$$

so that

$$\ln \frac{S_{det}(t_c)}{S_{det}(t)} = \kappa \int_t^{t_c} dt' h(t'). \tag{5.145}$$

This is the result that Sornette et al. obtain from the martingale argument. Inserting (5.143) into (5.145) and using the integral (obtained via integration by parts)

$$\int_0^\tau dx \frac{1}{x^\gamma} \cos(\omega \ln x + \psi)$$
$$= \frac{1-\gamma}{\omega^2 + (1-\gamma)^2} \frac{1}{\cos \varphi} \tau^{1-\gamma} \cos(\omega \ln \tau + \phi) \quad (0 < \gamma < 1),$$

in which

$$\tau = t_c - t, \qquad \tan \varphi = \frac{\omega}{1-\gamma}, \qquad \phi = \psi - \varphi,$$

we find

$$\ln S_{det}(t) = \ln S_{det}(t_c) - \left[C_0 \tau^\beta + C_1 \tau^\beta \cos(\omega \ln \tau + \phi) \right], \tag{5.146}$$

---

i.e., on an arbitrarily chosen factor used for rescaling. Since the physical properties must be independent of this factor, the oscillations should not have any significance [152, 194]. Usually, they are supposed to drop out. This assumption can be proved in specific examples [152].

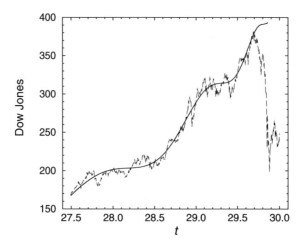

Fig. 5.20 Comparison of the Dow Jones Index prior to the crash of October 1929 (*dashed line*) with (5.148) (*solid line*). The data are the same as in Fig. 5.17. The time axis comprises 735 trading days ranging from July 1, 1927 (= 27.5), to December 31, 1929 (= 30). The parameters of (5.148) are: $S_{\text{det}}(t_c) = 571$, $C'_0 = 267$, $C'_1 = 14.3$, $\beta = 0.45$, $t_c = 30.23$ ($\approx$ March 1930), $\omega = 7.9$, and $\phi = 2.25$. These parameters are identical to those of [96] except that the sign of C'_0 had to be reversed and t_c and ϕ had to be chosen to be slightly different to produce a fit of comparable quality

in which

$$C_0 = \frac{\kappa B_0}{\beta}, \quad C_1 = \frac{\kappa B_1}{\cos\varphi} \frac{\beta}{\omega^2 + \beta^2} \quad (0 < \beta = 1 - \gamma < 1). \tag{5.147}$$

The choice $\gamma \in (0, 1)$ is motivated by the fact that it yields a finite price at t_c. If t is close to t_c, one can further approximate

$$\ln \frac{S_{\text{det}}(t)}{S_{\text{det}}(t_c)} \approx \frac{S_{\text{det}}(t)}{S_{\text{det}}(t_c)} - 1,$$

so that

$$S_{\text{det}}(t) \approx S_{\text{det}}(t_c) - \left[C'_0 \tau^\beta + C'_1 \tau^\beta \cos(\omega \ln \tau + \phi)\right], \tag{5.148}$$

with $C'_0 = S_{\text{det}}(t_c) C_0$ and $C'_1 = S_{\text{det}}(t_c) C_1$.

Qualitatively, (5.146) and (5.148) seem to be reasonable: They exhibit a strong (power-law) increase of the price which is superimposed by oscillations whose period shortens for $t \to t_c^-$, as observed in Fig. 5.17 for the Dow Jones Index. This approach has been tested using several applications [97] (see Fig. 5.20 for an example). Some comments can still be made:

- A priori, fits by (5.146) or (5.148) contain seven open parameters. This makes the extraction of meaningful results rather involved. A possible approach has been developed in [196] and extended in [97, 195] to also include the next-order correction to (5.146) and (5.148).

- Motivated by the analogy to critical phenomena, one would expect that out of the seven fit parameters β and ω should have universal values. The empirical evidence for this expectation is not very strong, but a bit more convincing for ω than for β. Typically, one finds $0.2 \lesssim \beta \lesssim 0.6$ and $6 \lesssim \omega \lesssim 8$ [96, 97].
- The critical time t_c should not be taken as the actual crash time, but rather as the most probable time for the crash, because the hazard rate $h(t)$ is interpreted as a probability. This has two consequences. First, there is always a finite chance that the crash does not really happen, although fits by (5.146) or (5.148) would have predicted it. This point has been demonstrated and criticized in [113].[10] Second, the empirical results for t_c tend to systematically overestimate the crash time in cases where the crashes actually occur. Both points limit the predictive power of the presented approach to some extent [96, 197].

5.5 Summary

On the financial market, thousands of traders daily purchase and sell assets. Although trading is regulated by rules which are either imposed by the stock exchange or negotiated bilaterally according to rational criteria on the OTC market, this deterministic character at the microscopic level vanishes when the individual transactions are summed. The resulting fluctuating balance of supply and demand confers a random character on the macroscopic price evolution.

This justifies the stochastic description of financial markets which has been pursued since the seminal work of Bachelier. The classical approach relies on the efficient market hypothesis: All necessary information for the future evolution is contained in the present prices. There is no particular advantage in analyzing the past. This amounts to a Markov assumption which underlies the model of geometric Brownian motion. Geometric Brownian motion postulates the following properties of the price dynamics:

- The fundamental financial variable is the (infinitesimal) increment of the returns $d \ln S$.
- The time evolution of the returns is a Wiener process with drift. It is determined by two parameters: the expected growth rate μ (mean value) and the volatility σ (standard deviation).
- Since the probability density of the returns is a normal distribution, that of the prices is log–normal.
- Trading is continuous, i.e., the time interval Δt between successive quotations tends to zero. Therefore, the number of quotations in any finite time T diverges: $N = T/\Delta t \to \infty$. The central limit theorem then implies that the price process is Gaussian on all time scales, i.e., it is self-similar.

---

[10]Some of the authors of [113] therefore propose an alternative description of crashes based on a Langevin approach [16]. In this model, a crash is interpreted as an activated process which occurs by the succession of many improbable events.

In contrast to that, the statistical analysis reveals differences between this model and the real price dynamics:

- Instead of the returns the (mathematically more convenient) price increments can also be chosen as variables for short times. The corresponding times are of the order of a month to about 100 days.
- Quotations are separated by finite time intervals so that $N = T/\Delta t$ is not a priori (infinitely) large for any T and the central limit theorem does not always apply.
- If T is of the order of a few minutes to hours, the distribution of the price increments (or returns) is strongly non-Gaussian and exhibits broad tails. A reasonable description is provided by an exponentially truncated Lévy distribution. Despite the finite variance, the Lévy character of this distribution yields an extremely slow convergence to a Gaussian with increasing T.

This analysis points out that there are several distinguishable time scales:

- the decorrelation time of price increments $\tau_{\Delta S}$, which is typically of the order of 15 to 30 min,
- the crossover time $T_\times$ to the Gaussian distribution, which is of the order of weeks to months, and
- the decorrelation time of the magnitude of the price increments (i.e., of the volatility), which is of the order of weeks to months or even longer.

These properties apply to liquid markets in normal times. But we have seen that they are all derived by assuming the stationarity of the underlying time series. This assumption is, however, questionable based on the behavior obtained. Deriving the underlying process from an analysis of the time series alone seems statistically impossible. So one has to assume a simple class of underlying process. An alternative is the construction of agent based trading models which incorporate the structure of the trading process at a stock exchange and insight into behavioral patterns of traders. Furthermore, there is evidence that crashes occur in times of atypical market fluctuations and need special treatment, perhaps in analogy to critical systems with discrete scale invariance.

5.6 Further Reading

This section lists and briefly comments the main references from which our discussion has considerably profited:

Finance Books The book of Wilmott, Dewynne and Howison [215] develops the theory of option pricing by means of partial differential equations. Our discussion of the Black-Scholes theory has been heavily influenced by their presentation. Hull's book [82] on 'Options, Futures, and Other Derivatives' is not as mathematically oriented as [215]. It gives a comprehensive introduction to the whole field, with numerous illuminating examples. The book by Deutsch [36] has been written by a practitioner in risk management. It offers a clear and detailed presentation of the background and tools of modern risk management. The mathematical aspects of finance, such as martingale theory and risk-neutral valuation, are rigorously developed in the books of Bingham and Kiesel [11] and Korn and Korn [109]. Despite mathematical rigor, both texts are very accessible.

Physics Books Our presentation of 'Models Beyond Geometric Brownian Motion' strongly profited from the papers of Bouchaud, Potters et al., of Mantegna, Stanley et al. and of McCauley et al. This research has found its way into three very commendable textbooks. Bouchaud and Potters [18] develop the theory of financial risk by explicitly taking deviations from Gaussian behavior into account. This non-traditional approach is very lucidly presented. The book by Mantegna and Stanley [131] provides a comprehensive overview of the statistical description of financial time series and of pertinent models, ranging from Lévy to GARCH processes. It is an excellent introduction to the field of econophysics. The book by McCauley [132] critically examines common assumptions of financial economics, like the stability and stationarity of markets, in view of the statistics of real financial time series and proceeds to formulations of option pricing and portfolio selection based on minimal assumptions about the underlying market dynamics. Furthermore, the book by Voit [207], resulting from lectures on the 'Statistical Mechanics of Financial Markets', provides a good introduction not only to the classical ideas about the financial market (random walk, Black-Scholes theory, etc.), but also to the diverse modern approaches at the interface between physics and finance. Also the classic textbook on stochastic methods by Gardiner [59] has been extended to a treatment of financial markets in its 4th edition.

Appendix A
Stable Distributions Revisited

This appendix is meant to complement the discussions of the central limit theorem (Sect. 2.1) and of stable distributions (Sect. 4.1). Its aim is two-fold. On the one hand, further criteria are presented for testing whether the probability density $p(\ell)$ belongs to the domain of attraction of the stable distribution $L_{\alpha,\beta}(x)$. On the other hand, useful asymptotic expansions of $L_{\alpha,\beta}(x)$ and (most of) the known closed-form expressions for specific α and β are compiled.

A.1 Testing for Domains of Attraction

Let $p(\ell)$ be the probability density of the independent and identically distributed random variables ℓ_n ($n = 1, \ldots, N$). Its distribution function $F(\ell)$ and characteristic function $p(k)$ are given by

$$F(\ell) = \int_{-\infty}^{\ell} \mathrm{d}\ell'\, p(\ell'), \qquad p(k) = \int_{-\infty}^{+\infty} \mathrm{d}\ell\, p(\ell) \exp(\mathrm{i}k\ell). \tag{A.1}$$

The density $p(\ell)$ is said to belong to the 'domain of attraction' of the limiting probability density $L_{\alpha,\beta}(x)$ if the distributions of the normalized sums

$$\hat{S}_N = \frac{1}{B_N} \sum_{n=1}^{N} \ell_n - A_N$$

converge to $L_{\alpha,\beta}(x)$ in the limit $N \to \infty$. In Sect. 4.1, we quoted a theorem that all possible limiting distributions must be stable, i.e., invariant under convolution (see Theorem 4.1). This stability criterion leads to the class of Lévy (stable) distributions $L_{\alpha,\beta}(x)$ which are characterized by two parameters, α and β (see Theorem 4.2). The Gaussian is a specific example of $L_{\alpha,\beta}(x)$ with $\alpha = 2$.

Theorem A.1 *A probability density $p(\ell)$ belongs to the domain of attraction of the Gaussian if and only if*

$$\lim_{y \to \infty} \frac{y^2 \int_{|\ell|>y} d\ell\, p(\ell)}{\int_{|\ell|<y} d\ell\, \ell^2 p(\ell)} = 0, \tag{A.2}$$

and to the domain of $L_{\alpha,\beta}(x)$ with $0 < \alpha < 2$ if and only if

1.
$$\lim_{\ell \to \infty} \frac{F(-\ell)}{1 - F(\ell)} = \frac{c_-}{c_+},$$

2.
$$\lim_{\ell \to \infty} \frac{1 - F(\ell) + F(-\ell)}{1 - F(r\ell) + F(-r\ell)} = r^\alpha \quad \forall r > 0,$$

where c_+ and c_- are proportional to the amplitudes of the asymptotic behavior of $p(\ell)$ for large $\pm\ell$ (see (4.16)).

Alternatively, one can also test for the domains of attraction by means of the characteristic function due to the following theorem:

Theorem A.2 *The density $p(\ell)$ belongs to the domain of attraction of the Gaussian if and only if*

$$\lim_{r \to 0} \frac{\mathrm{Re}[\ln p(rk)]}{\mathrm{Re}[\ln p(r)]} = k^2 \quad (any\ k \geq 0) \tag{A.3}$$

converges uniformly with respect to k, and to the domain of $L_{\alpha,\beta}(x)$ with $0 < \alpha < 2$ if and only if the following conditions are satisfied for $k \geq 0$ and $r \to 0$:

1. $0 < \alpha < 2, \alpha \neq 1$:

$$\lim_{r \to 0} \frac{\mathrm{Re}[\ln \tilde{p}(rk)]}{\mathrm{Re}[\ln \tilde{p}(r)]} = k \tag{A.4}$$

and

$$\lim_{r \to 0} \frac{\mathrm{Im}[\ln \tilde{p}(rk)]}{\mathrm{Re}[\ln \tilde{p}(r)]} = \beta r^\alpha \tan\left(\frac{\pi}{2}\alpha\right). \tag{A.5}$$

2. $\alpha = 1$:

$$\lim_{r \to 0} \frac{\mathrm{Re}[\ln \tilde{p}(rk)]}{\mathrm{Re}[\ln \tilde{p}(r)]} = k \tag{A.6}$$

and

$$\lim_{r \to 0} \frac{\mathrm{Im}[\ln \tilde{p}(rk) - r \ln \tilde{p}(k)]}{\mathrm{Re}[\ln \tilde{p}(r)]} = \frac{2\beta}{\pi} k \ln k. \tag{A.7}$$

In these equations, $\tilde{p}$ stands for

$$\tilde{p}(k) = p(k) \quad for\ \alpha \leq 1,$$
$$\tilde{p}(k) = e^{-i\langle x \rangle k} p(k) \quad for\ \alpha > 1.$$

Both theorems are taken from [62].

A.2 Closed-Form Expressions and Asymptotic Behavior

Although the canonical representation completely defines the expression of all possible stable distributions by (see (4.11))

$$\ln L_{\alpha,\beta}(k) = i\gamma k - c|k|^\alpha \left(1 + i\beta \frac{k}{|k|}\omega(k,\alpha)\right),$$

where γ is an arbitrary real constant, $c \geq 0$, $0 < \alpha \leq 2$, $-1 \leq \beta \leq 1$, and the function $\omega(k,\alpha)$ is given by

$$\omega(k,\alpha) = \begin{cases} \tan(\pi\alpha/2) & \text{for } \alpha \neq 1, \\ (2/\pi)\ln|k| & \text{for } \alpha = 1, \end{cases}$$

the inverse Fourier transform of $L_{\alpha,\beta}(k)$ can only be carried out in a limited number of cases. In the following, we compile (most of) these cases and present several useful asymptotic expansions. Our presentation is based on the excellent discussions in [17, 144, 147], where further details can also be found.

The Gaussian: $\alpha = 2$ If $\alpha = 2$, $\omega(k,2) = 0$. Then $L_{2,\beta}(x)$ becomes independent of β and is given by

$$L_2(x) = \frac{1}{2\pi} \int_{-\infty}^{+\infty} dk\, e^{-ik(x-\gamma)-ck^2} = \frac{1}{\sqrt{4\pi c}} \exp\left(-\frac{(x-\gamma)^2}{4c}\right). \tag{A.8}$$

This shows that γ merely shifts the origin and c is only a scale factor if the imaginary term of (4.11) vanishes. Therefore, we set $\gamma = 0$ and $c = 1$ for the following discussion of the symmetric cases, where $\beta = 0$.

The Symmetric Cases: $\beta = 0$ If $\beta = 0$, $L_{\alpha,0}(x)$ is an even function of x. Explicit expressions are known for $\alpha = 1/2, 2/3, 1$.

$\alpha = 1/2$:

$$\begin{aligned} L_{1/2,0}(x) &= \frac{1}{2\pi} \int_{-\infty}^{+\infty} dk\, \exp(-ikx - |k|^{1/2}) \\ &= \frac{1}{\pi} \int_0^\infty dk\, \cos(kx)\exp(-k^{1/2}) \\ &= \frac{1}{2\pi x} \int_0^\infty dk\, k^{-1/2}\sin(kx)\exp(-k^{1/2}) \\ &\stackrel{y=\sqrt{kx}}{=} \frac{1}{\pi x^{3/2}} \int_0^\infty dy\, \sin(y^2)\exp\left(-\frac{y}{x^{1/2}}\right), \end{aligned} \tag{A.9}$$

where we used an integration by parts to obtain the penultimate expression. The integral in (A.9) is given by (see [1], p. 303, Eq. (7.4.23))

$$\left(\frac{2}{\pi}\right)^{1/2} \int_0^\infty dy \sin(y^2) \exp\left(-\frac{y}{x^{1/2}}\right)$$
$$= \left[\frac{1}{2} - C\left(\sqrt{\frac{1}{2\pi x}}\right)\right] \cos\left(\frac{1}{4x}\right) + \left[\frac{1}{2} - S\left(\sqrt{\frac{1}{2\pi x}}\right)\right] \sin\left(\frac{1}{4x}\right), \quad \text{(A.10)}$$

where $x > 0$ and $C(x)$ and $S(x)$ are Fresnel integrals

$$C(x) = \int_0^x dy \cos\left(\frac{1}{2}\pi y^2\right), \quad S(x) = \int_0^x dy \sin\left(\frac{1}{2}\pi y^2\right). \quad \text{(A.11)}$$

The solution of the integral in (A.10) is tabulated in [1], where it is abbreviated by $g(x)$ (see Table 7.8, p. 323 and Eq. (7.3.6), p. 300). The function $g(x)$ starts at $1/2$ for $x = 0$ and decreases monotonously towards zero if x tends to infinity. Therefore, the full solution for the stable distribution reads

$$L_{1/2,0}(x) = \frac{1}{(2\pi)^{1/2} x^{3/2}} g\left([1/2\pi x]^{1/2}\right). \quad \text{(A.12)}$$

$\alpha = 2/3$: The stable law was found by Zolotarev and can be related to the Whittaker function $W_{u,v}(x)$ ([1], p. 505) by [147]

$$L_{2/3,0}(x) = \frac{1}{\pi} \int_0^\infty dk \cos(kx) \exp(-k^{2/3})$$
$$= \left(\frac{3}{\pi}\right)^{1/2} \frac{1}{x} \exp\left(-\frac{2}{27x^2}\right) W_{1/2,1/6}\left(\frac{4}{27x^2}\right). \quad \text{(A.13)}$$

$\alpha = 1$: This yields the Cauchy distribution

$$L_{1,0}(x) = \frac{1}{2\pi} \int_{-\infty}^{+\infty} dk \, e^{-ikx - |k|} = \frac{1}{\pi} \frac{1}{1 + x^2}. \quad \text{(A.14)}$$

In addition to these explicit formulas, the asymptotic expansions for the general case of α are known if $x \to 0$ and $x \to \infty$. A series expansion for small x can be obtained from

$$L_{\alpha,0}(x) = \frac{1}{2\pi} \int_{-\infty}^{+\infty} dk \exp(-ikx - |k|^\alpha) = \frac{1}{\pi} \int_0^\infty dk \cos(kx) \exp(-k^\alpha)$$
$$= \frac{1}{\pi} \sum_{n=0}^\infty \frac{(-1)^n}{(2n)!} x^{2n} \int_0^\infty dk \, k^{2n} \exp(-k^\alpha)$$
$$\stackrel{y=k^\alpha}{=} \frac{1}{\pi\alpha} \sum_{n=0}^\infty \frac{(-1)^n}{(2n)!} \Gamma\left(\frac{1+2n}{\alpha}\right) x^{2n}. \quad \text{(A.15)}$$

A.2 Closed-Form Expressions and Asymptotic Behavior

This series diverges for all x if $0 < \alpha < 1$ and converges for all x if $1 < \alpha < 2$. In the divergent range, it may further be considered as an asymptotic expansion around $x = 0$. If $x = 0$, only the $n = 0$ term of (A.15) survives and one obtains

$$L_{\alpha,0}(0) = \frac{1}{\pi\alpha}\Gamma\left(\frac{1}{\alpha}\right). \tag{A.16}$$

On the other hand, Wintner found the following large-x expansion [144, 147]:

$$L_{\alpha,0}(x) = \frac{\alpha}{\pi}\sum_{n=0}^{\infty}\frac{(-1)^n}{n!}\frac{\Gamma([1+n]\alpha)}{x^{\alpha(n+1)+1}}\sin\left[\frac{1}{2}\pi(n+1)\alpha\right] \tag{A.17}$$

$$= \frac{1}{\pi}\sum_{n=1}^{\infty}\frac{(-1)^{n+1}}{n!}\frac{\Gamma(1+n\alpha)}{x^{\alpha n+1}}\sin\left(\frac{1}{2}\pi n\alpha\right), \tag{A.18}$$

in which we used

$$\Gamma\big(1+(n+1)\alpha\big) = (n+1)\alpha\Gamma\big((n+1)\alpha\big)$$

in the last step. The series (A.17) and (A.18) converge for $x > 0$ when $0 < \alpha < 1$, but diverge for all x if $1 < \alpha < 2$. Even in the first case, the convergence is very slow if x is small. The leading term of this expansion is

$$L_{\alpha,0}(x) \sim \frac{\Gamma(1+\alpha)\sin(\pi\alpha/2)}{\pi}\frac{1}{x^{\alpha+1}} \quad (x \to \infty). \tag{A.19}$$

The Asymmetric Cases: $\beta = \pm 1$ If $0 < \alpha < 1$, the stable distributions are concentrated in the intervals $(-\infty, 0]$ and $[0, \infty)$ (choosing $\gamma = 0$) for $\beta = 1$ and $\beta = -1$, respectively. We want to illustrate this general property by two examples where closed-form expressions can be obtained, $\alpha = 1/3$ and $\alpha = 1/2$:

$\alpha = 1/3$: Let $K_{1/3}(x)$ be the modified Bessel function of order $1/3$. This function is real and positive for $x > 0$ and tends to zero for $x \to \infty$ [1]. If $0 < x < \infty$, Zolotarev found that $L_{\alpha,\beta}(x)$ with $\alpha = 1/3$, $\beta = -1$ and $c = 1$ can be related to $K_{1/3}(x)$ by the following expression [17, 147]:

$$L_{1/3,-1}(x) = \frac{1}{3^{1/4}\pi}\left(\frac{2}{3x}\right)^{3/2}K_{1/3}\left(\frac{3^{1/4}4}{9}\left(\frac{2}{3x}\right)^{1/2}\right). \tag{A.20}$$

$\alpha = 1/2$: For this choice, $\tan(\pi\alpha/2) = 1$. Choose, again, $c = 1$. Then the stable distribution reads

$$\begin{aligned}
L_{1/2,-1}(x) &= \frac{1}{2\pi}\int_{-\infty}^{+\infty}dk\,\exp\left[-ikx - |k|^{1/2}\left(1-i\frac{k}{|k|}\right)\right] \\
&= \frac{1}{\pi}\int_0^{\infty}dk\,\cos(kx - k^{1/2})\exp(-k^{1/2}) \\
&\stackrel{y=k^{1/2}}{=} \frac{2}{\pi}\int_0^{\infty}dy\,y\cos(y^2 x - y)\exp(-y).
\end{aligned} \tag{A.21}$$

Decomposing the cosine as

$$\cos(a - b) = \cos a \cos b + \sin a \sin b,$$

we obtain two integrals

$$\int_0^\infty dy\, y \cos(y^2 x) \cos(y) e^{-y}, \qquad \int_0^\infty dy\, y \sin(y^2 x) \sin(y) e^{-y},$$

both of which have the same value (see [67], p. 532, Eq. (3.965)), for instance,

$$\int_0^\infty dy\, y \cos(y^2 x) \cos(ay) e^{-ay} = \frac{a}{4}\sqrt{\frac{\pi}{2x^3}} e^{-a^2/2x} \quad (x > 0). \tag{A.22}$$

If $x > 0$, both integrals have the same magnitude and the same sign, whereas they differ in sign when $x < 0$ and thus cancel each other. If $x = 0$, both integrals vanish (see [67], p. 524, Eqs. (3.944–6.)). Therefore, we find

$$L_{1/2,-1}(x) = \begin{cases} 0 & \text{for } x \leq 0, \\ \frac{1}{(2\pi)^{1/2}} \frac{1}{x^{3/2}} \exp(-\frac{1}{2x}) & \text{for } x > 0. \end{cases} \tag{A.23}$$

For the general case $0 < \alpha < 1$, one can derive a series expansion [17, 52, 144]

$$L_{\alpha,-1}(x) = -\frac{1}{\pi}\sum_{n=1}^\infty \frac{(-c')^n}{n!} \frac{\Gamma(1+n\alpha)}{x^{\alpha n+1}} \sin(\pi n\alpha), \tag{A.24}$$

where $c' = c/\cos(\pi\alpha/2)$, and show that $L_{\alpha,-1}(x)$ may be written as an inverse Laplace transform [144]

$$L_{\alpha,-1}(x) = \frac{1}{2\pi i}\int_{\delta-i\infty}^{\delta+i\infty} ds\, e^{sx} \exp(-c' s^\alpha). \tag{A.25}$$

Appendix B
Hyperspherical Polar Coordinates

Let $\mathbf{r}$ denote a vector in d-dimensional Euclidean space. Its components,

$$r_1, r_2, \ldots, r_d,$$

can be expressed in hyperspherical polar coordinates,

$$r, \theta_1, \ldots, \theta_{d-2}, \phi,$$

as

$$
\begin{aligned}
r_1 &= r \sin\theta_{d-2} \sin\theta_{d-3} \cdots \sin\theta_1 \cos\phi, \\
r_2 &= r \sin\theta_{d-2} \sin\theta_{d-3} \cdots \sin\theta_1 \sin\phi, \\
r_3 &= r \sin\theta_{d-2} \sin\theta_{d-3} \cdots \sin\theta_2 \cos\theta_1, \\
&\vdots \\
r_{d-2} &= r \sin\theta_{d-2} \sin\theta_{d-3} \cos\theta_{d-4}, \\
r_{d-1} &= r \sin\theta_{d-2} \cos\theta_{d-3}, \\
r_d &= r \cos\theta_{d-2},
\end{aligned}
\tag{B.1}
$$

where $r = |\mathbf{r}| \geq 0$ and

$$
\begin{aligned}
0 &\leq \theta_i \leq \pi, \quad i = 1, \ldots, d-2, \\
0 &\leq \phi \leq 2\pi.
\end{aligned}
\tag{B.2}
$$

In these coordinates, the d-dimensional volume element is given by

$$\mathrm{d}^d \mathbf{r} = r^{d-1} (\sin\theta_{d-2})^{d-2} (\sin\theta_{d-3})^{d-3} \cdots \sin\theta_1 \, \mathrm{d}r \, \mathrm{d}\theta_{d-2} \cdots \mathrm{d}\theta_1 \, \mathrm{d}\phi. \tag{B.3}$$

The total surface area, S_d, of a d-dimensional unit sphere can be calculated either by integrating (B.3) over all angles or by the following trick: Since

$$\int_{-\infty}^{+\infty} d^d r \exp(-r^2) = S_d \int_0^\infty dr\, r^{d-1} \exp(-r^2)$$
$$\stackrel{r^2=x}{=} \frac{1}{2} S_d \int_0^\infty dx\, x^{(d/2)-1} e^{-x} = \frac{1}{2} S_d \Gamma(d/2),$$

but also

$$\int_{-\infty}^{+\infty} d^d r \exp(-r^2) = \prod_{\alpha=1}^{d} \left[\int_{-\infty}^{+\infty} dr_\alpha \exp(-r_\alpha^2) \right] = \pi^{d/2},$$

one finds

$$S_d = \frac{2\pi^{d/2}}{\Gamma(d/2)}, \qquad (B.4)$$

where $\Gamma(x)$ denotes the Gamma function [1]. To obtain the same result by direct integration, we need the following integrals (m is a positive integer) [67]:

$$\int_0^\pi (\sin x)^{2m} dx = \frac{(2m-1)!!}{(2m)!!} \pi, \qquad (B.5)$$

$$\int_0^\pi (\sin x)^{2m+1} dx = 2\frac{(2m)!!}{(2m+1)!!} \pi. \qquad (B.6)$$

The double factorial notation means

$$(2m)!! = 2m(2m-2)\cdots 4\cdot 2, \quad 0!! = 1, \qquad (B.7)$$
$$(2m+1)!! = (2m+1)(2m-1)\cdots 3\cdot 1. \qquad (B.8)$$

Quite generally, we have

$$S_d = 2\pi \int_0^\pi d\theta_{d-2}(\sin\theta_{d-2})^{d-2} \cdots \int_0^\pi d\theta_1 \sin\theta_1$$
$$= \left(\int_0^\pi d\theta_{d-2}(\sin\theta_{d-2})^{d-2} \right) S_{d-1}. \qquad (B.9)$$

This recurrence relation is sometimes needed in the main text. If d is even, we then obtain

$$S_d = 2\pi \int_0^\pi d\theta_{d-2}(\sin\theta_{d-2})^{d-2} \cdots \int_0^\pi d\theta_1 \sin\theta_1$$
$$= 2\pi \left(\frac{(d-3)!!}{(d-2)!!} \pi \right) \left(2\frac{(d-4)!!}{(d-3)!!} \right) \left(\frac{(d-5)!!}{(d-4)!!} \pi \right) \cdots \left(2\frac{0!!}{1!!} \right)$$
$$= \frac{(2\pi)^{d/2}}{(d-2)!!} = \frac{2\pi^{d/2}}{\Gamma(d/2)}, \qquad (B.10)$$

B Hyperspherical Polar Coordinates

since
$$(d-2)!! = 2^{(d/2)-1}[(d/2)-1]! = 2^{(d/2)-1}\Gamma(d/2).$$

Similarly, if d is odd,
$$\begin{aligned}S_d &= 2\pi\left(2\frac{(d-3)!!}{(d-2)!!}\right)\left(\frac{(d-4)!!}{(d-3)!!}\pi\right)\cdots\left(2\frac{0!!}{1!!}\right)\\ &= 2\pi^{d/2}\left(\frac{2^{(d-1)/2}}{\sqrt{\pi}(d-2)!!}\right) = \frac{2\pi^{d/2}}{\Gamma(d/2)},\end{aligned} \quad (\text{B.11})$$

in which the relation [1]
$$\Gamma\left(m+\frac{1}{2}\right) = \frac{\sqrt{\pi}(2m-1)!!}{2^m}$$

was exploited in the last step.

With the help of (B.4), one can calculate the surface area $S_d(R)$ and the volume $V_d(R)$ of a hypersphere with radius R from (B.3). The results are

$$S_d(R) = \frac{2\pi^{d/2}R^{d-1}}{\Gamma(d/2)}, \qquad V_d(R) = \frac{2\pi^{d/2}R^d}{d\Gamma(d/2)}. \quad (\text{B.12})$$

Appendix C
The Weierstrass Random Walk Revisited

Imagine a random walk in a d-dimensional space whose 'steps' can be isotropically performed in any direction and have a variable length distributed according to

$$p(\boldsymbol{\ell}) = \frac{1}{S_d \ell^{d-1}} \frac{N-1}{N} \sum_{j=0}^{\infty} \frac{1}{N^j} \delta\bigl(|\boldsymbol{\ell}| - b^j a\bigr) =: \frac{1}{S_d \ell^{d-1}} p_1(\ell), \qquad \text{(C.1)}$$

where $S_d = 2\pi^{d/2}/\Gamma(d/2)$ is the surface area of a unit sphere in d dimensions. The prefactor in (C.1) guarantees that $p(\boldsymbol{\ell})$ is normalized. Equation (C.1) generalizes the one-dimensional Weierstrass walk of Sect. 4.2 to higher space dimensions. It is sometimes called the 'fractal Rayleigh-Pearson walk' [81].

As in Sect. 4.2, the properties of the walk can be elaborated by calculating the characteristic function. Using hyperspherical polar coordinates (see Appendix B),

$$\ell, \theta_1, \ldots, \theta_{d-2}, \phi,$$
$$(\ell = |\boldsymbol{\ell}| \geq 0, 0 \leq \theta_i \leq \pi \text{ for } i = 1, \ldots, d-2, 0 \leq \phi \leq 2\pi)$$

so that

$$\mathrm{d}^d \boldsymbol{\ell} = \ell^{d-1} (\sin \theta_{d-2})^{d-2} (\sin \theta_{d-3})^{d-3} \cdots \sin \theta_1 \, \mathrm{d}\ell \, \mathrm{d}\theta_{d-2} \cdots \mathrm{d}\theta_1 \, \mathrm{d}\phi,$$

we can express the characteristic function as

$$p(\boldsymbol{k}) = \int_{-\infty}^{+\infty} \mathrm{d}^d \boldsymbol{\ell} \, p(\boldsymbol{\ell}) \exp(\mathrm{i} \boldsymbol{k} \cdot \boldsymbol{\ell})$$

$$= \frac{S_{d-1}}{S_d} \int_0^{\infty} \mathrm{d}\ell \, p_1(\ell) \int_0^{\pi} \mathrm{d}\theta (\sin \theta)^{d-2} \exp(\mathrm{i}|\boldsymbol{k}||\boldsymbol{\ell}| \cos \theta)$$

$$= \frac{S_{d-1}}{S_d} \int_0^{\infty} \mathrm{d}\ell \, p_1(\ell) \int_0^{\pi/2} \mathrm{d}\theta \, 2(\sin \theta)^{d-2} \cos(|\boldsymbol{k}||\boldsymbol{\ell}| \cos \theta)$$

$$= \Gamma\!\left(\frac{d}{2}\right) \int_0^{\infty} \mathrm{d}\ell \left(\frac{2}{k\ell}\right)^{(d/2)-1} J_{(d/2)-1}(k\ell) p_1(\ell) \quad (d > 1), \qquad \text{(C.2)}$$

W. Paul, J. Baschnagel, *Stochastic Processes*, DOI 10.1007/978-3-319-00327-6,
© Springer International Publishing Switzerland 2013

in which we have used the following integral representation of the Bessel function $J_{(d/2)-1}(x)$ (see [67], p. 962, Eqs. (8.411–4.)):

$$J_{(d/2)-1}(x) = \frac{2}{\sqrt{\pi}\,\Gamma([d-1]/2)}\left[\frac{x}{2}\right]^{(d/2)-1}\int_0^{\pi/2} d\theta\,(\sin\theta)^{2[(d/2)-1]}\cos(x\cos\theta), \tag{C.3}$$

which is valid as long as $d > 1$. Inserting (C.1) into (C.2) we obtain

$$p(k) = \frac{N-1}{N}\sum_{j=0}^{\infty}\frac{1}{N^j}\left[\Gamma\left(\frac{d}{2}\right)\left(\frac{2}{kb^j a}\right)^{(d/2)-1} J_{(d/2)-1}(kb^j a)\right]. \tag{C.4}$$

Equation (C.4) leaves us with the problem of summing the infinite series. This seems to be a formidable task. However, there is a cunning way to proceed, by expressing the term in the square brackets as a *Mellin transform*.

Interpolation: Mellin Transforms Mellin transforms are integral transformations such as Fourier or Laplace transforms. The Mellin transform $f(p)$ of a function $f(x)$ is defined by

$$f(p) = \int_0^\infty dx\, x^{p-1} f(x). \tag{C.5}$$

The transform exists if

$$\int_0^\infty |f(x)| x^{q-1} dx < \infty$$

for some $q > 0$. Then, the inverse Mellin transform is given by

$$f(x) = \frac{1}{2\pi i}\int_{c-i\infty}^{c+i\infty} dp\, \frac{1}{x^p} f(p) \quad (x > 0), \tag{C.6}$$

with $c > q$. Consider now a series of the form

$$\phi(k) = \sum_{j=0}^{\infty}\frac{1}{N^j} f(kb^j a). \tag{C.7}$$

If we replace $f(kb^j a)$ by its inverse Mellin transform,

$$f(kb^j a) = \frac{1}{2\pi i}\int_{c-i\infty}^{c+i\infty} dp\, \frac{1}{(kb^j a)^p} f(p) \quad (k > 0), \tag{C.8}$$

we obtain a geometric series which can be summed to give

$$\phi(k) = \frac{1}{2\pi i}\int_{c-i\infty}^{c+i\infty} dp\, \frac{1}{1 - N^{-1}b^{-p}}\frac{1}{(ka)^p} f(p), \tag{C.9}$$

C The Weierstrass Random Walk Revisited

provided the order of summation and integration may be interchanged. The hope is that these manipulations lead to a contour integral which is tractable by the residue theorem.

Let us see whether we can successfully apply these ideas to the d-dimensional Weierstrass random walk. The function $f(x)$ in (C.7) corresponds to the term in square brackets in (C.4). However, the convergence properties of this term are not sufficient for our purpose, since

$$\Gamma\left(\frac{d}{2}\right)\left(\frac{x}{2}\right)^{1-d/2} J_{(d/2)-1}(x) \xrightarrow{x\to 0} 1.$$

It is therefore much better to consider

$$f(x) = \Gamma\left(\frac{d}{2}\right)\left(\frac{x}{2}\right)^{1-d/2} J_{(d/2)-1}(x) - 1, \tag{C.10}$$

which has the inverse Mellin transform

$$f(x) = \frac{1}{2}\frac{1}{2\pi i}\int_{c-i\infty}^{c+i\infty} dp \left(\frac{2}{x}\right)^p \frac{\Gamma(d/2)\Gamma(p/2)}{\Gamma([d-p]/2)} \tag{C.11}$$

for $-2 < c = \text{Re}(p) < 0$. The characteristic function can now be written as

$$p(\boldsymbol{k}) = \frac{N-1}{N}\sum_{j=0}^{\infty}\frac{1}{N^j}\left[1 + f\left(kb^j a\right)\right]$$

$$= 1 + \frac{N-1}{2N(2\pi i)}\sum_{j=0}^{\infty}\int_{c-i\infty}^{c+i\infty} dp \left(\frac{1}{Nb^p}\right)^j \left(\frac{2}{ka}\right)^p \frac{\Gamma(d/2)\Gamma(p/2)}{\Gamma([d-p]/2)}. \tag{C.12}$$

Now, we would like to interchange the order of summation and integration. This is possible if we restrict c to the interval $\max(-2, -\ln N/\ln b) < c < 0$, since

$$c > -\alpha := -\frac{\ln N}{\ln b} \Rightarrow \frac{1}{|Nb^p|} < 1 \tag{C.13}$$

and

$$|\Gamma(x+iy)| \sim \sqrt{2\pi}|y|^{x-1/2}\exp\left[-\frac{1}{2}\pi|y|\right].$$

when $|y| \to \infty$, which implies absolute convergence, so the order of summation and integration can be interchanged. Therefore, we obtain

$$p(\boldsymbol{k}) = 1 + \frac{N-1}{2N(2\pi i)}\int_{c-i\infty}^{c+i\infty} dp \frac{\Gamma(p/2)}{1-N^{-1}b^{-p}}\left[\frac{2}{ka}\right]^p \frac{\Gamma(d/2)}{\Gamma([d-p]/2)} \tag{C.14}$$

with $\max(-2, -\alpha) < c < 0$. In the integration interval, the integrand has simple poles at

- $p = -2n = -2, -4, \ldots$ (due to the Γ function),
- $p = p_m = -\alpha + (2\pi i/\ln b)m$, $m = 0, \pm 1, \pm 2, \ldots$ (due to $(1 - N^{-1}b^{-p})^{-1}$).

This gives rise to the following residues:

- Since the poles of the Γ function are simple, we have

$$\lim_{p \to -2n}\left[(p+2n)\Gamma\left(\frac{p}{2}\right)\right] = \lim_{\epsilon \to 0}\left[2\epsilon\Gamma(-n+\epsilon)\right]$$
$$= \lim_{\epsilon \to 0}\left(2\epsilon\frac{(-1)^n\Gamma(1+\epsilon)}{(n-\epsilon)\cdots(1-\epsilon)\epsilon}\right)$$
$$= 2\frac{(-1)^n}{n!}$$

so that the residue of the integrand is

$$\lim_{p \to -2n}\left[\frac{(p+2n)\Gamma(p/2)}{1-N^{-1}b^{-p}}\left(\frac{2}{ka}\right)^p\frac{\Gamma(d/2)}{\Gamma([d-p]/2)}\right]$$
$$= \frac{2(-1)^n}{n!}\frac{1}{1-N^{-1}b^{2n}}\left(\frac{ka}{2}\right)^{2n}\frac{\Gamma(d/2)}{\Gamma(n+d/2)}.$$

- The poles of $(1-N^{-1}b^{-p})^{-1}$ are also simple. This gives (with L'Hospital's rule)

$$\lim_{p \to p_m}\left(\frac{p-p_m}{1-N^{-1}b^{-p}}\right) = \frac{N}{b^{-p_m}\ln b} = \frac{1}{\ln b},$$

which implies that the residue of the integrand is

$$\lim_{p \to p_m}\left[\frac{(p-p_m)\Gamma(p/2)}{1-N^{-1}b^{-p}}\left(\frac{2}{ka}\right)^p\frac{\Gamma(d/2)}{\Gamma([d-p]/2)}\right]$$
$$= \frac{(ka/2)^\alpha}{\ln b}\frac{\Gamma(d/2)\Gamma(p_m/2)}{\Gamma([d-p_m]/2)}\exp\left(-2\pi i m\frac{\ln(|ka|/2)}{\ln b}\right).$$

If we now choose the rectangle with the corners $p = c \pm (2M+1)\pi i/\ln b$ and $p = -(2K+1) \pm (2M+1)\pi i/\ln b$ as the integration contour $\mathcal{C}$, the integral of (C.14) can be expressed as

$$\int_{c-i\infty}^{c+i\infty} = \oint_\mathcal{C} - \int_{c+(2M+1)\pi i/\ln b}^{-(2K+1)+(2M+1)\pi i/\ln b} - \int_{-(2K+1)+(2M+1)\pi i/\ln b}^{-(2K+1)-(2M+1)\pi i/\ln b}$$
$$- \int_{-(2K+1)-(2M+1)\pi i/\ln b}^{c-(2M+1)\pi i/\ln b}.$$

C The Weierstrass Random Walk Revisited

In the limits $M \to \infty$ and $K \to \infty$, the last three integrals on the right-hand vanish [81], whereas the contour integral is given by the residue theorem

$$\frac{1}{2}\frac{1}{2\pi i}\oint_C dp \frac{\Gamma(p/2)}{1-N^{-1}b^{-p}}\left(\frac{2}{ka}\right)^p \frac{\Gamma(d/2)}{\Gamma([d-p]/2)}$$

$$= \sum_{n=1}^{\infty} \frac{(-1)^n}{n!} \frac{1}{1-N^{-1}b^{2n}} \left(\frac{ka}{2}\right)^{2n} \frac{\Gamma(d/2)}{\Gamma(n+d/2)}$$

$$+ \frac{(ka/2)^\alpha}{2\ln b} \sum_{m=-\infty}^{\infty} \frac{\Gamma(d/2)\Gamma(p_m/2)}{\Gamma([d-p_m]/2)} \exp\left(-2\pi i m \frac{\ln(|ka|/2)}{\ln b}\right).$$

(C.15)

Inserting these results into (C.14) we finally obtain for the following characteristic function of the d-dimensional Weierstrass random walk:

$$p(k) = \frac{N-1}{N} \sum_{n=0}^{\infty} \frac{(-1)^n}{n!} \frac{1}{1-N^{-1}b^{2n}} \left(\frac{ka}{2}\right)^{2n} \frac{\Gamma(d/2)}{\Gamma(n+d/2)}$$

$$+ \frac{N-1}{N} \frac{(ka/2)^\alpha}{2\ln b} \sum_{m=-\infty}^{\infty} \frac{\Gamma(d/2)\Gamma(p_m/2)}{\Gamma([d-p_m]/2)} \exp\left[-2\pi i m \frac{\ln(|ka|/2)}{\ln b}\right].$$

(C.16)

The One-dimensional Case Since (C.3) is only valid for $d > 1$, the solution of the one-dimensional Weierstrass walk is not a special limit of (C.16), although the correct result is obtained by formally setting $d = 1$ in (C.3) and exploiting the following properties of the Γ function (see [1], p. 256, Eqs. (6.1.17–18)):

$$\Gamma(2z) = (2\pi)^{-1/2} 2^{2z-1/2} \Gamma(z)\Gamma(z+1/2),$$

$$\Gamma(z+1/2)\Gamma(1/2-z) = \pi/\cos\pi z.$$

A rigorous analysis of the Weierstrass walk has to start from (see (4.26))

$$p(k) = \frac{N-1}{N} \sum_{j=0}^{\infty} \frac{1}{N^j} \cos(kb^j a),$$

and, after this, proceed analogously to the previous discussion with a different function $f(x)$ than used in (C.10). If $1/2 < \alpha < 2$, a suitable choice is

$$f(x) = \cos x - 1 = \frac{1}{2\pi i} \int_{c-i\infty}^{c+i\infty} dp \frac{\Gamma(p)}{x^p} \cos\left(\frac{\pi p}{2}\right), \qquad (C.17)$$

whereas one should take, for $0 < \alpha \leq 1/2$,

$$f(x) = \cos x \exp(-\epsilon x) - 1 \quad (\epsilon > 0)$$

$$= \frac{1}{2\pi i} \int_{c-i\infty}^{c+i\infty} dp \frac{\Gamma(p)}{x^p (1+\epsilon^2)^{p/2}} \cos[p \arctan(1/\epsilon)], \qquad (C.18)$$

where the additional exponential factor is needed to warrant absolute convergence. The final result is obtained in the limit $\epsilon \to 0$. It reads, for $0 < \alpha < 2$,

$$p(k) = \frac{N-1}{N} \frac{(ka)^\alpha}{\ln b} \sum_{m=-\infty}^{\infty} \Gamma(p_m) \cos\left(\frac{\pi}{2} p_m\right) \exp\left(-2\pi i n \frac{\ln|ka|}{\ln b}\right)$$

$$+ \frac{N-1}{N} \sum_{n=0}^{\infty} \frac{(-1)^n}{(2n)!} \frac{(ka)^{2n}}{1-N^{-1}b^{2n}}, \qquad (C.19)$$

where p_m is given by $p_m = -\alpha + (2\pi i / \ln b)m$, as before. A more detailed discussion can be found in [81].

Appendix D
The Exponentially Truncated Lévy Flight

Consider N independent random variables, $\{\ell_n\}_{n=1,\ldots,N}$, which are identically distributed according to an exponentially truncated Lévy flight. This means that the probability density $p(\ell)$ behaves asymptotically for large ℓ as

$$p(\ell) \sim \begin{cases} c_- e^{-\lambda|\ell|}|\ell|^{-(1+\alpha)} & \ell < 0, \\ c_+ e^{-\lambda\ell}\ell^{-(1+\alpha)} & \ell > 0, \end{cases} \quad (D.1)$$

where $0 < \alpha < 2$ and λ is a positive cut-off parameter. Using this definition, we want to derive the distribution $L^t_{\alpha,\beta}(x, N)$ for the sum of the random variables ℓ_n in the limit of large N. Following [108], this limit distribution can be defined in terms of the characteristic function[1]

$$L^t_{\alpha,\beta}(k, N) = \int_{-\infty}^{+\infty} dx\, L^t_{\alpha,\beta}(x, N) e^{ikx}, \quad (D.2)$$

by

$$\ln L^t_{\alpha,\beta}(k, N) = i\gamma' k + N \int_{-\infty}^{+\infty} d\ell \left(e^{ik\ell} - 1\right) p(\ell), \quad (D.3)$$

if $0 < \alpha < 1$, and by

$$\ln L^t_{\alpha,\beta}(k, N) = i\gamma'' k + N \int_{-\infty}^{+\infty} d\ell \left(e^{ik\ell} - 1 - ik\ell\right) p(\ell), \quad (D.4)$$

if $1 < \alpha < 2$. Here, γ' and γ'' are arbitrary real constants. The different integrands for the two regimes of α are necessary to remove the singularity at the origin. The

---

[1] Note that the sign in $\exp(ik\ell)$ of (D.2) is opposite to the convention of [108], who uses $\exp(-ik\ell)$ in the definition of the characteristic function. The consequence of the different choices is that our result for β (see (D.15)) is just -1 times the result of [108]. This is the reason why the β values can differ in sign in the literature.

integrals are then convergent. As explained in [62], this regularization only changes the value of γ' and γ''. Since these constants merely shift the origin in real space, the different regularizations do not affect the (essential) result for the probability distribution.

If z^* denotes the conjugate of the complex variable z, we immediately see from (D.3) or (D.4) that

$$\ln L^{\mathrm{t}}_{\alpha,\beta}(|k|, N) = \left[\ln L^{\mathrm{t}}_{\alpha,\beta}(-|k|, N)\right]^*. \tag{D.5}$$

Therefore, it suffices to consider $k > 0$. Let us now separately consider the cases $0 < \alpha < 1$ and $1 < \alpha < 2$.

Case: $0 < \alpha < 1$ For this range of α, the integral in (D.3) is convergent when (D.1) is inserted, and we can write

$$\ln L^{\mathrm{t}}_{\alpha,\beta}(k, N) = \mathrm{i}\gamma'k + Nc_- I_-(k, \alpha) + Nc_+ I_+(k, \alpha), \tag{D.6}$$

where

$$I_+(k, \alpha) = \int_0^\infty d\ell \left(e^{\mathrm{i}k\ell} - 1\right) \frac{e^{-\lambda\ell}}{\ell^{1+\alpha}} \tag{D.7}$$

and

$$\begin{aligned} I_-(k, \alpha) &= \int_{-\infty}^0 d\ell \left(e^{\mathrm{i}k\ell} - 1\right) \frac{e^{-\lambda|\ell|}}{|\ell|^{1+\alpha}} \\ &= \int_0^\infty d\ell \left(e^{-\mathrm{i}k\ell} - 1\right) \frac{e^{-\lambda\ell}}{\ell^{1+\alpha}} \\ &= I_+^*(k, \alpha). \end{aligned} \tag{D.8}$$

Therefore, we only have to calculate the integral $I_+(k, \alpha)$. Substituting $y = \lambda \ell$ and defining $\kappa = k/\lambda$, an integration by parts yields

$$\begin{aligned} I_+(k, \alpha) &= \lambda^\alpha \int_0^\infty dy \left(e^{\mathrm{i}\kappa y} - 1\right) \frac{e^{-y}}{y^{1+\alpha}} \\ &= \frac{\lambda^\alpha}{\alpha} \left[\int_0^\infty dy\, y^{(1-\alpha)-1} e^{-y} + (\mathrm{i}\kappa - 1) \int_0^\infty dy\, y^{(1-\alpha)-1} e^{-(1-\mathrm{i}\kappa)y}\right] \\ &= \frac{\lambda^\alpha}{\alpha} \Gamma(1-\alpha)\left[1 + (\mathrm{i}\kappa - 1)(1+\kappa^2)^{-(1-\alpha)/2} e^{\mathrm{i}(1-\alpha)\arctan\kappa}\right], \end{aligned} \tag{D.9}$$

where we performed an integration by parts in the penultimate step and then used the following integral (see [67], p. 364, Eqs. (3.381–5.)):

$$\int_0^\infty dy\, y^{\nu-1} e^{-(p+\mathrm{i}q)y} = \Gamma(\nu)(p^2+q^2)^{-\nu/2} e^{-\mathrm{i}\nu \arctan(q/p)}, \tag{D.10}$$

D The Exponentially Truncated Lévy Flight

which is valid for $p > 0$ and $\text{Re}\,\nu > 0$. With $p = 1$ and $\nu = 1 - \alpha$, both conditions are satisfied in the present case. If we take into account that

$$\arctan \kappa = \frac{1}{2i} \ln \frac{1 + i\kappa}{1 - i\kappa}$$

so that

$$\frac{i\kappa - 1}{\sqrt{1 + \kappa^2}} \exp(i \arctan \kappa) = -1,$$

the integral $I_+(k, \alpha)$ can finally be written as

$$I_+(k, \alpha) = \frac{\lambda^\alpha}{\alpha} \Gamma(1 - \alpha) \left[1 - (1 + \kappa^2)^{\alpha/2} \exp(-i\alpha \arctan \kappa)\right]. \tag{D.11}$$

Inserting this result into (D.6) and using (D.8), we obtain

$$\ln L^t_{\alpha,\beta}(k, N) = i\gamma' k + c_0 - c(N) \frac{(k^2 + \lambda^2)^{\alpha/2}}{\cos(\pi\alpha/2)} \cos\left(\alpha \arctan \frac{|k|}{\lambda}\right)$$
$$\times \left[1 + i\beta \tan\left(\alpha \arctan \frac{|k|}{\lambda}\right)\right], \tag{D.12}$$

where $c(N)$ is given by

$$c(N) = N \frac{\pi(c_+ + c_-) \cos(\pi\alpha/2)}{\alpha \Gamma(\alpha) \sin(\pi\alpha)} = N \frac{\pi(c_+ + c_-)}{2\alpha \Gamma(\alpha) \sin(\pi\alpha/2)}, \tag{D.13}$$

c_0 is a normalization constant (i.e., $L^t_{\alpha,\beta}(k = 0, N) = 1$)

$$c_0 = \frac{\lambda^\alpha}{\cos(\pi\alpha/2)} c(N), \tag{D.14}$$

and β has the following value:

$$\beta = \frac{c_- - c_+}{c_+ + c_-}. \tag{D.15}$$

Equation (D.12) is valid for $k > 0$. However, the solution for $k < 0$ is directly related to (D.12) via (D.5). Taking the complex conjugate of (D.12) only changes the sign of the imaginary part in parentheses, so the full solution for all k can be written

$$\ln L^t_{\alpha,\beta}(k, N) = i\gamma' k + c_0 - c(N) \frac{(k^2 + \lambda^2)^{\alpha/2}}{\cos(\pi\alpha/2)} \cos\left(\alpha \arctan \frac{|k|}{\lambda}\right)$$
$$\times \left[1 + i\beta \frac{k}{|k|} \tan\left(\alpha \arctan \frac{|k|}{\lambda}\right)\right]. \tag{D.16}$$

Case: $1 < \alpha < 2$ The analysis for this range of α-values proceeds similarly to the previous one if we start from (D.4) which we write

$$\ln L_{\alpha,\beta}^{t}(k, N) = i\gamma'' k + N c_- J_-(k, \alpha) + N c_+ J_+(k, \alpha), \tag{D.17}$$

where

$$J_+(k, \alpha) = \int_0^\infty d\ell \left(e^{ik\ell} - 1 - ik\ell \right) \frac{e^{-\lambda \ell}}{\ell^{1+\alpha}} \tag{D.18}$$

and

$$J_-(k, \alpha) = \int_{-\infty}^0 d\ell \left(e^{ik\ell} - 1 - ik\ell \right) \frac{e^{-\lambda|\ell|}}{|\ell|^{1+\alpha}}$$

$$= \int_0^\infty d\ell \left(e^{-ik\ell} - 1 + ik\ell \right) \frac{e^{-\lambda \ell}}{\ell^{1+\alpha}}$$

$$= J_+^*(k, \alpha). \tag{D.19}$$

As before, we only have to calculate $J_+(k, \alpha)$. Now, we introduce the new variable $y = \lambda \ell$, define $\kappa = k/\lambda$ and integrate by parts. This gives

$$J_+(k, \alpha) = \frac{\lambda^\alpha}{\alpha} \left(i\kappa \Gamma(2 - \alpha) + (i\kappa - 1) \int_0^\infty dy \left(e^{i\kappa y} - 1 \right) \frac{e^{-y}}{y^\alpha} \right)$$

$$= \frac{\lambda^\alpha}{\alpha} \left(i\kappa \Gamma(2 - \alpha) + (i\kappa - 1) I_+(k, \alpha - 1) \right)$$

$$= \frac{\lambda^\alpha}{\alpha} \Gamma(2 - \alpha) \left[i\kappa + \frac{i\kappa - 1}{\alpha - 1} \left(1 + \frac{i\kappa - 1}{(1 + \kappa^2)^{(2-\alpha)/2}} e^{i(2-\alpha)\arctan \kappa} \right) \right], \tag{D.20}$$

in which (D.9) was inserted in the last step. Since

$$\frac{(i\kappa - 1)^2}{1 + \kappa^2} \exp(2i \arctan \kappa) = 1,$$

we find

$$J_+(k, \alpha) = ik \left(\lambda^{\alpha-1} \frac{\Gamma(2-\alpha)}{\alpha - 1} \right) + I_+(k, \alpha), \tag{D.21}$$

which is, up to first term, the same result as for $0 < \alpha < 1$. However, the factor in parentheses is constant and may be incorporated in a redefinition of γ''. Therefore,

D The Exponentially Truncated Lévy Flight

we can write down the solution analogous to (D.16)

$$\ln L^t_{\alpha,\beta}(k,N) = i\gamma'' k + c_0 - c(N)\frac{(k^2+\lambda^2)^{\alpha/2}}{\cos(\pi\alpha/2)}\cos\left(\alpha\arctan\frac{|k|}{\lambda}\right)$$
$$\times\left[1 + i\beta\frac{k}{|k|}\tan\left(\alpha\arctan\frac{|k|}{\lambda}\right)\right]. \qquad (D.22)$$

This shows that the limit distribution of the exponentially truncated Lévy flight has the same form for both $0 < \alpha < 1$ and $1 < \alpha < 2$, barring the constants γ' and γ''. Since these constants only determine the origin in real space, they can be chosen to be equal to zero. Therefore, we have derived (4.98)–(4.101).

Appendix E
Put–Call Parity

A European option is a contract between two parties in which

- the seller (= writer) of the option
- grants the buyer (= holder) of the option
- the right to purchase (= *call option*) from the writer or to sell (= *put option*) to him an underlying with spot price $S(t)$
- for the strike price K
- at the expiry date T in the future.

At first sight, put and call options seem to be very different. However, they are intimately related to one another. An elegant way to show this is the arbitrage-pricing technique.

Suppose we have the following portfolio: we have bought one asset and one put option with prices S and $\mathcal{P}$, respectively, and we have sold one call for price $\mathcal{C}$. Both the put and the call option are chosen to have the same strike price. The portfolio has the value

$$\Pi(t) = S(t) + \mathcal{P}(S,t) - \mathcal{C}(S,t). \tag{E.1}$$

At expiry, the options yield (see (5.9) and (5.10))

$$\mathcal{C}(S,T) = \max\bigl(S(T) - K, 0\bigr),$$
$$\mathcal{P}(S,T) = \max\bigl(K - S(T), 0\bigr),$$

so that the payoff of the portfolio is

$$\text{if } S \geq K: \quad S + 0 - (S - K) = K,$$
$$\text{if } S \leq K: \quad S + (K - S) - 0 = K.$$

The portfolio gives a guaranteed gain K at time T. What is its value for $t < T$?

Since the outcome of the portfolio at expiry is fixed, the portfolio yields a risk-free profit. The only way to obtain such a riskless profit is to deposit an amount $Ke^{-r(T-t)}$ in a bank at time t and to wait until $t = T$. At time T, the initial deposit has grown to K by the addition of (continuously compounded) interest. Therefore, the value of the portfolio at an earlier time t must equal its discounted payoff, i.e.,

$$\Pi(t) = Ke^{-r(T-t)}, \tag{E.2}$$

otherwise arbitrage would be possible.

To prove this assertion, let us assume that $\Pi(t) = \epsilon Ke^{-r(T-t)}$ with $\epsilon \neq 1$. Two cases are possible, $\epsilon < 1$ and $\epsilon > 1$:

- $\Pi(t) < Ke^{-r(T-t)}$: In this case, an investor could borrow the amount $\Pi(t)$ from a bank and buy the portfolio. At time T, he owes the bank $\Pi(t)e^{r(T-t)}$, but receives the payoff K from the portfolio. Therefore, his total balance reads

$$K - \Pi(t)e^{r(T-t)} = K - \left(\epsilon Ke^{-r(T-t)}\right)e^{r(T-t)}$$
$$= K(1-\epsilon) > 0 \quad (\epsilon < 1).$$

- $\Pi(t) > Ke^{-r(T-t)}$: In this case, the investor could sell the portfolio at price $\Pi(t)$ and invest the portion $Ke^{-r(T-t)}$ of this receipt in the bank, while keeping

$$\Pi(t) - Ke^{-r(T-t)} = (\epsilon - 1)Ke^{-r(T-t)} > 0 \quad (\epsilon > 1).$$

At time T, his bank deposit has risen to K, and this is exactly the amount that he owes the buyer of the portfolio.

In both cases, the investor gains a riskless profit, which is not possible in an arbitrage-free market. Thus, (E.2) must be correct. Inserting (E.1) into (E.2), we obtain a relation between European call and put options, the so-called *put–call parity*

$$\mathcal{P}(S,t) = \mathcal{C}(S,t) - S(t) + Ke^{-r(T-t)}. \tag{E.3}$$

Appendix F
Geometric Brownian Motion

Geometric Brownian motion is a special variant of an Itô process,

$$dS = a_1(S,t)dt + \sqrt{a_2(S,t)}dW(t), \tag{F.1}$$

where the coefficients $a_1(S,t)$ and $\sqrt{a_2(S,t)}$ are proportional to the random variable S (see (2.119)–(2.121)). Let the constants of proportionality be denoted by μ (>0) and σ (>0). Thus, the motion is defined by the following stochastic differential equation (Langevin equation):

$$dS = \mu S(t)dt + \sigma S(t)dW(t) \quad (S(0) > 0). \tag{F.2}$$

The solution of (F.2) yields a log–normal distribution for the transition probability from (S',t') to (S,t),

$$p(S,t|S',t') = \frac{1}{\sqrt{2\pi(\sigma S)^2(t-t')}} \exp\left(-\frac{[\ln(S/S') - (\mu - \sigma^2/2)(t-t')]^2}{2\sigma^2(t-t')}\right), \tag{F.3}$$

from which the mean and the variance can be calculated using (1.33). The result is

$$E[S] = S'e^{\mu(t-t')}, \qquad \text{Var}[S] = (S')^2 e^{2\mu(t-t')}\left[e^{\sigma^2(t-t')} - 1\right]. \tag{F.4}$$

The aim of this appendix is to derive (F.3). To this end, we exploit the fact that (F.1) is equivalent to the Fokker–Planck equation

$$\frac{\partial}{\partial t} p(S,t|S',t')$$
$$= -\frac{\partial}{\partial S}[a_1(S,t)p(S,t|S',t')] + \frac{1}{2}\frac{\partial^2}{\partial S^2}[a_2(S,t)p(S,t|S',t')]. \tag{F.5}$$

Inserting the coefficients from (F.2) yields

$$\frac{\partial}{\partial t}p = (\sigma^2 - \mu)p + (2\sigma^2 - \mu)S\frac{\partial}{\partial S}p + \frac{1}{2}(\sigma S)^2\frac{\partial^2}{\partial S^2}p. \tag{F.6}$$

Equation (F.6) must be solved subject to the following boundary conditions:

$$\begin{aligned} t = t': \quad & p(S, t'|S', t') = \delta(S - S'), \\ S = 0: \quad & p(0, t|S', t') = 0, \\ S \to \infty: \quad & p(S, t|S', t') \to 0. \end{aligned} \tag{F.7}$$

The first condition simply states that the random variable S has the value S' at $t = t'$. The second can be understood by considering (F.2). If S vanishes at some time, it stays zero for all future times, and the motion stops. On the other hand, if $S(0) > 0$, as we assumed, it can never become zero at a later time if evolving according to (F.2). Therefore, $p(S, t|S', t')$ has to vanish for $S = 0$. Finally, the third condition expresses the fact that S cannot grow beyond any bound in a finite time interval $[t, t']$.

Given these boundary conditions, the solution of (F.6) proceeds in the same way as that of the Black-Scholes theory for call options (see Sect. 5.2.2). The main steps are as follows: First, we eliminate the dependence of the prefactors on S by introducing a new 'spatial' variable. Furthermore, we rescale all variables to make the equation dimensionless. This is achieved by using

$$p(S, t|S', t') = \frac{1}{S'} f(x, \tau), \quad S = S'e^x, \quad t = t' + \frac{\tau}{(\sigma^2/2)}, \tag{F.8}$$

which allows us to rewrite (F.6) and (F.7) as

$$\frac{\partial f}{\partial \tau} = \frac{\partial^2 f}{\partial x^2} + [3 - \kappa]\frac{\partial f}{\partial x} + [2 - \kappa]f \quad (\kappa = 2\mu/\sigma^2), \tag{F.9}$$

$$\begin{aligned} \tau = 0: \quad & f(x, 0) = \delta(e^x - 1), \\ x \to -\infty: \quad & f(x, \tau) \to 0, \\ x \to +\infty: \quad & f(x, \tau) \to 0. \end{aligned} \tag{F.10}$$

The second step consists in converting (F.9) into a diffusion equation by another change of variables. Let us make the ansatz

$$f(x, \tau) = e^{ax + b\tau} g(x, \tau) \quad (a \text{ and } b \text{ are real and arbitrary}), \tag{F.11}$$

and insert it into (F.9) and (F.10). This gives

$$\frac{\partial g}{\partial \tau} = \frac{\partial^2 g}{\partial x^2} + [2a + (3 - \kappa)]\frac{\partial g}{\partial x} + [a^2 + (3 - \kappa)a + 2 - \kappa - b]g. \tag{F.12}$$

F Geometric Brownian Motion

In order to obtain a diffusion equation, the second and third terms of (F.12) must vanish. Since a and b are arbitrary, this can be achieved by choosing

$$a := \frac{1}{2}(\kappa - 3), \qquad b := a^2 + (3-\kappa)a + 2 - \kappa = -\frac{1}{4}(\kappa-1)^2. \tag{F.13}$$

Thus, we obtain a diffusion equation,

$$\frac{\partial g}{\partial \tau} = \frac{\partial^2 g}{\partial x^2}, \tag{F.14}$$

with the following boundary conditions:

$$\tau = 0: \quad g(x,0) = e^{-(\kappa-3)x/2}\delta(e^x - 1)$$

$$\Rightarrow g(x,0)e^{-\alpha x^2} \xrightarrow{|x|\to\infty} 0 \,(\alpha > 0), \tag{F.15}$$

$$\tau > 0: \quad g(x,\tau)e^{-\alpha x^2} \xrightarrow{|x|\to\infty} 0 \,(\alpha > 0),$$

where α is an arbitrary, real, positive constant. This means that g vanishes sufficiently fast for $|x| \to \infty$, so a unique solution exists.

As in the case of the Black-Scholes theory, the solution of (F.14) can be obtained by the Green's function method (see (5.34)–(5.41)). The final result reads

$$g(x,\tau) = \frac{1}{\sqrt{4\pi\tau}} \int_{-\infty}^{+\infty} dy\, g(y,0) \exp\left(-\frac{(x-y)^2}{4\tau}\right)$$

$$\stackrel{u=e^y}{=} \frac{1}{\sqrt{4\pi\tau}} \int_0^\infty du\, \delta(u-1) \frac{e^{-\frac{1}{2}(\kappa-3)\ln u}}{u} \exp\left(-\frac{(x-\ln u)^2}{4\tau}\right)$$

$$= \frac{1}{\sqrt{4\pi\tau}} \exp\left(-\frac{x^2}{4\tau}\right). \tag{F.16}$$

If (F.16) is substituted into (F.11) and (F.13) is taken into account, we find

$$f(x,\tau) = \frac{1}{\sqrt{4\pi\tau}} \exp\left(\frac{1}{2}(\kappa-3)x - \frac{1}{4}(\kappa-1)^2\tau\right) \exp\left(-\frac{x^2}{4\tau}\right)$$

$$= e^{-x}\left[\frac{1}{\sqrt{4\pi\tau}} \exp\left(-\frac{(x-(\kappa-1)\tau)^2}{4\tau}\right)\right]. \tag{F.17}$$

If we restore the original variables,

$$p(S,t|S',t') = \frac{1}{S'} f(x,\tau), \quad S = S'e^x, \quad t = t' + \frac{\tau}{(\sigma^2/2)},$$

and remember $\kappa = 2\mu/\sigma^2$, so that

$$(\kappa - 1)\tau = (\mu - \sigma^2/2)(t - t'),$$

the desired result (F.3) is obtained.

References

1. M. Abramowitz, I.A. Stegun, *Handbook of Mathematical Functions* (Dover, New York, 1970)
2. L. Arnold, *Stochastic Differential Equations: Theory and Applications* (Wiley, New York, 1974)
3. L. Bachelier, Théorie de la spéculation. Doctoral dissertation, Faculté des Sciences de Paris (1900). Ann. Sci. Ec. Norm. Super. **3**(17), 21 (1900). Reprinted in 1995. Translated into English in P. Cootner, The Random Character of the Stock Market (MIT Press, Cambridge, 1964)
4. P. Bak, *How Nature Works: The Science of Self-Organized Criticality* (Springer, New York, 1996)
5. P. Bak, M. Paczuski, M. Shubik, Price variations in a stock market with many agents. Physica A **246**, 430 (1997)
6. U. Balucani, M. Zoppi, *Dynamics of the Liquid State* (Oxford University Press, Oxford, 2003)
7. F. Bardou, J.-P. Bouchaud, A. Aspect, C. Cohen-Tannoudji, *Lévy Statistics and Laser Cooling* (Cambridge University Press, Cambridge, 2001)
8. Committee on Banking Supervision, *International Convergence of Capital Measurement and Capital Standards* (Bank for International Settlements, Basel, 2006)
9. K.E. Bassler, J.L. McCauley, G.H. Gunaratne, Nonstationary increments, scaling distributions and variable diffusion processes in financial markets. Proc. Natl. Acad. Sci. USA **104**, 17297 (2007)
10. K. Binder, D.W. Heermann, *Monte Carlo Simulation in Statistical Physics*, 5th edn. (Springer, Berlin, 2010)
11. N.H. Bingham, R. Kiesel, *Risk-Neutral Valuation* (Springer, London, 1998)
12. J.J. Binney, N.J. Dowrick, A.J. Fisher, M.E.J. Newman, *The Theory of Critical Phenomena* (Clarendon Press, Oxford, 1993)
13. F. Black, How we came up with the option formula. J. Portf. Manag. **15**, 4 (1989)
14. F. Black, M. Scholes, The pricing of options and corporate liabilities. J. Polit. Econ. **72**, 637 (1973)
15. J.-P. Bouchaud, Elements for a theory of financial risk. Physica A **263**, 415 (1999)
16. J.-P. Bouchaud, R. Cont, A Langevin approach to stock market fluctuations and crashes. Eur. Phys. J. B **6**, 543 (1998)
17. J.-P. Bouchaud, A. Georges, Anomalous diffusion in disordered media: statistical mechanisms, models and physical applications. Phys. Rep. **195**, 127 (1990)
18. J.-P. Bouchaud, M. Potters, *Theory of Financial Risk—From Data Analysis to Risk Management* (Cambridge University Press, Cambridge, 2000)

19. J.-P. Bouchaud, M. Potters, M. Meyer, Apparent multifractality in financial time series. Eur. Phys. J. B **13**, 595 (2000)
20. L. Breiman, *Probability* (SIAM, Philadelphia, 1992)
21. G. Brown, C. Randall, If the skew fits. Risk **April**, 62 (1999)
22. R.M. Bryce, K.B. Sprague, Revisiting detrended fluctuation analysis. Sci. Rep. **2**, 315 (2012)
23. W. Buffet, Omaha World-Herald, October 28, 1993. Quoted from S. Reynolds, Thoughts of Chairman Buffet: Thirty Years of Wisdom from the Sage of Omaha (Harper Collins, New York, 1998).
24. A. Bunde, S. Havlin, Percolation I, in *Fractals and Disordered Systems*, ed. by A. Bunde, S. Havlin (Springer, Berlin, 1996)
25. J.M. Chambers, C.L. Mallows, B.W. Stuck, A method for simulating stable random variables. J. Am. Stat. Assoc. **71**, 340 (1976)
26. S. Chandrasekhar, Stochastic problems in physics and astronomy. Rev. Mod. Phys. **15**, 1 (1943). See also in [209]
27. P. Cizeau, Y. Liu, M. Meyer, C.-K. Peng, H.E. Stanley, Volatility distribution in the S&P500 stock index. Physica A **245**, 441 (1997)
28. W.T. Coffey, Yu.P. Kalmykov, J.T. Waldron, *The Langevin Equation* (World Scientific, Singapore, 2005)
29. R. Cont, J.-P. Bouchaud, Herd behavior and aggregate fluctuations in financial markets. Macroecon. Dyn. **4**, 170 (2000)
30. R. Cont, M. Potters, J.-P. Bouchaud, Scaling in stock market data: stable laws and beyond, in *Scale Invariance and Beyond*, ed. by B. Dubrulle, F. Grander, D. Sornette (Springer, Berlin, 1997)
31. R. Cont, Empirical properties of asset returns: stylized facts and statistical issues. Quant. Finance **1**, 223 (2001)
32. M.G. Daniels, J.D. Farmer, L. Gillemot, G. Iori, E. Smith, Quantitative model of price diffusion and market friction based on trading as a mechanistic random process. Phys. Rev. Lett. **90**, 108102 (2003)
33. T.G. Dankel Jr., Mechanics on manifolds and the incorporation of spin into Nelson's stochastic mechanics. Arch. Ration. Mech. Anal. **37**, 192 (1970)
34. L. de Haan, A. Ferreira, *Extreme Value Theory: An Introduction* (Springer, Berlin, 2006)
35. J. Des Cloizeaux, G. Jannink, *Polymers in Solution—Their Modelling and Structure* (Oxford University Press, Oxford, 1990)
36. H.-P. Deutsch, *Derivatives and Internal Models 4* (Palgrave Macmillan, Houndmills, 2009)
37. D. Dohrn, F. Guerra, Nelson's stochastic mechanics on Riemannian manifolds. Lett. Nuovo Cimento **22**, 121 (1978)
38. D. Dohrn, F. Guerra, P. Ruggiero, Spinning particles and relativistic particles in the framework of Nelson's stochastic mechanics, in *Feynman Path Integrals*, ed. by S. Albeverio. Lecture Notes in Physics, vol. 106 (Springer, Berlin, 1979)
39. J.L. Doob, The Brownian movement and stochastic equations. Ann. Math. **43**, 351 (1942). See also in [209]
40. E.B. Dynkin, *Markov Processes*, vol. 1 (Springer, Berlin, 1965)
41. E.B. Dynkin, *Markov Processes*, vol. 2 (Springer, Berlin, 1965)
42. M. Ebert, W. Paul, Trading leads to scale-free self-organization. Physica A **391**, 6033 (2012)
43. A. Einstein, Über die von der molekularkinetischen Theorie der Wärme geförderte Bewegung von in ruhenden Flüssigkeiten suspendierten Teilchen. Ann. Phys. **17**, 549 (1905)
44. P. Embrechts, C. Klüppelberg, T. Mikosch, *Modelling Extremal Events for Insurance and Finance* (Springer, Berlin, 2008)
45. M. Emery, *Stochastic Calculus in Manifolds* (Springer, Berlin, 1989)
46. E.F. Fama, Efficient capital markets: a review of theory and empirical work. J. Finance **25**, 383 (1970)
47. J.D. Farmer, Market force, ecology and evolution. Ind. Corp. Change **11**, 895 (2002)
48. J. Feder, *Fractals* (Plenum, New York, 1989)

49. J.A. Feigenbaum, P.G.O. Freund, Discrete scale invariance in stock markets before crashes. Int. J. Mod. Phys. B **10**, 3737 (1996)
50. J.A. Feigenbaum, P.G.O. Freund, Discrete scale invariance and the second black Monday. Mod. Phys. Lett. B **12**, 57 (1998)
51. W. Feller, *An Introduction to Probability Theory and its Applications*, vol. 1 (Wiley, New York, 1966)
52. W. Feller, *An Introduction to Probability Theory and its Applications*, vol. 2 (Wiley, New York, 1966)
53. G.R. Fleming, P. Hänggi (eds.), *Activated Barrier Crossing: Applications in Physics, Chemistry and Biology* (World Scientific, Singapore, 1993)
54. P.J. Flory, *Statistical Mechanics of Chain Molecules* (Hanser, Munich, 1988)
55. E. Frey, K. Kroy, Brownian motion: a paradigm of soft matter and biological physics. Ann. Phys. (Leipz.) **14**, 20 (2005)
56. R. Frigg, C. Werndl, Entropy—a guide for the perplexed, in *Probabilities in Physics*, ed. by C. Beisbart, S. Hartmann (Oxford University Press, Oxford, 2011)
57. M. Fuchs, The Kohlrausch law as a limit solution to mode coupling equations. J. Non-Cryst. Solids **172–174**, 241 (1994)
58. X. Gabaix, P. Gopikrishnan, V. Plerou, H.E. Stanley, A theory of power-law distributions in financial market fluctuations. Nature **423**, 267 (2003)
59. C.W. Gardiner, *Stochastic Methods: A Handbook for the Natural and Social Sciences* (Springer, Berlin, 2009)
60. R.J. Glauber, Time-dependent statistics of the Ising model. J. Math. Phys. **4**, 294 (1963)
61. J. Glimm, A. Jaffe, *Quantum Physics—A Functional Integral Point of View*, 2nd edn. (Springer, New York, 1987)
62. B.W. Gnedenko, A.N. Kolmogorov, *Limit Distributions for Sums of Independent Random Variables* (Addison-Wesley, Reading, 1954)
63. W. Götze, Aspects of structural glass transitions, in *Liquids, Freezing and the Glass Transition, Part 1*, ed. by J.P. Hansen, D. Levesque, J. Zinn-Justin (North-Holland, Amsterdam, 1990)
64. W. Götze, *Complex Dynamics of Glass-Forming Liquids—A Mode-Coupling Theory* (Oxford University Press, Oxford, 2009)
65. P. Gopikrishnan, V. Plerou, L.A. Nunes Amaral, M. Meyer, H.E. Stanley, Scaling of the distribution of fluctuations of financial market indices. Phys. Rev. E **60**, 5305 (1999)
66. J.-F. Gouyet, *Physics and Fractal Structures* (Springer, Paris, 1995)
67. I.S. Gradshteyn, I.M. Ryzhik, A. Jeffrey (eds.), *Table of Integrals, Series and Products* (Academic Press, San Diego, 1994)
68. R. Graham, Macroscopic theory of activated decay of metastable states. J. Stat. Phys. **60**, 675 (1990)
69. P. Gregory, *Bayesian logical data analysis for the physical sciences: a comparative approach with mathematica support* (Cambridge University Press, Cambridge, 2005)
70. F. Guerra, Structural aspects in stochastic mechanics and stochastic field theory, in *New Stochastic Methods in Physics*, ed. by C. De Witt-Morette, K.D. Elworthy, (1981). See also in Phys. Rep. **77**, 263 (1981)
71. F. Guerra, L.M. Morato, Quantization of dynamical systems and stochastic control theory. Phys. Rev. D **27**, 1774 (1983)
72. J. Hager, L. Schäfer, θ-point behavior of dilute polymer solutions: can one observe the universal logarithmic corrections predicted by field theory? Phys. Rev. E **60**, 2071 (1999)
73. P. Hänggi, Escape from a metastable state. J. Stat. Phys. **42**, 105 (1986)
74. P. Hänggi, H. Grabert, P. Talkner, H. Thomas, Bistable systems: master equation versus Fokker-Planck modeling. Phys. Rev. A **29**, 371 (1984)
75. P. Hänggi, P. Talkner, M. Berkovec, Reaction-rate theory: fifty years after Kramers. Rev. Mod. Phys. **62**, 251 (1990)
76. P. Hänggi, H. Thomas, Stochastic processes: time evolution, symmetries and linear response. Phys. Rep. **88**, 207 (1982)

77. J.W. Haus, K. Kehr, Diffusion in regular and disordered lattices. Phys. Rep. **150**, 263 (1987)
78. R. Hilfer (ed.), *Applications of Fractional Calculus in Physics* (World Scientific, Singapore, 1999)
79. R. Hilfer, Threefold introduction to fractional derivatives, in *Anomalous Transport*, ed. by R. Klages, G. Radons, I.M. Sokolov (VCH, Weinheim, 2008)
80. J. Honerkamp, *Statistical Physics* (Springer, Berlin, 1998)
81. B.D. Hughes, E.W. Montroll, M.F. Shlesinger, Fractal random walks. J. Stat. Phys. **28**, 111 (1982)
82. J.C. Hull, *Options, Futures, and Other Derivatives* (Prentice Hall International, Englewood Cliffs, 1997)
83. N. Ikeda, S. Watanabe, M. Fukushima, H. Kunita, *Itô's Stochastic Calculus and Probability Theory* (Springer, Tokyo, 1996)
84. K. Ilinski, Critical crashes? Int. J. Mod. Phys. C **10**, 741 (1999)
85. K. Imafuku, O. Ichiro, Y. Yamanaka, Tunneling time based on the quantum diffusion process approach. Phys. Lett. A **204**, 329 (1995)
86. E. Ising, Beitrag zur Theorie des Ferromagnetismus. Z. Phys. **31**, 253 (1925)
87. K. Itô, H.P. McKean Jr., *Diffusion Processes and Their Sample Paths* (Springer, Berlin, 1996)
88. W. Janke, Monte Carlo simulations of spin systems, in *Computational Physics*, ed. by K.H. Hoffmann, M. Schreiber (Springer, Berlin, 1996)
89. R. Jameson, Playing the name game. Risk **October**, 38 (1998)
90. R. Jameson, Getting the measure of the beast. Risk **November**, 39 (1998)
91. E.T. Jaynes, Information theory and statistical mechanics. Phys. Rev. **106**, 620 (1957)
92. E.T. Jaynes, Information theory and statistical mechanics. II. Phys. Rev. **108**, 171 (1957)
93. E.T. Jaynes, *Probability Theory: The Logic of Science* (Cambridge University Press, Cambridge, 2003)
94. A. Johansen, O. Ledoit, D. Sornette, Crashes as critical points. Int. J. Theor. Appl. Finance **3**, 219 (2000)
95. A. Johansen, D. Sornette, Stock market crashes are outliers. Eur. Phys. J. B **1**, 141 (1998)
96. A. Johansen, D. Sornette, Critical crashes. Risk **January**, 91 (1999)
97. A. Johansen, D. Sornette, O. Ledoit, Predicting financial crashes using discrete scale invariance. J. Risk **1**, 5 (1999)
98. G. Jona-Lasinio, F. Martinelli, E. Scoppola, The semi-classical limit of quantum mechanics: a qualitative theory via stochastic mechanics, in *New Stochastic Methods in Physics*, ed. by C. De Witt-Morette, K.D. Elworthy, (1981). See also in Phys. Rep. **77**, 313 (1981)
99. N.L. Johnson, S. Kotz, N. Balakrishnan, *Continuous Univariate Distributions*, vol. 1 (Wiley, New York, 1994)
100. M. Kac, Random walk and the theory of Brownian motion. Am. Math. Mon. **54**, 369 (1947). See also in [209]
101. N.G. van Kampen, *Stochastic Processes in Physics and Chemistry* (North-Holland, Amsterdam, 1985)
102. N.G. van Kampen, Brownian motion on a manifold. J. Stat. Phys. **44**, 1 (1986)
103. I. Karatzas, S.E. Shreve, *Brownian motion and stochastic calculus* (Springer, Berlin, 1999)
104. A. Khintchine, Korrelationstheorie der stationären stochastischen Prozesse. Math. Ann. **109**, 604 (1934)
105. J. Klafter, M.F. Shlesinger, G. Zumofen, Beyond Brownian motion. Phys. Today **February**, 33 (1996)
106. P.E. Kloeden, E. Platen, *Numerical Solution of Stochastic Differential Equations* (Springer, Berlin, 1995)
107. A. Kolmogorov, *Grundbegriffe der Wahrscheinlichkeitsrechnung*. Ergebnisse der Math., vol. 2 (Springer, Berlin, 1933)
108. I. Koponen, Analytic approach to the problem of convergence of truncated Lévy flights towards the Gaussian stochastic process. Phys. Rev. E **52**, 1197 (1995)
109. R. Korn, E. Korn, *Optionsbewertung und Portfolio-Optimierung* (Vieweg, Wiesbaden, 1999)

References

110. H.A. Kramers, Brownian motion in a field of force and the diffusion model of chemical reactions. Physica **7**, 284 (1940)
111. R. Kutner, Hierachical spatio-temporal coupling in fractional wanderings. (I) Continuous-time Weierstrass flights. Physica A **264**, 84 (1999)
112. R. Kutner, M. Regulski, Hierachical spatio-temporal coupling in fractional wanderings. (II) Diffusion phase diagram for Weierstrass walks. Physica A **264**, 107 (1999)
113. L. Laloux, M. Potters, R. Cont, J.-P. Aguilar, J.-P. Bouchaud, Are financial crashes predictable? Europhys. Lett. **45**, 1 (1999)
114. D.P. Landau, K. Binder, *A Guide to Monte Carlo Simulations in Statistical Physics* (Cambridge University Press, Cambridge, 2009)
115. R. Landauer, T. Martin, Barrier interaction time in tunneling. Rev. Mod. Phys. **66**, 217 (1994)
116. D.A. Lavis, G.M. Bell, *The Statistical Mechanics of Lattice Systems*, vol. 1 (Springer, Berlin, 1999)
117. D.A. Lavis, G.M. Bell, *The Statistical Mechanics of Lattice Systems*, vol. 2 (Springer, Berlin, 1999)
118. K. Lindenberg, B. West, *The Nonequilibrium Statistical Mechanics of Open and Closed Systems* (VCH, Weinheim, 1991)
119. Y. Liu, P. Gopikrishnan, P. Cizeau, M. Meyer, C.-K. Peng, H.E. Stanley, The statistical properties of the volatility of price fluctuations. Phys. Rev. E **60**, 1390 (1999)
120. J.D. MacBeth, L.J. Merville, An empirical examination of the Black-Scholes call option pricing model. J. Finance **34**, 1173 (1979)
121. E. Madelung, Quantentheorie in Hydrodynamischer Form. Z. Phys. **40**, 322 (1926)
122. N. Madras, G. Slade, *The Self-Avoiding Walk* (Birkhäuser, Boston, 1995)
123. B.B. Mandelbrot, The variation of certain speculative prices. J. Bus. **36**, 394 (1963)
124. B.B. Mandelbrot, *The Fractal Geometry of Nature* (Freeman, New York, 1983)
125. B.B. Mandelbrot, *Fractals and Scaling in Finance* (Springer, New York, 1997)
126. B.B. Mandelbrot, J.W. van Ness, Fractional Brownian motions, fractional noises and applications. SIAM Rev. **10**, 422 (1968)
127. R.N. Mantegna, H.E. Stanley, Stochastic process with ultraslow convergence to a Gaussian: the truncated Lévy flight. Phys. Rev. Lett. **73**, 2946 (1994)
128. M.E. Shlesinger, Comment on 'Stochastic process with ultraslow convergence to a Gaussian: the truncated Lévy flight'. Phys. Rev. Lett. **74**, 4959 (1995)
129. R.N. Mantegna, H.E. Stanley, Scaling behaviour in the dynamics of an economic index. Nature **376**, 46 (1995)
130. R.N. Mantegna, H.E. Stanley, Econophysics: scaling and its breakdown in finance. J. Stat. Phys. **89**, 469 (1997)
131. R.N. Mantegna, H.E. Stanley, *An Introduction to Econophysics: Correlations and Complexity in Finance* (Cambridge University Press, Cambridge, 2000)
132. J.L. McCauley, *Dynamics of Markets: Econophysics and Finance* (Cambridge University Press, Cambridge, 2009)
133. J.L. McCauley, K.E. Bassler, G.H. Gunaratne, On the analysis of times series with nonstationary increments, in *Handbook of Research on Complexity*, ed. by B. Rossner (Edward Elgar, Cheltenham Glos, 2009)
134. J.L. McCauley, K.E. Bassler, G.H. Gunaratne, Martingales, detrending data, and the efficient market hypothesis. Physica A **387**, 202 (2008)
135. J.L. McCauley, G.H. Gunaratne, K.E. Basser, Martingale option pricing. Physica A **380**, 351 (2007)
136. M. McClendon, H. Rabitz, Numerical simulations in stochastic mechanics. Phys. Rev. A **37**, 3479 (1988)
137. B.M. McCoy, T.T. Wu, *The Two-Dimensional Ising Model* (Cambridge University Press, Cambridge, 1973)
138. R. Merton, Theory of Rational Option Pricing. Bell J. Econ. Manag. Sci. **4**, 141 (1973)
139. A. Messiah, *Quantum Mechanics*, vol. 1 (North-Holland, Amsterdam, 1969)
140. A. Messiah, *Quantum Mechanics*, vol. 2 (North-Holland, Amsterdam, 1969)

141. H. Metiu, K. Kitahara, J. Ross, Statistical mechanical theory of the kinetics of phase transitions, in *Fluctuation Phenomena*, ed. by E.W. Montroll, J.L. Lebowitz (North-Holland, Amsterdam, 1987)
142. N. Metropolis, A.W. Rosenbluth, M.N. Rosenbluth, A.H. Teller, E. Teller, Equation of state calculations by fast computing machines. J. Chem. Phys. **21**, 1087 (1953)
143. R. Metzler, J. Klafter, The random walk's guide to anomalous diffusion: a fractional dynamics approach. Phys. Rep. **339**, 1 (2000)
144. E.W. Montroll, J.T. Bendler, On Lévy (or stable) distributions and the Williams-Watts model of dielectric relaxation. J. Stat. Phys. **34**, 129 (1984)
145. E.W. Montroll, M.F. Shlesinger, On the wonderful world of random walks, in *Nonequilibrium Phenomena II: From Stochastics to Hydrodynamics*, ed. by J.L. Lebowitz, E.W. Montroll (Elsevier, Amsterdam, 1984)
146. E.W. Montroll, G.H. Weiss, Random walks on lattices II. J. Math. Phys. **6**, 167 (1965)
147. E.W. Montroll, B.J. West, On an enriched collection of stochastic processes, in *Fluctuation Phenomena*, ed. by E.W. Montroll, J.L. Lebowitz (North-Holland, Amsterdam, 1987)
148. S. Moss de Oliveira, P.M.C. de Oliveira, S.D. Evolution, *Money, War, and Computers* (Teubner, Stuttgart, 1999)
149. E. Nelson, Derivation of the Schrödinger equation from Newtonian mechanics. Phys. Rev. **150**, 1079 (1966)
150. E. Nelson, *Quantum fluctuations* (Princeton University Press, Princeton, 1985)
151. E. Nelson, *Dynamical Theories of Brownian Motion* (Princeton University Press, Princeton, 1967)
152. Th. Niemeijer, J.M.J. van Leeuwen, Renormalization theory for Ising-like spin systems, in *Phase Transitions and Critical Phenomena*, vol. 6, ed. by C. Domb, M.S. Green (Academic Press, London, 1976)
153. Nobel Prize in Economic Sciences (1970). Available at http://www.nobel.se/economics/laureates/1970/index.html
154. Nobel Prize in Economic Sciences (1997). Available at http://www.nobel.se/economics/laureates/1997/index.html
155. H.C. Öttinger, *Stochastic Processes in Polymeric Fluids* (Springer, Berlin, 1996)
156. L. Onsager, Crystal statistics I: a two-dimensional model with order disorder transition. Phys. Rev. **65**, 117 (1944)
157. I. Oppenheim, K. Shuler, G.H. Weiss, *The Master Equation* (MIT Press, Cambridge, 1977)
158. M.F.M. Osborne, *The Stock Market and Finance from a Physicist's Viewpoint* (Crossgar Press, Minneapolis, 1995)
159. A. Pagan, The econometrics of financial markets. J. Empir. Finance **3**, 15 (1996)
160. V. Pareto, *Cours d'Economique Politique* (F. Rouge, Lausanne, 1897)
161. M. Pavon, A new formulation of stochastic mechanics. Phys. Lett. A **209**, 143 (1995)
162. M. Pavon, Hamilton's principle in stochastic mechanics. J. Math. Phys. **36**, 6774 (1995)
163. M. Pavon, Lagrangian dynamics for classical, Brownian and quantum mechanical particles. J. Math. Phys. **37**, 3375 (1996)
164. W. Paul, D.W. Heermann, K. Binder, Relaxation of metastable states in mean-field kinetic Ising systems. J. Phys. A, Math. Gen. **22**, 3325 (1989)
165. K. Pearson, *The history of statistics in the 17th and 18th centuries* (Macmillan, London, 1978)
166. C.-K. Peng, S.V. Buldyrev, S. Havlin, M. Simons, H.E. Stanley, A.L. Goldberger, Mosaic organization of DNA nucleotides. Phys. Rev. E **49**, 1685 (1994)
167. C.-K. Peng, J. Mietus, J.M. Hausdorff, S. Havlin, H.E. Stanley, A.L. Goldberger, Long-Range anticorrelations and non-Gaussian behavior of the heartbeat. Phys. Rev. Lett. **70**, 1343 (1993)
168. A.M. Perelomov, *Generalized Coherent States and Their Applications* (Springer, Berlin, 1986)
169. V. Plerou, H.E. Stanley, Stock return distributions: tests of scaling and universality from three distinct stock markets. Phys. Rev. E **77**, 037101 (2008)

170. B. Podobnik, D. Horvatic, A. Petersen, H.E. Stanley, Cross-correlations between volume change and price change. Proc. Natl. Acad. Sci. USA **106**, 22079 (2009)
171. M. Potters, R. Cont, J.-P. Bouchaud, Financial markets as adaptive systems. Europhys. Lett. **41**, 239 (1998)
172. T. Preis, S. Golke, W. Paul, J.J. Schneider, Multi-agent-based order book model of financial markets. Europhys. Lett. **75**, 510 (2006)
173. T. Preis, S. Golke, W. Paul, J.J. Schneider, Statistical analysis of financial returns for a multiagent order book model of asset trading. Phys. Rev. E **76**, 016108 (2007)
174. W.H. Press, B.P. Flannery, S.A. Teukolsky, W.T. Vetterling, *Numerical Recipes—The Art of Scientific Computing* (Cambridge University Press, Cambridge, 1989)
175. B. Prum, J.C. Fort, *Stochastic Processes on a Lattice and Gibbs Measures* (Kluwer, Dordrecht, 1991)
176. L.E. Reichl, *A Modern Course in Statistical Physics* (VCH, Weinheim, 2009)
177. E.J. Janse van Rensburg, Monte Carlo methods for the self-avoiding walk. J. Phys. A, Math. Theor. **42**, 323001 (2009)
178. H. Risken, *The Fokker-Planck Equation: Methods of Solution and Applications* (Springer, Berlin, 1996)
179. M. Rubinstein, R.H. Colby, *Polymer Physics* (Oxford University Press, London, 2003)
180. D. Ruelle, *Statistical Mechanics—Rigorous Results* (Benjamin, New York, 1969)
181. S.G. Samko, A.A. Kilbas, O.I. Marichev, *Fractional Integrals and Derivatives—Theory and Applications* (Gordon & Breach, New York, 1993)
182. P.A. Samuelson, Rational theory of warrant pricing. Ind. Manage. Rev. **6**, 13 (1965)
183. I. Schneider (ed.), *Die Entwicklung der Wahrscheinlichkeitsrechnung von den Anfängen bis 1933* (Wissenschaftliche Buchgesellschaft, Darmstadt, 1986)
184. T. Schwiertz, Computersimulation von Zufallsprozessen: Extremwertstatistik, Rekurrenzen und die Bewertung pfadabhängiger Derivate. Diploma thesis, University of Mainz (2004)
185. C.E. Shannon, A mathematical theory of communication. Bell Syst. Tech. J. **27**, 379 (1948). See also Bell Syst. Tech. J. **27**, 623 (1948)
186. D. Shirvanyants, S. Panyukov, Q. Liao, M. Rubinstein, Long-Range correlations in a polymer chain due to its connectivity. Macromolecules **41**, 1475 (2008)
187. M.F. Shlesinger, Fractal time in condensed matter. Annu. Rev. Phys. Chem. **39**, 269 (1988)
188. M.F. Shlesinger, B.D. Hughes, Analogs of renormalization group transformations in random processes. Physica A **109**, 597 (1981)
189. M.F. Shlesinger, G.M. Zaslavsky, U. Frisch (eds.), *Lévy flights and related topics in physics* (Springer, Berlin, 1995)
190. E. Smith, J.D. Farmer, L. Gillemot, S. Krishnamurthy, Statistical theory of the continuous double auction. Quant. Finance **3**, 481 (2003)
191. M. von Smoluchowski, *Abhandlungen über die Brownsche Bewegung und verwandte Erscheinungen* (Akademische Verlagsgesellschaft, Leipzig, 1923)
192. A.D. Sokal, Monte Carlo methods for the self-avoiding walk, in *Monte Carlo and Molecular Dynamics Simulations in Polymer Science*, ed. by K. Binder (Oxford University Press, London, 1995)
193. O. Malcai, O. Biham, P. Richmond, S. Solomon, Theoretical analysis and simulations of the generalized Lotka-Volterra model. Phys. Rev. E **66**, 031102 (2002)
194. D. Sornette, Discrete scale invariance and complex dimensions. Phys. Rep. **297**, 239 (1998)
195. D. Sornette, A. Johansen, Large financial crashes. Physica A **245**, 411 (1997)
196. D. Sornette, A. Johansen, J.-P. Bouchaud, Stock market crashes, precursors and replicas. J. Phys., I **6**, 167 (1996)
197. D. Sornette, Critical market crashes. Phys. Rep. **378**, 1 (2003)
198. D. Sornette, Dragon-Kings: mechanisms, statistical methods and empirical evidence. Eur. Phys. J. Spec. Top. **205**, 1 (2012)
199. H.E. Stanley, V. Plerou, X. Gabaix, A statistical physics view of financial fluctuations: evidence for scaling and universality. Physica A **387**, 3967 (2008)
200. D. Stauffer, Can percolation theory be applied to the stock market? Ann. Phys. **7**, 529 (1998)

201. D. Stauffer, Agent based models of the financial market. Rep. Prog. Phys. **70**, 409 (2007)
202. P. Talkner, Mean first passage time and the lifetime of a metastable state. Z. Phys. B **68**, 201 (1987)
203. G.E. Uhlenbeck, L.S. Ornstein, On the theory of Brownian motion. Phys. Rev. **36**, 823 (1930). See also in [209]
204. J.B. Davies, S. Sandstrom, A. Shorrocks, E.N. Wolff, *The World Distribution of Household Wealth* (UNU-WIDER, Helsinki, 2006). http://www.wider.unu.edu
205. N. Vandewalle, M. Ausloos, Ph. Boveroux, A. Minguet, How the financial crash of October 1997 could have been predicted. Eur. Phys. J. B **4**, 139 (1998)
206. N. Vandewalle, Ph. Boveroux, A. Minguet, M. Ausloos, The crash of October 1987 seen as a phase transition: amplitude and universality. Physica A **255**, 201 (1998)
207. J. Voit, *The Statistical Mechanics of Financial Markets* (Springer, Berlin, 2001)
208. C.W. Wang, G.E. Uhlenbeck, On the theory of Brownian motion II. Rev. Mod. Phys. **17**, 323 (1945). See also in [209]
209. N. Wax (ed.), *Selected Papers on Noise and Stochastic Processes* (Dover, New York, 1954)
210. A. Wehrl, General properties of entropy. Rev. Mod. Phys. **50**, 221 (1978)
211. G.H. Weiss, *Aspects and Applications of the Random Walk* (North-Holland, Amsterdam, 1994)
212. U. Weiss, *Quantum Dissipative Systems* (World Scientific, Singapore, 1993)
213. B.J. West, W. Deering, Fractal physiology for physicists: Lévy statistics. Phys. Rep. **246**, 1 (1994)
214. D. Williams, *Probability with Martingales* (Cambridge University Press, Cambridge, 1997)
215. P. Wilmott, J. Dewynne, S. Howison, *Option Pricing—Mathematical Models and Computation* (Oxford Financial Press, Oxford, 1993)
216. J.P. Wittmer, A. Cavallo, H. Xu, J.E. Zabel, P. Polińska, N. Schulmann, H. Meyer, J. Farago, A. Johner, S.P. Obukhov, J. Baschnagel, Scale-free static and dynamical correlations in melts of monodisperse and flory-distributed homopolymers. J. Stat. Phys. **145**, 1017 (2011)
217. V.M. Yakovenko, J. Barkley Rosser Jr., Colloquium: statistical mechanics of money, wealth, and income. Rev. Mod. Phys. **81**, 1703 (2009)
218. K. Yasue, Stochastic calculus of variations. J. Funct. Anal. **41**, 327 (1981)
219. W.-X. Zhou, Determinants of immediate price impacts at the trade level in an emerging order-driven market. New J. Phys. **14**, 023055 (2012)
220. R. Zwanzig, *Nonequilibrium Statistical Mechanics* (Oxford University Press, Oxford, 2001)

Index

A
Action, 115, 123
Agent based modeling of financial markets, *see* traders
American options, *see* options
Arbitrage
 definition, 176
 hypothesis of the arbitrage-free market, 176
Arbitrage-pricing technique, 176
 put–call parity, 259, 260
Arbitrageurs, *see* traders
Arrhenius behavior, 98
Asymmetry parameter β
 of a fractal-time Poisson process, 153
 of a truncated Lévy flight, 156
 of stable distributions, 135
Autocorrelation function, 40, 78, 83

B
Bachelier, 2, 163, 168, 205, 206
Bank, use of the term, 165
Bid–ask spread, 167, 215
Binomial distribution, 4
Black-Scholes theory
 and risk neutrality, 185, 186
 assumptions, 174–176
 comparison with Bachelier's theory, 205, 206
 derivation of differential equation, 176–178
 equivalent martingale probability for, 186, 187
 history of, 163, 174
 independence of drift, 178, 184, 185
 interpretation of pricing formulas, 183, 184
 solution, 179–183
Boolean algebra, 23
Borel algebra, 19

Brownian motion, 75, 90, 126
 geometric, 168–170, 261–263
 hydrodynamics of, 110
 hypothesis of universal, 115
 in a potential, 92
 reflection principle, 91
 stopping time, 90

C
Caldeira-Leggett model, 84–90
Call options
 boundary conditions for call price, 178
 call price of Black-Scholes theory, 182
 definition, 171
 hedging by, 173
 payoff at expiry, 174
Canonical representation of stable distributions, 134
Cauchy distribution, 131, 240
Central limit theorem (CLT), 8, 33, 131, 159, 160
 domain of attraction, 136, 237, 238
 violation by correlations, 161
 violation by diverging moments, 160
Chapman-Kolmogorov equation, 44, 46, 47, 58
Characteristic exponent α
 of a fractal-time Poisson process, 153
 of a truncated Lévy flight, 156
 of stable distributions, 135
 of the Weierstrass random walk, 140
Characteristic function, 31, 64
Coherent states, 123
Conditional
 expectation, 33, 40
 probability, 32
Conservative diffusion process, 114, 116, 118, 123

Cont-Bouchaud model, *see* traders
Contingent claim, *see* derivatives
Continuity equation, 112
    quantum mechanical, 120
Continuous trading, *see* quotations
Continuous-time random walk (CTRW), 72–75
    separable, 72, 150
Convergence to
    Gaussian distribution, 136, 237, 238
    stable (Lévy) distributions, 135, 237, 238
Covariance
    functional, 126
    matrix, 32
Crashes
    and discrete scale invariance, 230
    as a result of the cooperativity between traders, 226, 228
    Dow Jones Index (1929, 1987), 225
    empirical properties, 222–225
    hazard rate, 230, 231
    log–periodic oscillations, 225, 230–233
    Sornette-Johansen model, 228–233
Credit risk, *see* risk
Cross-correlation function, 40
Cumulants, 31
Cumulative distribution, 30
    and extreme value theory, 36

D

Delivery date, 171
Delivery price, 171
Delta-hedge, *see* hedging
Demand, 214, 218, 226, 230
Derivatives, 170
Diffusion
    coefficient, 8, 10, 47, 109
    Fickian diffusion, 6
    free diffusion, 6
    non-stationary, 99
    process, 51, 59, 75, 110, 121, 126
    unbounded, 98
Discounting, *see* numéraire
Discrete scale invariance
    and crashes, 230
    and fractal-time Poisson process, 151
    and Weierstrass random walk, 141
Distribution
    binomial, 4
    canonical, 28
    Cauchy, 131, 240
    cumulative, 30
        and extreme value theory, 36
    Fréchet, 37, 38
    Gaussian, 8, 13, 239
    Gumbel, 37, 38
    leptokurtic, 194
    limiting, 35, 90
    log–normal distribution in financial markets, 170
    marginal, 32
    n-point distribution function, 39
    non-degenerate, 36
    of asset prices, 194–204
    Poisson, 11
    stable (Lévy) distributions, 131–136
    waiting-time, 72, 74
    Weibull, 37, 38
Diversification, 170
Dividends, 175
Doob's theorem, 81
Dow Jones Index
    crashes in 1929 and 1987, 225
    first published, 166
    modeling prior to crash of 1929, 232
Drift
    and hazard rate of crashes, 231
    as a market parameter, 169
    coefficient, 47
    elimination in a risk-neutral world, 184–189
    negligible for short maturities, 205
    relation to hazard rate of crashes, 231
    velocity, 10

E

Economics, 2
Econophysics, 164, 235
Efficient market hypothesis, 209
    as an assumption of the Black-Scholes theory, 174
    definition, 167
    giving rise to a Markov assumption, 168, 175
Ensemble
    canonical, 20, 27
        and maximum entropy principle, 27
    microcanonical, 27, 179
Entropy
    information entropy, 26
    microcanonical, 28
    statistical entropy, 27
Equivalent martingale measure
    and historic price distribution, 189, 207
    of Black-Scholes theory, 186
Euclidean quantum fields, 126
European options, *see* options

Index

Event, 17
    simple event, 17, 22
    simple sets, 19
Exchange, *see* stock exchange
Exercise price of an option, *see* strike price of an option
Exotic options, *see* options
Expectation value, 4, 20, 117
    conditional, 33, 40
    of binomial distribution, 5
Extreme value
    distributions, 35
    Fréchet distribution, 37
    Gumbel distribution, 37
    maximal excursion, 90
    theory, 36
    von Mises representation of the distributions, 38
    Weibull distribution, 37

F
Fair game, 42
First passage time, 94, 99
    distribution, 95
    mean, 95–98, 110, 118, 121
Fluctuation–dissipation theorem, 78, 89
Fluctuations, 77
Fokker-Planck equation, 47, 59, 76, 94, 109, 111
    backward, 48, 95
    forward, 48
Forwards
    comparison with futures, 171
    definition, 171
    hedging by, 173
Fractal dimension
    definition, 146
    of a fractal-time Poisson process, 153
    of the random walk, 147
    of the Weierstrass random walk, 147
Fractal Rayleigh-Pearson walk, 247
Fractal-time Poisson process
    as an asymmetric stable distribution, 153
    asymptotic behavior, 153
    characteristic exponent α, 153
    definition, 151
    fractal dimension, 153
    log–periodic oscillations, 153
    solution, 151–153
Fractal-time random walks, 150–155
    crossover to normal diffusion, 155
    existence of moments, 150
    subdiffusive behavior, 154, 155
    waiting-time distribution, 150–153

Fractional derivative, 143
Fractional diffusion equation, *see* Weierstrass random walk
Free energy
    generalized, 106
    thermodynamic, 106
Free lunch, *see* arbitrage
Freely jointed chain, 71
Friction coefficient, 75, 88
Fundamentalist, *see* traders
Futures
    comparison with forwards, 171
    comparison with options, 171
    definition, 171
    history of, 171

G
GARCH, 204
Gaussian, 34, 81, 123
    colored noise, 90
    distribution, 8, 13, 239
    propagator, 80, 181
    stochastic field, 125
    white noise, 39, 50
Generating function, 65, 67, 74
Geometric Brownian motion, 209
    as a model of financial markets, 168–170
    as a special case of an Itô process, 169, 261
    transition probability, 170, 261
Green's function, 80

H
Hamilton-Jacobi equations, 115, 117
Hamiltonian, 20, 101
Hazard rate of crashes, 230, 231
    relation to drift, 231
Hedging
    by call options, 173
    by forwards, 173
    definition, 170
    delta-hedge, 177, 183, 184
    risk-less hedging and Legendre transformation, 178
Heteroskedasticity, 204
Holder of an option, 171
Hurst exponent, 211, 217
Hyperspherical polar coordinates, 243–245

I
Iff, 32
Iid, 33
Imitation, *see* traders
Interest rate
    risk-free, 168
    stochastic, 175

Ising model, 101
  mean-field version, 105
Itô calculus, 53
Itô formula, 58, 111
  application to option prices, 176
  application to stock prices, 169, 187

K

Kohlrausch function, 154
  relation to stable distributions, 154
Kramers
  equation, 76
  problem, 92, 97, 118, 121
Kramers-Moyal expansion, 47, 59, 108
Kurtosis, 13, 20, 35
  implied, 191, 209
  of Gaussian distribution, 13
  of Poisson distribution, 13
  of price distribution, 197, 202–204
  of truncated Lévy flights, 157

L

Langevin equation, 39, 47, 49, 51, 57
  generalized, 88
Lévy distributions, *see* stable distributions
Lévy flights
  and Polya's problem, 149, 150
  and Weierstrass random walk, 146
  as a model for price distributions, 195
  definition, 147
  in higher dimensions, 147–150
  truncated, 155–159
Lévy walks, 159
Limiting distributions and stability, 134
Liquid market, 167
Log–normal distribution, 13
Log–periodic oscillations
  and critical phenomena, 230
  crashes, 225, 230–233
  of a fractal-time Poisson process, 153
  of the Weierstrass random walk, 141
Long position, 171

M

Madelung fluid, 116
Margins, 171
Market
  arbitrage-free, 175
  definition, 164
  financial, 164, 235
  microscopic modeling of, 225–233
Market capitalization, 166
Market friction, 167
Market risk, *see* risk

Markov
  chain, 103
  process, 43, 81, 92, 125
    and efficient market hypothesis, 168
Martingale, 41, 54, 77, 127
  and geometric Brownian motion, 186–188
  and stationarity of financial time series, 213, 214
  equivalent martingale measure and risk neutrality, 185
Master equation, 9, 44–47, 58, 63, 103, 108
Maximum entropy principle, 26
Mean acceleration, 113
Mean backward derivative, 112, 114
Mean forward derivative, 111, 114
Mean velocity, 112
Mean-square displacement, 82, 83
Mean-square limit, 53, 56
Measure, 19, 103, 105
  equivalent, 29
  Gibbs measure, 20, 30
  Lebesgue measure, 19, 20, 29
  Radon-Nikodým derivative of, 29
Median, 14
  of Gaussian distribution, 14
  of log–normal distribution, 14
Mellin transforms, 248
Memory effect, 88
Metastable state, 92
Moneyness, 184
Monte Carlo simulations, 102, 104
  detailed balance, 103
  importance sampling, 103
  random numbers, 104

N

n-point function, 39, 43, 58
Noise
  colored, 90
  Gaussian white noise, 39
  white, 50
Noise trader, *see* traders
Non-anticipating function, 54, 57, 110
Non-specific (systemic) risk, *see* risk
Normal distribution, 8, 13
Numéraire
  definition, 187
  discounting by, 187

O

Operational risk, *see* risk
Options
  American, 172
  at-the-money, 184, 190

Index

Options (*cont.*)
  call option, 171
  European, 171
  exotic, 172
  expiry date, 172
  history of, 172
  in-the-money, 184, 190
  out-of-the-money, 184, 190
  plain vanilla, 172
  price of, 172, 173
  pricing by Black-Scholes theory, 173–184
  put option, 171
Order book, 215
  ask price, 215
  bid price, 215
  bid–ask spread, 215
  continuous double auction, 215
  depth, 216
  limit order, 215
  market makers, 216
  market order, 215
  price impact function, 219
  price-time priority, 216
  tick size, 215
  volume, 216
Ornstein-Uhlenbeck process, 77, 81, 110, 114, 124, 127
Osmotic velocity, 113, 118
Over-the-counter (OTC), definition and properties, 165

P

Partition function
  canonical, 20, 28
Payoff of an option, 174
Percolation model, *see* price distribution
Periodic boundary conditions, 63
Plain vanilla options, *see* options
Poisson
  distribution, 11
  process, 151
Polya problem, 66, 99
Portfolio
  definition, 170
  risk-less, 179, 184
Present value, 183
Price
  decorrelation time of increments, 196
  geometric Brownian motion versus empirical price dynamics, 233
  spot price, 164

Price distribution
  and percolation model, 227
  and self-organized criticality, 228
  and time-dependent volatility, 199–202
  empirical, 194–198
  for geometric Brownian motion, 170
  in the Black-Scholes model, 186
  modeling by a Lévy flight, 195, 202
  modeling by a truncated Lévy flight, 196
  non-Lévy-like power law, 197, 202
Probability
  conditional, 32
  current, 119, 121
  density, 30
  of return, 67
Probability theory, 17
  and logic, 21
  history of, 1
  mathematical structure, 17
  prior information, 22, 23
  rational expectation, 22
Process
  self-similar, 211
Put options
  boundary conditions for put price, 178
  definition, 171
  payoff at expiry, 174
  put price of Black-Scholes theory, 183
Put–call parity, 176, 259, 260

Q

Quantum
  fields, 122
  mechanics, 110
Quotations
  continuous trading, 174
  decorrelation time of increments, 196
  typical time interval between, 192

R

Random force, 75, 88
Random variable, 19
  independent and identically distributed (iid), 33
  statistically independent random variables, 32
  uncorrelated random variables, 32
Random walk, 3, 63, 126
  Rayleigh-Pearson walk, 3, 69, 71
  versus random flight, 147
Rate equations, *see* master equation
Recurrence, 69
  Lévy flights, 150
Renewal equation, 67, 99

Return
  definition, 169
  versus prices, 193, 194
Risk, 36, 165
  credit risk, 165
  elimination of market risk in Black-Scholes theory, 177–179
  hedging market risk by call options, 173
  hedging market risk by forwards, 173
  market risk, 166
  non-specific (systemic) risk, 170
  operational risk, 166
  reduction by diversification, 170
  risk management, 165
  specific (non-systemic) risk, 170
Risk-neutral valuation, 184–189

S

σ-algebra, 19, 30
  Borel algebra, 19
Sample space, 17
  as probability space, 19
Schrödinger equation, 21, 116, 117, 119
Self-averaging, 5, 28
Self-avoiding walk, 72
Self-organized criticality, see price distribution
Self-similarity
  meaning of, 146
  self-similar process, 211
  Weierstrass random walk, 137, 138
Sharpe ratio, 205
Short position, 171
Skewness, 13, 20, 35
  of Poisson distribution, 13
Smoluchowski equation, 94
Sornette-Johansen model, see crashes
Specific (non-systemic) risk, see risk
Speculation bubble, 229
Spot price, see price
Stable distributions, 8, 131–136, 237–242
  and Kohlrausch function, 154
  and Weierstrass random walk, 141
  application to human heart beats, 155
  as limiting distributions, 134
  asymmetric ($\beta = \pm 1$) cases and Laplace transform, 242
  asymmetry parameter β, 135
  asymptotic behavior, 135
  asymptotic expansions, 240–242
  canonical representation, 134
  characteristic exponent α, 135
  closed-form expressions of asymmetric ($\beta = \pm 1$) cases, 241, 242
  closed-form expressions of symmetric ($\beta = 0$) cases, 239, 240
  definition, 132
  domain of attraction, 135
  existence of moments, 135
  scaling property, 155
  stability of the Cauchy distribution, 134
  stability of the Gaussian distribution, 132, 133
Standard & Poor's Composite Index (S&P500)
  definition, 166
  fit by Lévy flight, 195
  fit by truncated Lévy flight, 196
  time evolution (1950–1999), 167
Stationarity
  and financial time series, 209–214
  and Ornstein-Uhlenbeck process, 81
  non-stationarity of the Wiener process, 84
  of a Markov process, 44
  of a stochastic process, 40, 50, 90
Statistical physics, 20
  canonical ensemble, 20, 27, 102, 105
  equipartition theorem, 78
  Gibbs measure, 20
  kinetic theory of gases, 2
  microcanonical ensemble, 27
  partition function, 20, 28, 102, 106
  statistical mechanics, 2
Statistically independent random variables, 32
Stochastic differential equations, 57, 59, 76, 110, 123, 126
Stochastic integrals, 53, 54
  Itô, 54
  Stratonovich, 54
Stochastic processes
  continuous time, 39
  discrete time, 39
  history of, 1
  Markov, 43, 81
  martingale, 41
  mathematical structure, 17
  Ornstein-Uhlenbeck process, 77, 81, 110, 114, 124, 127
  stationary, 40, 50, 81, 90
Stochastic variable, see random variable
Stock exchange
  as a clearing house, 171
  tasks and rules, 164
Stock index, 166
Stopping time, 90
Stretched exponential function, see Kohlrausch function
Strike price of an option, 172
Structure function, 64, 74

Stylized facts of financial markets, 212
Subdiffusion
  due to diverging moments, 160
  of fractal-time random walks, 154, 155
Superdiffusion
  due to correlations, 161
  due to diverging moments, 160
  of the Weierstrass random walk, 142–147
Supply, 214, 218, 226, 230
Surface area
  of a d-dimensional hypersphere, 245
  of a d-dimensional unit sphere, 244

T
Thermally activated, 93
Thermodynamic limit, 28
Time reversal, 111
Time scales, 93
  equilibration time, 93
  escape time, 93
Time scales of financial markets, 234
  crossover to Gaussian behavior, 197
  price increments, 196, 200
  quotations, 192
  trading day, 197, 198, 200
  trading month, 197, 198, 200
  volatility, 200
Time series
  stationary, 209
  stochastic, 209
Traders
  agent based modeling of financial markets, 214–221
  arbitrageurs, 176
  Cont-Bouchaud model, 226–228
  fundamentalist, 215
  noise trader (trend chaser), 215
  self-determination versus imitation, 215
  use of the term, 165
Transience, 69
  Lévy flights, 150
Transition
  probability, 44, 58, 68, 103
  rate, 45, 108
Trend chaser, *see* traders
Truncated Lévy flights, 155–159
  as a model for price distributions, 196
  characteristic function, 156, 253–257
  crossover from Lévy to Gaussian behavior, 157–159
  exponential truncation, 156
  kurtosis of, 157

Tunnel effect, 118
  tunnel splitting, 119
  tunneling time, 120, 121

U
Uncertainty relation, 116
Uncorrelated random variables, 32
Underlying, 170
  arbitrarily divisible, 175, 177

V
Variance, 5, 20
  of binomial distribution, 5
  of Gaussian distribution, 13
Venn diagram, 24
Viscous drag, 75
Volatility
  as a market parameter, 169
  as a measure of risk, 170
  dominance for short maturities, 205
  historic, 190
  implied, 189–191, 208
  influence on price distribution, 199–202
  intra-day pattern, 199
  stochastic, 175, 198–202, 204
  volatility clustering, 199
  volatility smile, 191
    as a precursor to Gaussian behavior, 205–209
  volatility surface, 191
Volume element in hyperspherical polar coordinates, 243
Volume of a d-dimensional hypersphere, 245

W
Waiting-time distribution, 72, 74
  of fractal-time random walks, 150–153
Weierstrass function, 138
Weierstrass random walk, 136–150
  and Lévy flights, 146
  and stable distributions, 141
  asymptotic behavior, 141
  characteristic exponent α, 140
  criterion for self-similarity, 137
  definition, 137
  discrete scale invariance, 141
  examples from simulations, 144
  fractal dimension, 147
  fractional diffusion equation, 143
  log–periodic oscillations, 141
  self-similar structure, 137, 138
  solution in arbitrary dimensions, 247–252

Weierstrass random walk (*cont.*)
    solution in one dimension, 139–142, 251
    superdiffusive behavior, 142–147
    two-dimensional, 144
Wiener process, 51, 53, 76, 92, 125, 127
    as a limit of Brownian motion, 84
    as Bachelier's market model, 168
    continuous sample path, 51
    covariance of, 54, 82
    non-stationarity, 84
    time reversed, 111
Writer of an option, 171

Printed by Printforce, the Netherlands